Applied Microbiology in Intensive Care Medicine

Nissar Shaikh • Arshad Hussain Chanda
Editors

Applied Microbiology in Intensive Care Medicine

Springer

Editors
Nissar Shaikh
Surgical Intensive Care Unit
Hamad Medical Corporation
Doha, Qatar

Arshad Hussain Chanda
Surgical Intensive Care Unit
Hamad Medical Corporation
Doha, Qatar

ISBN 978-981-97-4005-5   ISBN 978-981-97-4006-2 (eBook)
https://doi.org/10.1007/978-981-97-4006-2

This Springer imprint is published by the registered company Springer Nature Singapore Pte Ltd.
The registered company address is: 152 Beach Road, #21-01/04 Gateway East, Singapore 189721, Singapore

Contents

Necrotizing Fasciitis: Pillars for Improved Patients' Outcome

Nissar Shaikh, Umm E Amara, Umme Nashrah, Firdos Ummunnisa, Mohannad Al-Tarakji, and Abdelhakem Tabeb

Abstract Necrotizing fasciitis (NF) is a common soft tissue infection, and it remains a life- and limb-threatening clinical disease. Minor trauma or incomplete surgical intervention and comorbidities are the risk for NF infection. NSAID (non-steroidal anti-inflammatory drug) seems to mask the disease process, sometimes inhibiting pain. Significantly higher pain without obvious skin manifestations and more systemic toxicity should raise a strong suspicion of NF. LRINC (Laboratory Risk Indicator for Necrotizing Fasciitis) score is helpful in the early diagnosis of NF by excluding cellulitis. Further investigations are also useful to diagnose NF, and imaging studies are increasingly used to diagnose NF with increased frequency. NF is classified according to the area involved or the microbes causing the necrotizing fasciitis.

NF has a typical pathophysiology. Initially, the microbe gets lodged in fascial layers and causes necrosis and rapid progression within fascial layers causing systemic toxicity; this is called the horizontal spread of infection. The later phase is cutaneous manifestation due to vertical spread and necrotizing infection with thrombosis of vessels supplying the skin.

The management includes aggressive medical and surgical therapy. Medical therapy includes early proper antibiotics, resuscitation, correction of coagulopathy, and hydration. The immediate surgical intervention is the bold debridement of necrotic tissue till the wound starts to bleed. Organ supportive care ranges from

N. Shaikh (✉)
Surgical Intensive Care Unit, Hamad Medical Corporation, Doha, Qatar
e-mail: smaheboob@hamad.qa

U. E Amara
Apollo Institute of Medical Sciences, Hyderabad, Telangana, India

U. Nashrah
Deccan College of Medical Sciences, Hyderabad, Telangana, India

F. Ummunnisa
Dr. Halima Al Tamimi: Obstetrics and Gynecology Centre, Doha, Qatar

M. Al-Tarakji · A. Tabeb
Surgery Department, Hamad Medical Corporation, Doha, Qatar

N. Shaikh, A. H. Chanda (eds.), *Applied Microbiology in Intensive Care Medicine*, https://doi.org/10.1007/978-981-97-4006-2_1

hemodynamic support to invasive ventilation. As known NF has higher morbidity and mortality, around 20–40% requires amputations. Delayed diagnosis and delayed surgical intervention will increase NF patients' mortality significantly. Early medical consultation in patients with unhealthy wounds and fever is needed to detect NF early and to prevent the progression to multi-organ dysfunction and shock.

Keywords Emergency · Males · Necrotizing fasciitis · Pain · LRINC score · Antibiotics · Debridement · Amputation · Ventilation · Coagulopathy · Vasopressor · Shock

1 Introduction

Necrotizing fasciitis (NF) is the necrotizing infection of the fascia, with rapid spread causing systemic toxicity and shock, and if not diagnosed early and treated properly can cause death. It is the most serious infection of the musculoskeletal system. The combination of low incidence and high mortality makes this soft tissue infection difficult to observe and study. It is the most serious illness among soft tissue infections, and it is known from the days of Hippocrates. There are various synonyms such as hospital gangrene, suppurative fasciitis, necrotizing cellulitis, and Fournier's gangrene [1]. During the American Civil War, surgeon Joseph Jones depicted necrotizing fasciitis, and later on, Meleney first described the disease in 1924 in China, whereas Wilson gave the term necrotizing fasciitis in 1952 [1].

To diagnose necrotizing fasciitis early, and properly, it is essential that all acute care teams, emergency physicians, surgeons, anesthetists, intensive care physicians, and surgeons should be well versed, aware, and updated about necrotizing fasciitis to improve the outcome of these patients.

2 Epidemiology

Fortunately, necrotizing fasciitis is rare, and the reported estimated incidence ranges from 0.3 to 15 cases for every 100,000 population. As the diagnosis of necrotizing fasciitis is a challenge, this incidence may be underestimated and underreported. In the USA, the incidence of necrotizing fasciitis is reported to be among four per 100,000 population, whereas in the UK, it is around 500 cases per year, and in Qatar, it is around 10 cases per year [2]. Necrotizing fasciitis is more common in males with a ratio of 3:1 and frequently affects extremities around 54% [3].

3 Classification

Necrotizing fasciitis is classified either based on the area involved or microbiological growth. Abdominal necrotizing fasciitis is termed Meleney's necrotizing fasciitis; when NF affects the neck, it is cervical NF; when NF affects the genital and perineum, it is called Fournier's gangrene; and submandibular NF is called Ludwig's angina.

Classification according to microbes causing necrotizing fasciitis is described in Table 1.

Type 1 NF: Combination of microbial flora–derived bowel; it is more common and affects the perineum and genitalia.

Type 2 NF: Gram-positive bacteria, commonly *Streptococcus* or *Staphylococcus* or combined, and skin- or throat-derived; most of these patients present with shock.

Type 3 NF: This is a gram-negative mono-bacterial. Commonly get infected with marine bacteria from backwaters or stagnated water, particularly *Vibrio* and *Aeromonas*. These patients are very sick on presentation and in shock.

Type 4 NF: It is fungal necrotizing fasciitis, which occurs due to *Candida* in immunocompromised patients and zygomycetes in immunocompetent patients. It is more frequently seen in trauma patients [4].

4 Risk Factors

Diabetes is the most frequent risk factor. The other risk factors are males, hypertension, coronary artery disease, obesity, immunosuppression, malnutrition, and skin viral infections that can complicate necrotizing fasciitis. Surgical intervention, incomplete evacuation or source control, trauma, penetrating injuries, and intravenous drug use can lead to and complicate necrotizing fasciitis.

Details of risk factors are mentioned in Table 2.

Table 1 Showing microbiological classification of NF

NF type	Microbes
NF type 1	Polymicrobial: Aerobes and anaerobes
NF type 2	Mono microbial: Gram-positive
NF type 3	Gram-negative mono bacteria are commonly the main bacteria
NF type 4	Fungal

Table 2 Risk factors for necrotizing fasciitis

Male
Diabetes mellitus
Coronary artery disease
Hypertension
Kidney disease
Surgery
Penetrating injury
Minor invasive procedures
Intravenous drug users
Peripheral vascular disease

5 Pathophysiology

Pathophysiology of the necrotizing fasciitis is the rapid progression of necrotizing infections through the fascial layer causing systemic toxicity (horizontal spread). A vertical spread is a later phase where patient infection causes thrombosis of dermal vasculature causing skin manifestations.

Initially, the pathogen lodges into the fascial layer, usually after a trivial injury, and it proliferates and produces hyaluronidase enzymes which degrade and cause necrosis of the fascia and subcutaneous tissue, thus creating an environment of rapid bacterial growth and spread of infection. Further, leukocytes in deeper fascia and dermis cause suppuration and thrombosis of veins and arteries to the dermis.

In mono-bacterial NF, the organism produces a more virulent factor that causes significant tissue damage, while the streptococci and staphylococci generate surface antigens M1 and 3 and exotoxins A, B, and C streptolysin O.

The M protein increases the adherence ability of these bacteria which helps microbes' phagocytes, thus hijacking the acute phase and escaping the web of inflammatory changes. Bacteria cause loss of vascular integrity, inducing capillary leakage and tissue edema. Overall, the combined effect of the above in these patients causes septic shock and organ dysfunction [5].

6 Diagnosis

Early diagnosis of NF is vital for proper management. Initial patient presentations are confusing and benign. The NF is frequently misdiagnosed as cellulitis, and it alters the management leading to poor patient outcomes. The major suspicion of NF should arise when a patient presents with severe pain (out of proportion to clinical signs) but very few dermatological manifestations. A total of 70–100% of NF patients appear sick on presentation, fever is seen in 86%, and around 30% of NF patients present with hypotension (Fig. 1). The dermatological manifestations are

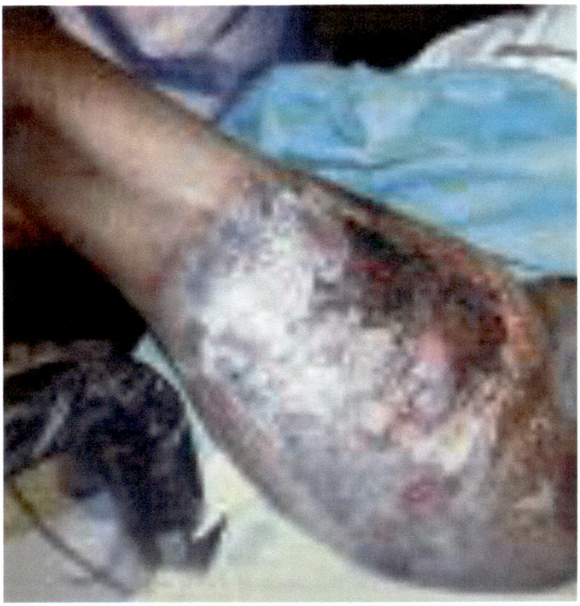

Fig. 1 Showing NF of the thigh

Table 3 Cutaneous grading of necrotizing fasciitis [6]

Stage 1	Erythema and swelling
Stage 2	Blisters, bulla warm skin
Stage 3	Crepitus, local anesthesia, skin necrosis

late to appear in NF. Wong et al. has proposed grading cutaneous signs and symptoms for NF patients [6] (Table 3).

The gold standard for the diagnosis of NF is a biopsy showing necrosis. Laboratory risk indicators for NF (LRINC) score can differentiate NF from cellulitis (Table 4). The parameters used in the LRINC score are basic blood workup results and high weightage given to CRP (C-reactive protein). Finger tests also can be performed at the bedside under local anesthesia, and NF can be diagnosed if there is a loss of fascial resistance. During surgery, NF is diagnosed and confirmed by loss of facial resistance and the presence of dishwash-like fluid in the tissue [4].

Imaging studies are also being increasingly used, although time should not be wasted on these imaging studies. An X-ray may show air in the subcutaneous tissue if the causative organisms are gas-forming. Ultrasound may show a subcutaneous collection. CT scan has more than 50% sensitivity in diagnosing NF. It will reveal facial edema and tissue swelling with or without gas. MRI is highly sensitive (more than 90%) and specific (50–80%) but may not be available in all hospitals, it is time-consuming, and it may delay the diagnosis and management of NF [2].

Table 4 LRINC score

Variable	Value	Score
C-reactive protein (mg/L)	≤150	0
	>150	4
Total white blood cell count	<15,000	0
	15–25,000	1
	>25,000	2
Hemoglobin (g/dL)	>13.5	0
	11–13.5	1
	<11	2
Sodium (mmol/L)	≥135	0
	<135	2
Creatinine (mg/dL)	≤1.6	0
	>1.6	2
Glucose (mg/dL)	≤180	0
	>180	1

7 Treatment

These patients should be evaluated and followed using the ABCDEF (A—assess and manage pain, B—both spontaneous breathing/Respiratory support/spontaneous awake trials (SBT/SAT), C—Circulatory support/choice of analgesia and sedation medication, D—delirium assess and manage, E—early mobility, F—family engagement) approach. Early aggressive medical and surgical management is key for improved outcomes in NF. Appropriate broad-spectrum antibiotics with fluid management are essential. Early bold debridement is key for a patient's better outcome, and the delay, particularly more than 24-h delay, not only increases mortality but also morbidities such as dehydration, toxemia, and biochemical disorders.

The NF patients need more than two or more debridements with multiple bedside looks and debridement. NF wound care is equally important. The wound should be washed and dressed with occlusive, absorptive dressing with bandage two to three times per day. It may be wise to consider cystostomy or colostomy to avoid contaminations of perineal or scrotal NF wounds. Nowadays, *Flexi-Seal* can be used to drain the stool with significant success while surgical intervention is in progress. It is necessary to support multiple organs when affected by NF. As polymicrobial type 1 NF is common, the initial antibiotics should be broad-spectrum, coverage including gram-positive and gram-negative bacteria and anaerobes. Clindamycin has extra advantages particularly in type 2 NF, as the efficacy of clindamycin is independent of inoculation size and bacterial growth, it is an effective suppressant of exotoxins in sub-inhibitor dose, and it facilitates phagocytosis and also reduces the penicillin-binding protein and has longer post antibacterial effects [7].

Intravenous immunoglobulins (IVIG) are useful in type 2 NF patients with toxic shock syndrome. Immunoglobulins neutralize toxins, superantigens, and bacterial activity as cytokine release and reduce the post-inflammatory cytokines. Around

80% of NF requires invasive ventilation, and 70% will be in septic shock requiring vasopressors, thus requiring intensive care therapy. NF patients need large amounts of fluids, blood, and blood products for resuscitation due to coagulopathy [7].

Hyperbaric oxygen therapy is also useful, once debridement is done and the patient is stable. It increases the oxygen saturation in the infected tissues by 1000 folds, has a bacteriostatic effect, improves polymorphonuclear function, stops the spread of the disease process, and enhances wound healing [8]. Adequate nutrition, with early enteral nutrition, is essential for a rapid better recovery of NF patients.

8 Complications

NF causes multiple organ dysfunction, and the most commonly affected is the hemodynamic system (70%) requiring vasopressor support. AKI (acute kidney injury) is also frequently affected and may require renal replacement therapy. The coagulation system is also commonly prolonged with DIC (disseminated intravascular coagulation), requiring blood, blood products, and fibrinogen therapy.

The regional complications are concerning. If the infection is not controlled, around 23–40% of NF requires amputation of the extremities. The reported risk factors for amputation are myonecrosis, unremitting shock, concurrent or associated vascular disease, and rapidly progressive infection with massive soft tissue necrosis [9].

We reported a few unusual complications of NF, such as pneumothorax, acute myocardial infarction, and ventricular tachycardia [7].

9 Prognosis

Neonatal and pediatric patients with NF have higher mortality than adults. The reported mortality of NF ranges from 20% to 75%. The most important factor for mortality is the delayed debridement of necrotic tissue which significantly increases morbidity and mortality.

Other risk factors which increase the NF patient's mortality are older age, diabetes mellitus, and two or more comorbidities, although other studies did not show any mortality difference with or without NF [1].

Type 2 and type 3 NF have a higher mortality compared to type 1 NF. The area involved with NF also dictates mortality: NF of the axilla, chest, and abdominal wall, as well as gluteal [3, 10].

10 Prevention

Keeping the skin intact and clean is the most important to avoid NF. Patients with skin infections with fever or sore throat should seek early medical attention.

11 Conclusion

NF is a rare soft tissue infection which is life- and limb-threatening if not treated early and properly. The common presentation is not septic in the early phase, and the common mistake is to confuse NF with cellulitis which alters the management.

NF should be suspected when the pain at the site is disproportionately higher than the dermatological manifestation. LRINC score helps in diagnosis by differentiating NF from cellulitis. Polymicrobial NF is more common. NF patients should be started immediately on broad-spectrum antibiotics in combination with clindamycin and taken for debridement. These patients may need multiple debridements. Organ supportive care is required in most of these patients. The reported mortality ranges from 20–70%, while the amputation rates in NF are 23–40%. Early diagnosis and aggressive management will improve the NF patient's outcome.

References

1. Shaikh N. Necrotizing fasciitis and diabetes mellitus. Int J Diabetes Dev Ctries. 2007;27:14.
2. Shaikh N, Khawaiter J, Al-Thani H. Necrotizing fasciitis: a surgical and medical emergency. Surg Sci. 2012;03:518–25.
3. Shaikh N, et al. Regional necrotizing fasciitis and its outcome. Qatar Med J. 2008;17(1):24–7.
4. Shaikh N, El-Menyar A, Mudali IN, Tabeb A, Al-Thani H. Clinical presentations and outcomes of necrotizing fasciitis in males and females over a 13-year period. Ann Med Surg. 2015;4:355–60.
5. Seal DV. Necrotizing fasciitis. Curr Opin Infect Dis. 2001;14:127.
6. Wang Y-S, Wong C-H, Tay Y-K. Staging of necrotizing fasciitis based on the evolving cutaneous features. Int J Dermatol. 2007;46:1036–41.
7. Shaikh N. Necrotizing fasciitis: a decade of surgical intensive care experience. Indian J Crit Care Med. 2006;10:225–9.
8. Weaver LK. Hyperbaric oxygen in the critically ill. Crit Care Med. 2011;39:1784–91.
9. Ozalay M, Ozkoc G, Akpinar S, Hersekli MA, Tandogan RN. Necrotizing soft-tissue infection of a limb: clinical presentation and factors related to mortality. Foot Ankle Int. 2006;27:598–605.
10. Jabbour G, El-Menyar A, Peralta R, Shaikh N, et al. Pattern and predictors of mortality in necrotizing fasciitis patients in a single tertiary hospital. World J Emerg Surg. 2016;11:40.

Hospital-Acquired Infections: Key for Prevention and Clinical Bundles

Abdelrahman Ali Balal, Umm E Amara, Umme Nashrah, Asad Malik, Arshad Hussain Chanda, and Nissar Shaikh

Abstract Hospital-acquired infections (HAIs), nosocomial infections, or healthcare-associated infections are not present in patients at the time of admission to the hospital, and these infections are acquired during hospitalization or as a consequence of treatment during hospitalization and manifest after 48 h of hospitalization. According to the literature, around 4% of hospitalized patients suffer from HAIs. The major HAIs in acute settings are catheter-related bloodstream infections (CRBSIs), ventilator-associated pneumonia (VAP), surgical site infections (SSIs), and catheter-associated urinary tract infections (CAUTIs). Other hospital-acquired infections which are not discussed here are MRSA (methicillin-resistant *Staphylococcus aureus*) infections, hospital-acquired pneumonia, and *Clostridium difficile* colitis. Over the last few decades, hospitals and medical institutions have been very careful and serious about hospital-acquired infections and established an infection tracking and surveillance system. Along with the treatment, robust prevention bundle strategies are applied to reduce the rate and incidence of hospital-acquired infections.

Care bundles are evidence-based measures. When implemented, it has been shown to improve healthcare and have a much better impact than isolated measures. The major hospital-acquired infection prevention bundles are CRBSI, CAUTI, VAP, and SSI prevention bundles. These bundles are simple, clear, and concise. It decreases the major hospital-acquired infection with improved morbidity, mortality, and hospital course if effectively implemented.

A. A. Balal
Surgical Intensive Care Unit, Hamad General Hospital, Doha, Qatar

U. E Amara
Apollo Medical College, Apollo Institute of Medical Sciences, Hyderabad, Telangana, India

U. Nashrah
Deccan College of Medical Sciences, Hyderabad, Telangana, India

A. Malik
Department of Medicine, Al-Wakrah Hospital, Hamad General Hospital, Doha, Qatar

A. H. Chanda (✉) · N. Shaikh
Surgical Intensive Care Unit, Hamad Medical Corporation, Doha, Qatar
e-mail: achanda@hamad.qa; smaheboob@hamad.qa

N. Shaikh, A. H. Chanda (eds.), *Applied Microbiology in Intensive Care Medicine*, https://doi.org/10.1007/978-981-97-4006-2_2

Keywords Hospital-acquired infections · Healthcare-associated infections Nosocomial infections · Surgical site infection · Ventilator-associated pneumonia Catheter-related bloodstream infection · Catheter-associated urinary tract infection Prevention bundle

1 Nosocomial or Hospital-Acquired Infections (HAIs)

Hospital-acquired infections or nosocomial infections affect hospitalized, under-treatment patients, and these infections are not present at the time of admission to the hospital. It occurs in any facility of the hospital, such as intensive care units (ICUs), wards, and long-term facilities. Nearly half of the HAIs occur outside of the ICU. Nosocomial infections mainly occur due to invasive procedures and major surgical interventions. Indwelling catheters, prosthetic devices, and invasive ventilation are other common sources of nosocomial infections. These nosocomial infections could be due to multidrug-resistant bacteria, viruses, and fungi [1].

2 Epidemiology

At any time, 7% of patients in developed countries and 10% of hospitalized patients in developing countries will acquire one HAI. In countries with limited resources, the HAI occurrence ranges from 5.7% to 19%, with an average of 10.1%. The overall HAI occurrence in intensive care settings was 47.9 per 1000 patients in developing countries. Surgical site infections were the leading cause in these countries with an incidence of 5.6/100 surgical procedures, and gram-negative bacilli represent the most common pathogen isolated in HAI patients [1]. The overall described rate of ventilator-associated pneumonia (VAP) was 16.8/1000 mechanical ventilator days, the catheter-associated urinary tract infection (CAUTI) was 5.5/1000 urinary catheter days, and the rate of catheter-related bloodstream infection (CRBSI) was reported to be 4.9/1000 central line days [2]. The major HAIs with significant impact are ventilator-associated pneumonia (VAP), catheter-related bloodstream infections (CRBSIs), surgical site infections (SSIs), and catheter-associated urinary tract infections (CAUTIs) [3]. All these HAIs are associated with invasive interventions or surgical procedures.

3 Risk Factors

The three major risk factors for hospital-acquired infections are related to host, agent, and environmental factors.

3.1 *The Host Factors*

The host risk factors for acquiring the infections during the hospital stay are elderly patients, malnutrition, malignancy, infection with HIV, severe burn and skin disease, diabetes mellitus, coma, open wounds, and hemodynamic instability. These patients have a poor defense against the microbes and get exposed to multidrug-resistant bacteria and other microbes.

3.2 *The Agent Factors*

Most HAIs are caused by bacteria, rarely by viruses and fungi. The bacteria and other microbes survive after contaminating various surfaces in the hospital.

3.3 *The Environmental Factors*

Environmental factors can affect both microbes and patients. These environmental factors include both animate and inanimate.

The inanimate environment includes medical instruments, equipment, diagnostic and therapeutic interventions, and environmental surfaces. The animate environment includes or refers to healthcare providers, other patients in the same unit, the patient's family, and visitors. Other environmental factors are cleanliness, sanitation, temperature, and humidity of the patient's environment.

4 Risk Factors for HAIs

The following risk factors for the respective hospital-acquired infections are presented in Table 1.

Table 1 Risk factors for the hospital-acquired infections

Various risk factors	Infections
• Catheter-related bloodstream infection (CRBSI)	Age, immunosuppression, femoral catheters, crowding of microbes that catheter insertion site
• Ventilator-associated pneumonia (VAP)	Age, comorbidities, brain injury, peptic ulcer prophylaxis, enteral feeding, subglottic aspiration
• Surgical site infection (SSI)	Age, diabetes mellitus, obesity, smoking, existing infection, trauma, shock, prolonged surgical infections
• Catheter-associated urinary tract infection (CAUTI)	Extremes of age, severity of the disease, comorbidities, prolonged duration of a urinary catheter, break in the closed system

The mode of transmission could be contact, airborne, or vector-borne. The better way of preventing and controlling these hospital-acquired infections is by following the clinical care bundle described below:

5 Surgical Site Infection Prevention Bundle

The skin is a natural barrier to infections. Surgery with initial infections causes a break in the skin continuum, making it more prone to infections. SSI is the term used for such infections at the surgery site. One to three percent of surgical incisions get complicated with infections, and typical SSI occurs within 30 days of the surgery. SSIs are most frequently reported after colorectal abdominal surgeries (3–30%).

SSIs are subdivided into three types: superficial, deep, and organ or SSI-space collection. Risk factors for SSI are described in Table 1. SSI prevention bundle is composed of seven or five points, and these points are adopted depending on the institution or hospital and tailored to the local requirements to control surgical site infections.

The commonly used SSI prevention bundle is a five-point bundle as shown in the following diagram (Fig. 1). Figure 2 shows the seven-point SSI bundle.

Each point in the SSI prevention bundle is important and should be understood.

In antibiotic prophylaxis, antibiotics should be chosen according to the surgery and local antibiogram. Antibiotics should be given as close as possible to the surgical incision time and continued for 24 h in the appropriate dosage. The preoperative anti-multidrug-resistant *Staphylococcus aureus* (MRSA) application is not shown to improve significantly the prevention of SSI, so it should be discouraged (particularly in pediatric age groups) [4]. It has been shown that the preoperative antiseptic (chlorhexidine) bath decreases gram-positive as well as gram-negative skin bacterial load, thus significantly reducing the chances of SSI in the postoperative period [5].

Hair should not be removed from the operative site unless these hair will interfere with the surgery. Hair should be removed by using clippers or a depilatory agent. The use of razors for hair removal has been shown to increase SSIs [6].

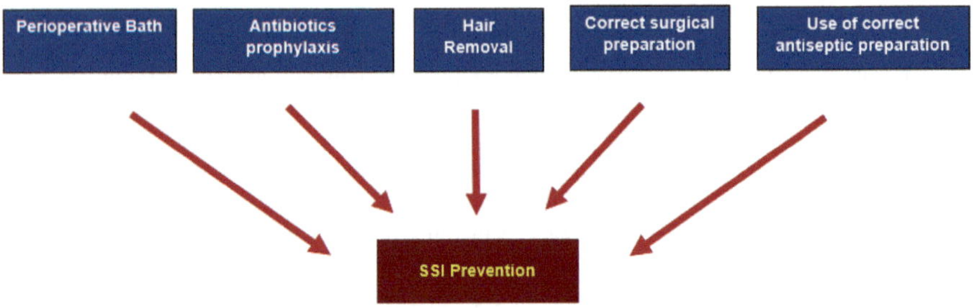

Fig. 1 Five-point SSI prevention bundle

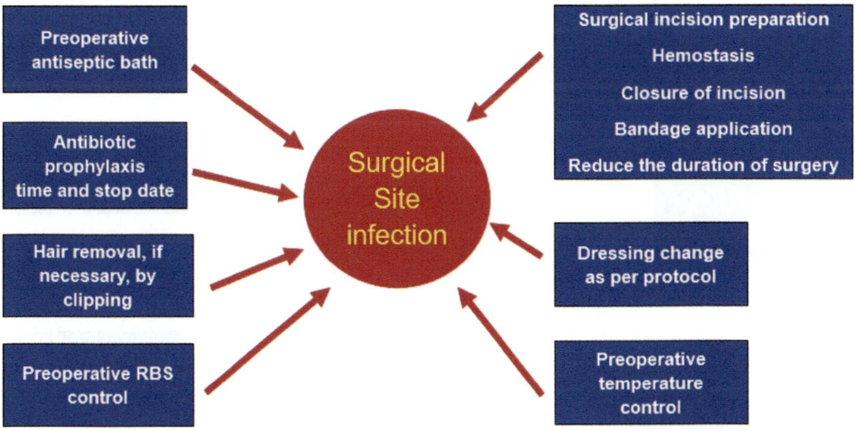

Fig. 2 Seven-point SSI prevention bundle

According to the National Institute for Clinical Excellence (NICE), the World Health Organisation (WHO) and the American College of Surgeons (ACS) recommend skin antiseptic preparations, unless contraindicated, such as cornea, mucosal membranes, and ear. We should use alcohol-based skin antiseptic solution for skin preparation. NICE, WHO, and ACS also recommended using alcohol-paced chlorhexidine gluconate (CHG) or iodine solution for skin preparation before surgery [6].

Preoperative glucose control is an essential element of the SSI bundle as by now it is clear that hyperglycemia in the preoperative period increases the risk of infections; hence, it is recommended to keep the random glucose sugar in the preoperative period between 6.1 and 8.3 mmol/L. Keeping RBS under 6.1 mmol/L increases the risk of hypoglycemia and therefore not recommended [6].

Preoperative hypothermia, defined as core body temperature less than 36 °C, is a consequence of anesthesia, surgery, and operative theater temperature. It increases morbidity and mortality in the postoperative period; hence, it is essential to prevent it. It comprises one of the important elements of SSI prevention [7].

6 Ventilator-Associated Pneumonia Prevention Bundle

When acutely ill ventilated patients are affected by ventilator-associated pneumonia, it increases the ventilatory days and length of ICU and hospital stays and increases the mortality. The use of a VAP bundle decreases the ventilatory days, length of stay, and mortality [8]. Components for prevention of VAP are mentioned in detail in Fig. 3.

The pathophysiology of VAP is described in following Fig. 4.

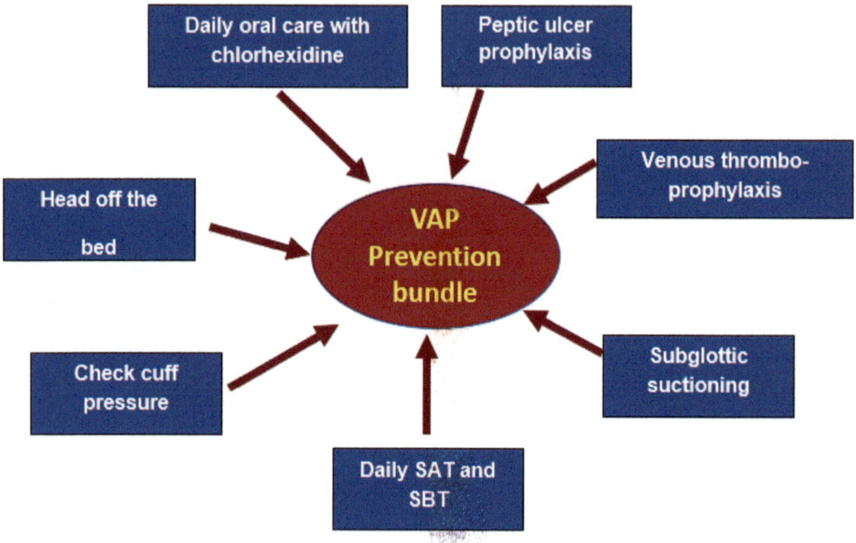

Fig. 3 The components of the VAP prevention bundle

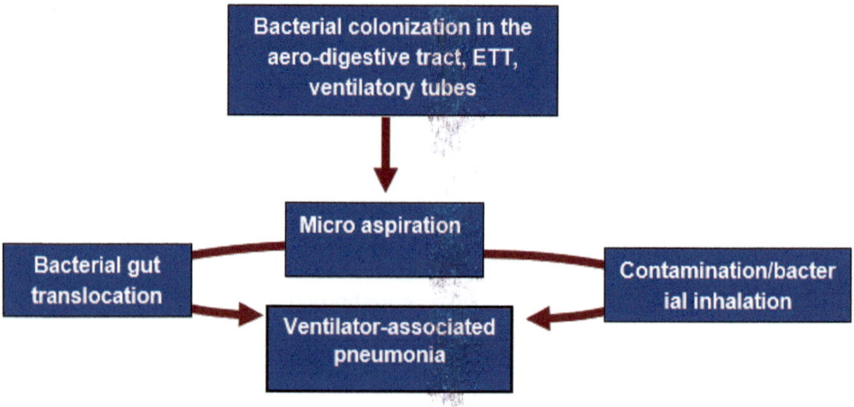

Fig. 4 Pathophysiology of VAP

6.1 Head off the Bed

Head off the bed (HOB) is the basic component of the VAP prevention bundle; the supine position increases microaspiration due to lower esophageal relaxation. Thus, the 25–30° head-up position helps in the prevention of aspiration, colonization, and VAP. HOB at 45° is more effective in prevention, but it is not practical and increases the incidence of pressure ulcers.

6.2 *Subglottic Suctioning of the Endotracheal Tube (ETT) and Cuff Pressure Measurement*

The ETT design modifications prevent bacterial colonization, prevent microaspiration, remove the biofilm from the ETT lumen, and lastly regulate the cuff pressure. The subglottic suction aspiration was reported to reduce the incidence of VAP significantly. The suction port above the ETT cuff is helpful in subglottic aspiration and can aspirate intermittently or continuously. There is no compelling evidence, but tapper shape cuff is supposed to prevent the patient from microchannels. Maintaining the ETT pressure >20 cm H_2O is essential to prevent subglottic secretions into the lungs; hence, continuous or 4 h of checking of the cuff pressure and monitoring is standard of care which is essential to prevent VAP [9].

6.3 *Oral Care*

Reflux causes colonization of the oral cavity with enteric bacteria and fungi. Oral care with chlorhexidine solutions was reported to reduce the incidence of VAP by up to 40% [9].

6.4 *SAT and SBT (Spontaneous Awakening and Spontaneous Breathing Trial)*

The SAT and SBT indirectly reduce VAP by reducing ventilatory days [9].

6.5 *Stress Ulcer and Deep Venous Thrombosis Prophylaxis*

Both interventions indirectly reduce morbidity and mortality by reducing gastrointestinal bleeding and the incidence of DVT [9].

7 Catheter-Related Bloodstream Infection Prevention Bundle (CRBSI)

A central venous catheter (CVC) is a commonly used device in intensive care setups. Despite its advantages, it can complicate into infections called CRBSI, and it is documented to increase morbidity or mortality in critically ill patients. CRBSI prevention bundle and its regular use are documented to be useful in prevention of the

Table 2 CRBSI prevention bundle

Insertion bundle	Maintenance bundle
1. Perform hand hygiene 2. Adhere to aseptic precautions 3. Use the maximum sterile barrier precautions 4. To minimize the infection and not infectious complications; choose the best site for insertion 5. Site of insertion preparation with chlorhexidine (more than 0.5%) and alcohol 6. Cover the insertion site with a sterile transparent, semipermeable dressing	1. Hand hygiene 2. Chlorhexidine bath 3. Clean access port with appropriate antiseptic before use 4. Use a sterile device to access the catheter 5. If the dressing gets wet, soiled, or removed, immediately replace it. Other change dressing after 48 h Semipermeable dressing every 7 days Use an antibiotic patch or chlorhexidine cushion at the insertion site 6. Change administration set of continuous infusions every 72 h Blood and blood products set 24 h Fat embolization every 24 h 7. Daily inspect and remove the line if not required or infected

CRBSI. CRBSI prevention bundle is composed of two sections: the insertion and maintenance bundle, as shown in following Table 2.

The use of a CRBSI bundle over a decade has reduced the incidence of CRBSI, thus reducing morbidity, mortality, and cost significantly [10].

Both CRBSI insertion and maintenance bundle components are simple and self-explanatory. The use of the best insertion site means the subclavian insertion and is found to have the least infectious and non-infectious complications, and using the chlorhexidine cleaning at the site of insertion of the line is reported to prevent the infectious complications [10].

8 Catheter-Associated Urinary Tract Infections (CAUTIs)

CAUTI (catheter-associated urinary tract infection) is one of the most common hospital-acquired infections; one in three catheterized patients will complicate into CAUTI. Prolonged catheterization, improper insertion, and handling of urinary catheters are the risks for CAUTI. CAUTI increases morbidity and mortality; hence, it is essential to prevent it. For the prevention of CAUTI in catheterized patients, one should follow the CAUTI prevention bundle. From the literature, it is dictated that the use of a CAUTI prevention bundle reduces the occurrence of CAUTI, ultimately decreasing urinary catheter infections and improving the patient's outcome [11]. Figure 5 shows the components of the CAUTI prevention bundle.

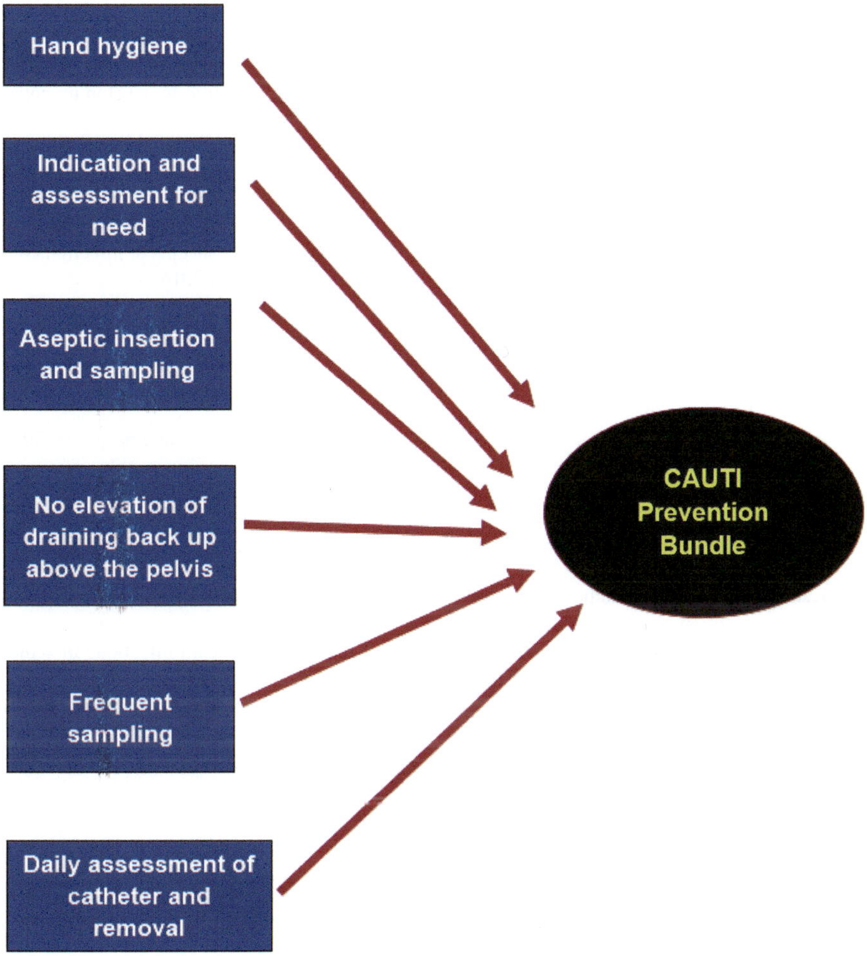

Fig. 5 CAUTI prevention bundle

9 Conclusion

Hospital-acquired or nosocomial infections are one of the significant causes of increased morbidity and mortality, hospital stay, and cost in acutely ill patients. These are CRBSIs due to central venous catheters. The ventilator-associated pneumonia occurs in intubated and ventilated patients. CAUTIs are one of the most frequent hospital-acquired infections in hospitalized patients, thus again causing increased morbidity or mortality. Surgical site infections are another nosocomial infection.

All these nosocomial infections are prevented by the maximum compliance of clinical bundles, mainly surgical site infection prevention, VAP prevention, CRBSI prevention, and CAUTI prevention bundles.

References

1. Allegranzi B, et al. Burden of endemic health-care-associated infection in developing countries: systematic review and meta-analysis. Lancet. 2011;377:228–41.
2. Rosenthal VD, et al. International Nosocomial Infection Control Consortium (INICC) report, data summary of 43 countries for 2007–2012. Device-associated module. Am J Infect Control. 2014;42:942–56.
3. Unahalekhaka A. Epidemiology of healthcare—associated infections (Chapter 3). In: Friedman C, editor. International Federation of Infection Control (IFIC) basic concepts of infection control. 3rd ed. Portadown: International Federation of Infection Control; 2016.
4. Woodward C, et al. Multicenter quality improvement project to prevent sternal wound infections in pediatric cardiac surgery patients. World J Pediatr Congenit Heart Surg. 2017;8:453–9.
5. Edmiston CE, Leaper D. Should preoperative showering or cleansing with chlorhexidine gluconate (CHG) be part of the surgical care bundle to prevent surgical site infection? J Infect Prev. 2017;18:311–4.
6. Anderson DJ, et al. Strategies to prevent surgical site infections in acute care hospitals: 2014 update. Infect Control Hosp Epidemiol. 2014;35:605–27.
7. Riley C, Andrzejowski J. Inadvertent perioperative hypothermia. BJA Educ. 2018;18:227–33.
8. Lim K-P, et al. Efficacy of ventilator-associated pneumonia care bundle for prevention of ventilator-associated pneumonia in the surgical intensive care units of a medical center. J Microbiol Immunol Infect Wei Mian Yu Gan Ran Za Zhi. 2015;48:316–21.
9. Kallet RH. Ventilator bundles in transition: from prevention of ventilator-associated pneumonia to prevention of ventilator-associated events. Respir Care. 2019;64:994–1006.
10. Buetti N, et al. Strategies to prevent central line-associated bloodstream infections in acute-care hospitals: 2022 update. Infect Control Hosp Epidemiol. 2022;43:553–69.
11. Saint S, et al. A program to prevent catheter-associated urinary tract infection in acute care. N Engl J Med. 2016;374:2111–9.

Antimicrobial Stewardship in ICU: A Success Story

Arshad Hussain Chanda, Mariyam Fatima, Asad Malik, Hussam Al Soub, and Nissar Shaikh

Abstract Antimicrobial resistance (AMR) has emerged as a global crisis. There is evidence of ever-emerging drug-resistant pathogens and very few new antimicrobial agents in development. The ICUs and the critical care setups are known as the epicenters for the development of antimicrobial resistance. One of the most important contributing factors to the development of multidrug-resistant organisms is the overuse and misuse of antibiotics in healthcare facilities. To combat this global crisis of AMR, antimicrobial stewardship (AMS) appears to be one of the major interventions, which appears to mitigate the problem. The goals of AMS are optimizing antibiotic use to limit the emergence of multidrug-resistant organisms (MDROs) and to reduce costs related to unnecessary antimicrobial use. The WHO, CDC, and scientific societies from North America and European countries have recommended guidelines to implement the "antibiotic stewardship program" (ASP) and key elements to establish the ASP. There are significant barriers to implementing the "ASP" in ICUs because of the specific criteria when treating critically ill patients. The barriers which are of concern for the ICUs are uncertainty of diagnosis, lack of rapid laboratory tools to diagnose the causative organism and its resistance pattern, lack of understanding about the antimicrobials and nonacceptance behavior by the ICU physicians, lack of time, lack of electronic medical records and personnel, etc. The ASP team comprises ICU physicians, infectious disease physicians, microbiologists, clinical pharmacists, and infection control practitioners. It led to early administration of the right drug and the right dose for the right duration, incorporating antimicrobial pharmacokinetics and pharmacodynamics (PK/PD) when prescribing

A. H. Chanda (✉) · N. Shaikh
Surgical Intensive Care Unit, Hamad Medical Corporation, Doha, Qatar
e-mail: achanda@hamad.qa

M. Fatima
Deccan College of Medical Sciences, Hyderabad, Telangana, India

A. Malik
Department of Medicine, Al-Wakrah Hospital, Hamad General Hospital, Doha, Qatar

H. Al Soub
Infectious Disease Section, Department of Medicine, Hamad General Hospital, Doha, Qatar

N. Shaikh, A. H. Chanda (eds.), *Applied Microbiology in Intensive Care Medicine*, https://doi.org/10.1007/978-981-97-4006-2_3

antibiotics and de-escalation according to the susceptibility and the optimal duration. Education, audit and feedback, guidelines, and a computerized antibiotic decision support system were identified as facilitators for the successful implementation of the ASP. The ASP implementation was associated with several positive outcomes like reduction of antibiotic use, the downward trend of *Clostridium difficile* infection incidence rate and costs, and decreased length of stay. Artificial intelligence may have a very important role in the coming days.

Keywords Antimicrobial stewardship · Antibiotic resistance · MDRO: multidrug-resistant organisms · Pharmacodynamics and pharmacokinetics · Antibiotics Sepsis and septic shock · Intensive care unit

1 Introduction

One of the top ten risks to world health is the rise in AMR among human and animal pathogens. It has been estimated that in 2019, about 4.95 million deaths were caused by bacterial AMR, including those directly related to it, while 1.27 million deaths were caused by drug-resistant illnesses worldwide [1]. According to projections, this number may increase to ten million global fatalities per year by 2050, similar to the number of deaths caused by cancer in 2020 [2]. ICUs are considered the epicenter for antibiotic resistance as severe infections are more frequent, and antibiotic use is more widespread in the ICU than in other hospital departments. Several conditions, such as a high risk of transmission, extensive surgical procedures, loss of physiological barriers, immunocompromised patients, prolonged hospital stay, and high antibiotic pressure, can encourage the development of resistance. Since very few drugs are being developed to address the issue of escalating antibiotic resistance, it is getting harder for ICU physicians to treat patients with MDR infections. Implementing an antimicrobial stewardship program (ASP) is a practical strategy to control the appropriate use of antibiotics since it prevents the development of MDR pathogens and adverse events in critically ill patients and increases cost-effectiveness.

Factors that contribute to the spread of AMR in the ICU include

- Higher severity of illness
- Presence of other medical conditions (multiple comorbidities)
- Invasive procedures—lines and devices
- Advanced age
- Dependence on others for daily living and/or cognitive decline
- Extended hospital stays (such as in nursing homes)
- Frequent interactions with medical facilities like hemodialysis units and IV infusions

In the ICU, the spread of AMR is by several factors. These include frequent interaction between medical staff caring for multiple patients, polluted environments, and shared equipment. The sources of infections can be reservoirs or vectors. Certain medical devices like CVCs, urinary catheters, and ETTs can also contribute to the spread of microorganisms by circumventing the host's natural defences [3]. Recent invasive procedures like surgery can also increase the risk of infection. Antibiotic use before admission to the ICU can create selection pressure that encourages the evolution of microorganisms resistant to many antibiotics.

2 Definition of Antimicrobial Stewardship

Antimicrobial stewardship (AMS) is a coherent set of actions that promote the responsible use of antimicrobials. This definition can be applied to actions at the individual level as well as the national and global levels and across human health, animal health, and the environment. Antimicrobial stewardship program (AMS program) is an organizational or system-wide healthcare strategy to promote the appropriate use of antimicrobials by implementing evidence-based interventions [4].

Antibiotic stewardship has been defined in a consensus statement from the Infectious Diseases Society of America (IDSA), the Society for Healthcare Epidemiology of America (SHEA), and the Pediatric Infectious Diseases Society (PIDS) as "coordinated interventions designed to improve and measure the appropriate use of [antibiotic] agents by promoting the selection of the optimal [antibiotic] drug regimen including dosing, duration of therapy, and route of administration" [5].

The Public Health Agency of Canada defines antimicrobial stewardship (AMS) [6] as, "Coordinated interventions designed to promote, improve, monitor, and evaluate the judicious use of antimicrobials so as to preserve their future effectiveness and to promote and protect human and animal health. Antimicrobial stewardship encompasses the five Rs of AMS: responsibility, reduction, refinement, replacement, and review."

3 "The Evolution of Antibiotic Stewardship: A Historical Perspective"

John E. McGowan Jr. and Dale N. Gerding in the USA wrote the first article to use the term "antimicrobial stewardship." In 1999, Ian Gould and Jos van der Meer pioneered antimicrobial stewardship. In 1998, they established the European Society of Clinical Microbiology and Infectious Diseases Study Group for Antimicrobial Stewardship, or ESGAP, which contributed to the global expansion of the term "antimicrobial stewardship [7]."

4 Milestones of Antimicrobial Stewardship

Antimicrobial stewardship was introduced to the hospital wards. Then, it was gradually introduced to the ICUs and critical care setups. Antimicrobial stewardship in ICUs progressed over a period of time and achieved several milestones. These could only be done after the realization that antimicrobial resistance in ICUs is increasing rapidly and contributes to poor patient outcomes. The inappropriate use of antibiotics and the rising cost of the treatment were other drivers to push forward antimicrobial stewardship. The major milestone in implementing antimicrobial stewardship in ICUs and critical care setups was the development of guidelines by various infectious disease, intensive care, critical care societies, and the World Health Organisation. These agencies recommend implementing antimicrobial stewardship in ICUs. The second decade of the twenty-first century saw some most important suggestions from the Infectious Diseases Society of America (IDSA) and the Society for Healthcare Epidemiology of America (SHEA) guidelines for developing an institutional program to enhance antimicrobial stewardship [8, 9] and the World Health Organization (WHO) guidelines on core components of antibiotic stewardship programs [10]. The most notable ones were the 2019 joint guideline by the Society of Critical Care Medicine (SCCM) and the Infectious Diseases Society of America (IDSA) on antimicrobial stewardship in ICUs, emphasizing the importance of antimicrobial stewardship strategies in the ICU. These guidelines stressed the recommendations for optimizing antimicrobial use, especially the appropriate empiric therapy, de-escalation, duration of treatment, and the need for local and national surveillance of antimicrobial resistance [11].

These milestones are evolving, reflecting that antimicrobial stewardship is also evolving with the multidisciplinary approach. There is an ongoing effort between various specialties to develop tools for fostering and implementing antimicrobial stewardship in ICUs.

5 Antimicrobial Stewardship in the Intensive Care Unit [12]

5.1 *The Goals of the Antimicrobial Stewardship Program*

The ICUs and the critical care setups are the high-risk areas in any hospital for using parental antibiotics and consume approximately ten times more than the wards. The cohorts of patients admitted to the ICUs are critically ill, with a high incidence of infection resulting in sepsis and septic shock. Certain practices are unique to the ICUs, like using empiric broad-spectrum antibiotics. The severity of the illness, which needs rapid interventions in the absence of definitive microbiological information like isolated organisms and their sensitivity, compels ICU physicians to adopt such practices. On one side, delay in antibiotics may lead to worsening of clinical condition; on the other, misuse of antibiotics may have multiple concerns. Thirty to sixty percent of antibiotics prescribed in ICUs are unnecessary, inappropriate, or suboptimal. These antibiotics are often used for long duration [8].

These antibiotics are often used for long duration. This practice leads to the development of multidrug-resistant organisms (MDROs), high incidence of adverse events like acute kidney injury, Clostridia diarrhea, and increased hospital cost due to a longer hospital stay and higher mortality.

In the brief background of the concerns aforementioned, the major goals of antimicrobial stewardship are to improve patient health outcomes, prevent or slow the emergence of antibiotic resistance, and reduce hospital costs. Figure 1 provides more details on these objectives.

The primary goals of an antimicrobial stewardship program as described by Ohl CA et al. [13] in 2011 are as follows:

- Prevent or slow the emergence of antimicrobial resistance
- Improve patient outcomes
- Reduce adverse drug events, including secondary infection (e.g., *C. difficile* antibiotic-associated diarrhea)
  - Reduce morbidity and mortality
- Reduce healthcare expenditures
  - Reduce antibiotic consumption
  - Reduce the length of stay
- Increase adherence/utilization of treatment guidelines
  - Optimize selection, dose, and duration of therapy
  - Decrease inappropriate use of antibiotics

In order to achieve the objectives of antimicrobial stewardship, a straightforward approach by Doron and Davidson (2011) states that ICU physicians should focus on

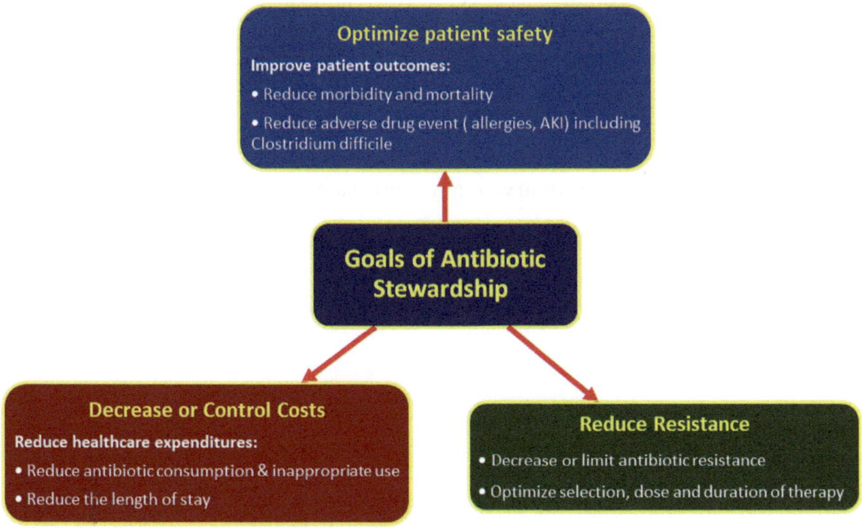

Fig. 1 The goals of the antimicrobial stewardship program in ICU

the “five Ds” to make sure that every patient receives the best possible antibiotic treatment [14].

1. The right **D**rug
2. The right **D**ose
3. The right **D**elivery method
4. Attention to **De**-escalation
5. The right **D**uration of administration

6 Establishing a Stewardship Program: In the General Hospital Setting

US hospitals must adopt antibiotic stewardship programs (ASPs) as per the guidelines released by the Centers for Disease Control and Prevention in 2014. These guidelines, known as the Core Elements, outline the necessary structural and operational elements. The CDC has revised its recommendations from 2014 to 2022, which include updates such as “Implementation of Antibiotic Stewardship Core Elements of Small and Critical Access Hospitals Supplemental Guidance” (2017), “Core Elements of Hospital Antibiotic Stewardship Programs Update” (2019), and “Priorities for Hospital Core Element Implementation Supplemental Guidance” (2022). These recommendations are focused on establishing stewardship programs in hospitals across any facility in a hospital (Table 1). Core elements of a hospital antimicrobial stewardship program as outlined by the US Centres for Disease Control and Prevention (CDC) [15] include Table 1.

Hospital antibiotic stewardship programs are essential in enhancing patient care and combating antimicrobial resistance. In 2021, almost 95% of American hospitals implemented antibiotic stewardship programs that adhered to all seven hospitals' Core Elements for Antibiotic Stewardship Programs.

The three most common interventions of the AMS are restrictive intervention which encompasses formulary restrictions, specialist preauthorization; secondly, the incentive prospective audit and feedback, education; and, last, the organizational multidisciplinary team approach, antimicrobial stewardship meeting [16].

Table 1 Core elements of a hospital antimicrobial stewardship program aimed at establishing stewardship programs in hospitals across any facility in a hospital

Core elements priorities for	Resources	Hospital core element implementation
1. Leadership commitment	Leadership support is critical to the success of antimicrobial stewardship programs Dedicate necessary human, financial, and information technology resources	Antibiotic stewardship physicians and/or pharmacist leader(s) have antibiotic stewardship responsibilities in their contracts, job descriptions, or performance reviews. Report activities and outcomes regularly to senior executives and the hospital board
2. Accountability	Appoint leaders or co-leaders, such as physicians and pharmacists, responsible for program management and outcomes	Program leader(s) responsible for program outcomes must be identified. Some programs are co-led by physicians and pharmacists
3. Pharmacy expertise	A pharmacist, ideally as the co-leader of the stewardship program, to help lead implementation efforts to improve antibiotic use	A pharmacist leader should be identified
4. Action—Implement interventions	Implement interventions, such as prospective audits and feedback or preauthorization, to improve antibiotic use, such as prospective audit and feedback or preauthorization, to improve antibiotic use	The antibiotic stewardship program has facility-specific treatment recommendations for common clinical condition(s) and performs prospective audit/feedback or preauthorization
5. Tracking	Monitor reports of antibiotic prescribing, the impact of interventions, and other important outcomes, like *C. difficile* infections and resistance patterns. Information on antibiotic use and resistance	The hospital submits antibiotic use data to the National Health Services on Antimicrobial Use option. One or more actions should be implemented by the program and should implement one or more actions, like prospective audits and feedback, preauthorization, and facility-specific guidelines for common infections
6. Reporting information on antimicrobial use and resistance regularly to hospital personnel and leadership	Antibiotic use reports	Antibiotic use reports are provided at least annually to target feedback to prescribers. In addition, the antibiotic stewardship program monitors adherence to facility-specific treatment recommendations for at least one common clinical condition
7. Education—Educate clinicians and patients about antimicrobial resistance and optimal use of antimicrobials	Guidelines Podcasts Seminars and symposiums On line courses	Every hospital should have a policy to educate, train, and renew the knowledge and provide adequate time and resources to the staff

7 Key Elements of AMS in the ICU

The core elements for establishing antimicrobial stewardship in ICUs include all the elements mentioned in the CDC guidelines for hospital antimicrobial stewardship. However, antimicrobial stewardship in the ICU necessitates considering several particular considerations. ICU physicians should practice antibiotic stewardship in all critical care units because it is one of their fundamental competencies. The intensive care unit (ICU) provides optimal patient care through close collaboration among healthcare professionals. However, treating infections in ICU patients poses a unique challenge that requires consultation with experts in clinical pharmacology, infectious diseases, and microbiology to determine the appropriate antibiotics. Intensivists must be actively involved in discussions regarding the hospital's antibiotic formulary and contribute to developing guidelines for common conditions frequently encountered in the intensive care unit. It is imperative that an intensivist becomes an integral part of the antimicrobial management team in the hospital. To improve communication and make better treatment decisions, the ICU ward round should include an infectious diseases physician, clinical microbiologist, or clinical pharmacist. This approach has been proven to be effective [17] (Fig. 2).

Fig. 2 Key elements of AMS in the ICU

Management of Vocal Fold Palsy in Thyroid Surgery

12

Jayakumar R Menon and Manju E. Issac

12.1 Introduction

The close anatomical relationship of both the recurrent laryngeal nerve (RLN) and external branch of the superior laryngeal nerve (ExSLN) to the thyroid gland makes it vulnerable to injury during thyroid surgery. Despite all development in the knowledge of anatomy, skills, and techniques and inclusion of intraoperative nerve monitoring (IONM), thyroid surgery remains number one among the iatrogenic causes for recurrent laryngeal nerve injury! This chapter is an attempt to cover the presentation and management of vocal fold paralysis following thyroid surgery.

A vocal fold palsy following thyroid surgery is a result of a direct injury or traction on the nerve or a thermal injury from a cautery. Sometimes it is quite a surprise to the experienced surgeon who is well sure about the integrity of the nerve. A temporary or permanent vocal fold paralysis following thyroid surgery varies from 0 to 5.8% [1]. It is noted that the risk of injury was increased in revision thyroid cases for benign or malignant diseases equally. Most studies reveal that the complications are more with malignant pathology, hyperthyroidism, thyroiditis, retrosternal goiter, and Grave's disease [2, 3].

12.2 Recurrent Laryngeal Nerve (RLN)

The RLN supplies motor input to all the intrinsic muscles of the larynx except the cricothyroid and the sensory innervations to the glottis, sub-glottis, proximal trachea, and esophagus.

COURSE: The vagus nerve courses downward in the neck after exiting the jugular foramen. The RLN axons congregate at the ventromedial aspect of the

J. R. Menon (✉) · M. E. Issac
Dr. Jayajumar's Laryngology Group, Thiruvananthapuram, Kerala, India

C. G. Nair, S. J. Abraham (eds.), *Surgical Management of Thyroid and Parathyroid Diseases*, https://doi.org/10.1007/978-981-97-3774-1_12

vagus nerve before branching. It travels in the carotid sheath and on the right side of the neck it passes over the subclavian artery and loops around it and travels cephalad in the neck towards the larynx. In less than 1% of cases on the right side, its branches directly from the vagus to the larynx at the level of the cricoid cartilage, the non-recurrent laryngeal nerve. This is explained by abnormal embryological development. Normally RLN is pulled caudad in the neck along the sixth arch. The left sixth arch remains as ductus arteriosus and later the ligamentum arteriosum. The right sixth arch resorbs and the right RLN passes inferior to the right fourth arch which becomes the subclavian artery. In rare situations where the right fourth arch also resorbs during embryonic development; the right subclavian artery arises from the descending aorta and right RLN branches from the vagus in the neck and enter the larynx directly without looping around a vascular structure. There are only a few reports of non-recurrent laryngeal nerve on the left side.

RLN ascends towards the neck from lateral to medial and courses upwards in the tracheoesophageal groove. The course of the nerve in the tracheoesophageal groove is important because several studies show that the course is variable, and one such variability is its anterolateral position where it poses a significant risk for iatrogenic injury.

There is often a branch of the inferior thyroid artery which is in close anatomical relationship with the anterior and posterior branches of RLN and can place the nerve at significant risk if one of these vessels bleed.

The retro-thyroid segment of RLN is also susceptible to surgical trauma. The relation of nerve to the posterior suspensory ligament of Berry can be lateral, medial, or through the ligament which is another site of injury. The Tubercle of Zuckerkandl is found in 63% and is another site of injury in the retro-thyroid segment [4].

12.3 Nerve Injury and Synkinesis

Understanding the basic issues in a paralyzed larynx is important in taking the right step regarding intervention. The loss of signal conduction due to demyelination without disruption of axons is termed neuropraxia which is temporary. But in the case of axonotmesis due to crush injury, there is disruption of axons and during recovery regenerating axons enter the corresponding endoneurial tubes. In neurotmesis, the structure of the nerve is disrupted, and regenerating axons enter the endoneurial conduits which end up in the wrong muscle. This is synkinesis which leads to the contraction of antagonistic muscles. As a result, there is no vocal fold movement during phonation or inspiration or for airway protection or the larynx exhibits paradoxical movement. Neurotmesis is evident on re-exploration as neuroma formation and scarring of the nerve can be seen in stretch injury. Impaired vocal fold motion due to synkinesis is a severe and major cause (66–88%) of all paralysis [5]. Loss of vocal fold motion can happen as a result of inadequate or dysfunctional reinnervation or denervation. Since RLN carries both adductor abductor fibers, regeneration is more problematic.

12.4 Unilateral Vocal Fold Paralysis

A neuropathological study of unilateral vocal fold palsy at a tertiary care center revealed that thyroidectomy continues to be the single most common surgical procedure responsible for unilateral vocal fold palsy [6]. The symptoms of unilateral vocal fold palsy are related to glottis insufficiency. Hoarseness of the voice in the immediate post-operative period needs attention to have a look at the movement of vocal folds with a fiberoptic laryngoscope. The voice quality varies from mild hoarseness in a compensated case to severe aphonia. Some patients develop swallowing difficulties such as cough on taking food especially liquids. They will have a weak and inefficient cough. Some patients may complain of being short of breath, but careful history reveals they are not having any airway obstruction but experiencing breathlessness during the conversation because of the inefficient laryngeal closure and air leakage. They experience vocal fatigue due to the same mechanism. The increased respiratory effort and laryngeal muscle activity to force glottal closure will result in fatigue. Exertional activities which require glottic closure like lifting weight, pulling, and pushing may be difficult for some patients.

The quality of the voice in a poorly compensated vocal fold paralysis is harsh and breathy due to the air escape. With time the quality improves due to the development of supraglottic compensatory strategies, especially in males. These patients compensate with false cord adduction or apposing epiglottis to arytenoids in an anterior-posterior dimension. These compensations will result in a rough, low-pitched voice. In females, there is a compensatory strategy of developing unnatural high-pitched voice referred to as "paralytic falsetto" which is a result of contraction of the ipsilateral cricothyroid muscle. Some patients exhibit symptoms of aspiration such as cough on taking food especially liquid.

Fiberoptic laryngoscope examination will demonstrate various patterns of vocal fold immobility (Figs. 12.1 and 12.2). Indirect laryngoscopy with a mirror and rigid scope; 70 or 90° can also help, but not as good as fiberoptic laryngoscope. Tongue

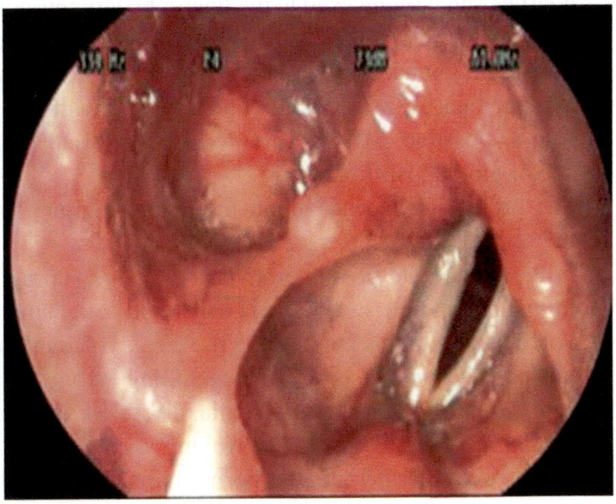

Fig. 12.1 Post thyroidectomy left vocal cord palsy

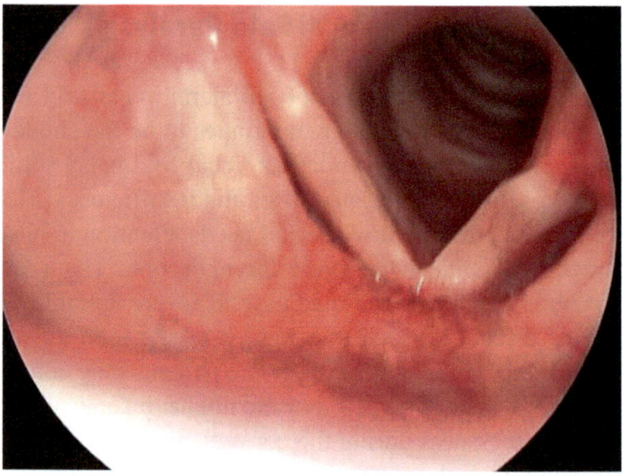

Fig. 12.2 Post thyroidectomy left EBSLN palsy

protrusion and grasping the tongue with gauze while performing rigid scopy affect the natural state of vocal fold mobility. The affected vocal fold exhibits different positions. The affected vocal fold may lie in any position depending upon the degree of re-innervation and synkinesis. It can either be an immobile vocal fold-paralysis, in abducted or paramedian position or a vocal fold with restricted adduction or abduction-paresis. When the affected vocal fold is at an adducted position hoarseness may improve over time because of the compensation of the unaffected vocal fold.

There are certain useful tasks applied during fiberoptic examination rather than checking vocal fold mobility alone. One such task is the "eee-sniff" maneuver in which the patient alternates between phonating the vowel "eee" and sniffing and by this a better judgment on the degree of impairment is possible. In patients with a compensatory supraglottic contraction or dysphonia plica ventricularis which can obscure the view of true vocal folds another phonatory task will help. This is called "unloading" described by Koufman removes the supraglottic phenomenon. The patient is instructed to hum through the nose or phonate with a sigh.

A strobe can be connected to the fiberoptic laryngoscope for a complete work-up. It may show incomplete closure or a large glottal gap or increased amplitude of vibration.

Maximum phonation time (MPT) is another test to evaluate glottic insufficiency. Here the patient is instructed to take a deep breath and phonate the vowel "ee" for as long as possible. MPT will be reduced in vocal fold palsy depending upon the degree of glottis insufficiency.

The prognosis depends upon the type of injury as well as the time elapsed since onset.

In the immediate post-operative period, the supraglottic swallowing technique (holding breath while swallowing) will avoid cough and make the swallowing better.

Voice therapy helps rehabilitate the weak voice in the postoperative period. And it can help the patient to avoid developing wrong compensatory techniques and subsequent muscle tension dysphonia. There are indirect and direct therapy procedures. Indirect therapy aims at improving vocal hygiene, counseling, and correcting posture. Direct therapy aims at normalizing the expiratory force, enhancing vocal projection, and optimizing medial compression of the affected vocal fold. Attempting high-pitched phonation and turning the head towards the side of paralysis and in opposite direction for paresis will enhance effective glottis closure and make the voice better. All these are employed in the scenario when the surgeon is sure about the integrity of the nerve. Methylcobalamin can be given. It has been claimed that it improves regeneration of the injured nerve. Intravenous steroids can be given to reduce the edema and speed up recovery.

While in an unfortunate event of a confirmed unilateral nerve injury such as transection of the nerve during surgery, non-selective reinnervation can be done which will retain the tone of the adductor muscles and help the vocal fold to attain adduction dynamically. Ansa cervicalis to strap muscles is the nerve of choice. The donor nerve is in the same field and it is a powerful adductor motor nerve. A segment of the desired length is dissected and released from the surrounding fascia and brought to the stump of recurrent laryngeal nerve above the point of injury near its entry point. Nerve to nerve anastomosis is done using 7-o nylon or PDS. Fat or a sheath of the vein is harvested to cover the area of microvascular anastomosis. Tissue fibrin glue is optional and can be used to reinforce the anastomotic site.

In unilateral paralysis, the aim is to medialize the paralyzed vocal fold to improve glottal closure. There are static procedures where we medialize the immobile vocal fold and dynamic procedures where nonselective reinnervation or selective reinnervation to the adductors are done (Fig 12.3). It can be combined with a static procedure like medialization. Non-selective re-innervation with medialization procedures like injection laryngoplasty or medialization thyroplasty is the procedure of choice when the vocal fold failed to normalize in 6–8 months.

In those who cannot wait for 6 months and in those with aspiration-related morbidity injection laryngoplasty is the choice. The medialization procedures are selected based on the position of paralyzed vocal fold from the midline as well as the time of vocal fold palsy. When the time of palsy is within 6–8 months an injection

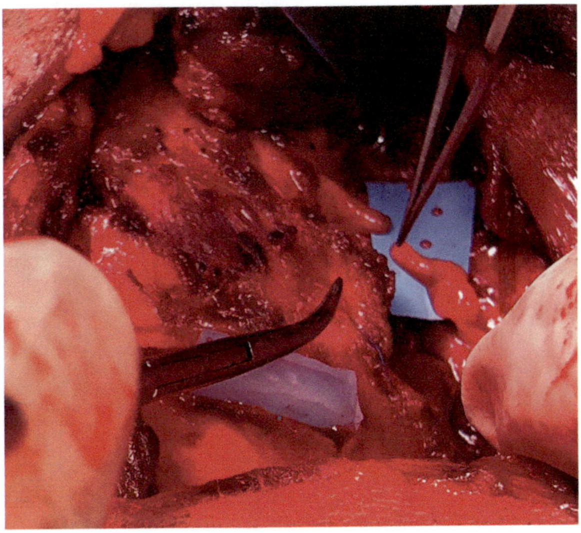

Fig. 12.3 Nonselective reinnervation ansa cervicalis and RLN. Thyroplasty implant also seen

laryngoplasty, i.e. endoscopic paraglottic fat injection which is done under GA is the choice. There are office procedures that can be done under direct vision of a flexible endoscope where hyalase or hydroxylapatite is injected into the paralyzed cord.

When the duration of paralysis is more than 6–8 months then a laryngeal framework surgery is the preferred treatment. Type 1 thyroplasty (medialization thyroplasty) where a silastic implant is carved based on measurement of the window made on the thyroid cartilage is inserted into the lower half of thyroid ala to medialize the affected vocal fold. The depth of medialization is assessed and the desired depth is measured and the implant is carved. There are modifications to this method where a titanium screw is used in some centers. Laryngeal framework surgeries are done under LA or MAC to assess the voice during surgery. Arytenoid rotation or adduction is combined in some situations where the paralyzed vocal fold is at an abducted position. A preoperative workup will decide whether the patient requires arytenoid adduction and medialization thyroplasty. A maximum phonation time of fewer than 6 s, the presence of aspiration, extremely breathy voice, and a negative medial compression test make the criteria for arytenoid adduction along with Type 1 or medialization thyroplasty (Figs. 12.4 and 12.5).

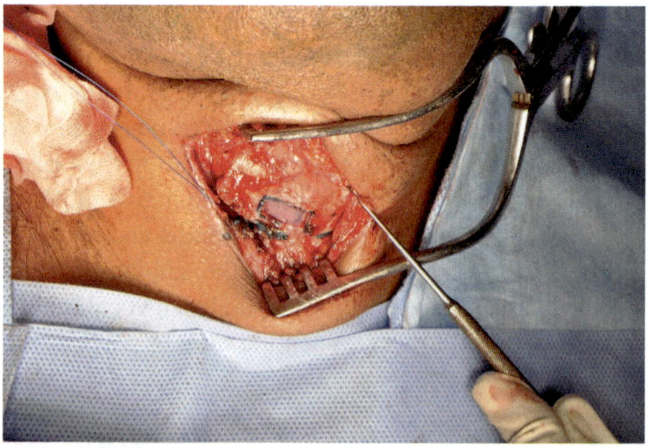

Fig. 12.4 Medialization thyroplasty right side—silastic implant

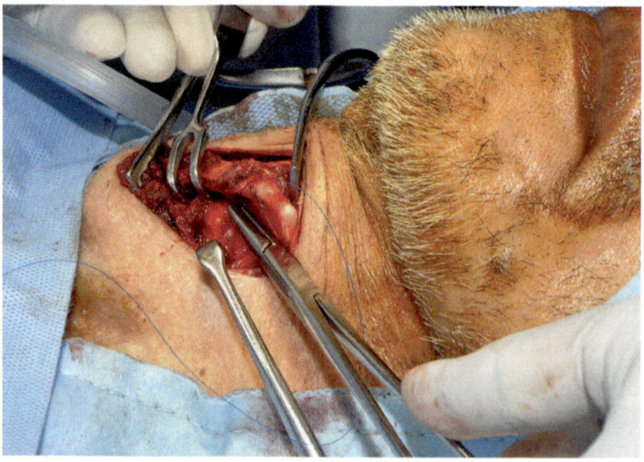

Fig. 12.5 Arytenoid rotation 4-0 prolene suture through the muscular process of left arytenoid

12.5 Bialteral Vocal Fold Palsy

Thyroidectomy continues to be the most common cause of bilateral vocal fold palsy [7]. The severity of clinical features depends upon the final position of the paralyzed vocal fold. In some, it will be a breathy voice and a weak cough to start with but eventually progress to snoring, noisy breathing, and saturation fall and end up in reintubation.

Bilateral vocal fold palsy following thyroidectomy requires knowledge to understand the emergency of different ranges of insufficient glottic spaces, experience to manage airway emergencies, and counseling the patient and bystander.

The initial part of the assessment requires flexible endoscopic evaluation of the status of the paralyzed vocal folds. This part is very important to correlate with the symptoms and other signs to decide on emergency management.

When the symptoms are mild and the vocal folds show some amount of movement or one side is having only a paresis one can still handle the situation with a propped-up position, O_2 mask, and C-PAP, intravenous steroids, methylcobalamin.

Moderate to severe symptoms not getting relieved with steroids and C-PAP should be planned for a temporary lateralization procedure like suture cordopexy of the least mobile or immobile cord.

Tracheostomy continues to be the most common treatment to provide airway without any compromise on voice and swallowing. The stigma associated with a tracheostomy is living with a breathing tube in the neck is the only disadvantage. Tracheostomy can be avoided in most cases since suture lateralization will provide enough space for the airway without much compromise on the voice. In the endoscopic suture lateralization procedure, 3-0 or 4-0 proline suture is taken through the thyroid cartilage, one above the vocal fold, another below the vocal fold, and endoscopically the sutures are then tied and pulled through the thyroid cartilage outer lamina and fixed over it. Here no tissue is removed. Suture lateralization provides enough glottis space without much compromise on voice. The only disadvantage is, it can cut through and reverse the effect of lateralization.

Arytnoidectomy is an irreversible method of lateralization where arytenoid cartilage is removed along with overlying mucosa. Coblator or laser can be used for this purpose. This is the recommended treatment for pediatric cases.

A permanent procedure like endoscopic posterior cordotomy (Kashima's procedure) assisted by laser or Coblator is required in an established case of BVCP who have waited for more than 6–8 months for the signs of recovery and in those to establish a permanent airway and to decannulate those patients who have been tracheostomized for critical glottis space (Fig 12.6). Patients with longstanding glottis stenosis can tolerate the insufficient glottic space. A permanent lateralization procedure like Kashima's surgery can be combined with adenoidectomy in some situations. The technique involves making a transverse cut just anterior to the vocal process with laser or Coblator. The tissue cut advanced along the same plane laterally till thyroid and cricoid inner perichondrium is reached. The ventricular band is partially removed for better visualization and to expose thyroid cartilage. Routine tracheostomy is not required in many cases. This can provide a better airway without much compromise on voice. The postoperative period is uneventful in most

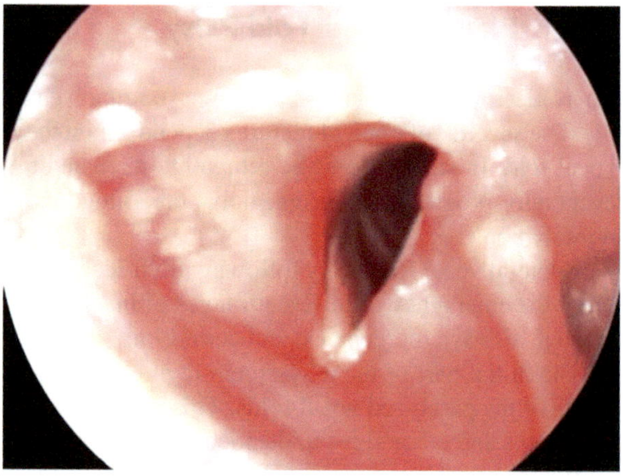

Fig. 12.6 Post thyroidectomy bilateral vocal cord palsy after Kashima's posterior cordotomy right side

cases. The cases without a tracheostomy can be monitored in the ICU for 24–48 h. Nebulization with adrenaline and intravenous antibiotics and steroids helps in reducing the edema at the operated site. Temporary aspiration especially to liquids has been noted in a few cases in the immediate postoperative period.

Most recently selective reinnervation for BVCP is getting appreciation since it can provide dynamic movement and lost tone to the vocal folds. The only disadvantage is the waiting period and in some patients a temporary or permanent lateralization procedure is required in the waiting period.

The phrenic nerve is the donor nerve for abductor muscle and the nerve to thyrohyoid or ansa can provide the adductor function to the adductors.

12.6 External Branch of Superior Laryngeal Nerve (ExSLN)

Because of difficulty in recognizing the subtle symptoms, injury to the superior laryngeal nerve is less documented. ExSLN or Nerve of Gali Curci supplies the cricothyroid muscle which tenses the vocal fold and helps in raising the vocal pitch. The nerve is at risk during upper pole dissection because of its close relationship with the superior thyroid vessels.

Anatomy: The superior laryngeal nerve (SLN) arises from the nodose ganglion of the vagus nerve and in its downward course at the level of the internal carotid artery it divides into internal and external branches. The internal branch runs inferior to the greater cornu of the hyoid, pierces the thyrohyoid membrane to innervate laryngeal mucosa. The external branch travels along the inferior constrictor muscle to reach the cricothyroid muscle. The nerve crosses superior thyroid vessels posteriorly near the superior pole of the thyroid gland.

Cernea et al. proposed their classification of ExSLN position in relation to the superior pole of the thyroid gland or the superior thyroid vessels and it is as follows [8]:

Type 1: The nerve crosses the superior vessels more than 1 cm above a plane tangent to the edge of the superior pole.

Type 2a: The nerve crosses the vessels within 1 cm above the upper limit of the superior pole.
Type 2b: This position is the most dangerous position for the nerve since it curves behind the superior pole below the upper edge.

In another study of the identification of ExSLN in large goiters, Cernia et al. demonstrated the higher frequency of Type 2b nerve in large goiters (54%). So, it is important that the identification of the nerve in these situations is very important to prevent its injury which is permanent and troublesome for the voice professionals [9].

There is another classification proposed by Kierner et al. in which there is a fourth category too where the nerve does not cross the trunk of superior thyroid vessels but runs dorsal to the artery until its ramification (14%) [10].

Friedman et al. proposed another classification based on the relation of ExSLN to inferior constrictor muscle at the junction of the cricothyroid muscle.

Type 1: The ExSLN descends its entire course superficially and laterally to the inferior constrictor muscle until it reaches the cricothyroid muscle.
Type 2: The nerve deepens into the lower portion of the inferior constrictor partially covered.
Type 3: The nerve runs its whole course covered and protected by inferior constrictor muscle [11].

Cricothyroid Muscle: The phonatory and respiratory functions of human cricothyroid muscle are controlled by different muscle bellies. Mu and Sanders, in their study on 30 autopsied adult humans, hemilarynges demonstrated the neuromuscular organization of cricothyroid muscle using microdissection, histology, and Sihler's stain. The result showed that the human cricothyroid was composed of three bellies. The rectus oblique and horizontal bellies and ExSLN were found to enter into the cricothyroid muscle as a single trunk (37.5%) or multiple as two to five branches (62.5%) and within cricothyroid muscle the ExSLN gave off three to seven branches to innervate rectus belly and one or two branches to supply oblique and horizontal bellies. ExSLN was found to give off branches to innervate ipsilateral Thyroarytenoid muscle (46%) and subglottic mucosa (67%) and connect with the RLN (25%) [12].

Wu et al. described the extension of the ExSLN that innervates the vocal cord [13]. In 12/27 (44%) of the specimens, a neural connection was found that from the medial surface of the cricothyroid muscle which enters into the lateral surface of the thyroarytenoid muscle.

This cranial nerve may be the nerve of the fifth branchial arch, a structure that has never been identified. Neuroanatomy of the larynx was explored in seven dogs to assess the communicating nerve from ExSLN to thyroarytenoid muscle by Nasri et al. and in three animals such innervations were identified [14].

The surgical damage is higher when the crossing of the ExSLN is lower, by transection, traction, or thermal damage. The cricothyroid artery supplies the ExSLN and the superior laryngeal artery supplies the internal branch of SLN [15]. A randomized controlled trial of visualization versus nerve monitoring of the ExSLN during thyroidectomy was conducted by Barczynski et al. and the use of IONM (INTRAOPERATIVE NERVE MONITORING) significantly improved the identification rate of the ExSLN during thyroidectomy as well as reduced the risk of early phonation changes after thyroidectomy [16].

ExSLN palsy is seen along with unilateral or bilateral RLN palsy or alone. Again, if it is ExSLN palsy alone it can present as unilateral or bilateral palsy.

12.7 Risk Factors for Surgical Injury

The situations of nerve damage are when a nerve gets entrapped during superior thyroid vessel ligature, stretching, transection, injury to the distal blood supply of the nerve, and resulting ischemia [17], or thermal damage by cautery. To prevent damage the following surgical techniques are proposed:

1. Individual ligature of the superior thyroid vessels near thyroid capsule
2. Visual identification of nerve prior to ligature
3. Use of electrical neural stimulation with IONM (intraoperative nerve monitoring)

12.8 Clinical Features

The symptoms of ExSLN palsy are inability to raise the pitch, lowered voice projection, and voice fatigue. Unilateral palsy is never diagnosed unless the patient is a professional voice user. The actual incidence of ExSLN palsy is not clear because of its minimal morbidity on the speaking voice and endoscopic findings are subtle and controversial. The main complaint will be an inability to raise the pitch.

The laryngoscopic findings are shortening of the aryepiglottic fold on side of paralysis, flabby vocal folds, hypomobility, and mucosal wave asymmetry, and phase asymmetry on applying strobe. Laryngeal behavior in unilateral superior laryngeal nerve paralysis was investigated in animal models and clinical cases by Tanaka et al. and concluded that a unilateral SLN palsy causes a rotation of the posterior glottis to the paralytic side when the unaffected cricothyroid muscle is markedly activated [18]. Rubin et al. demonstrated the hypomobility of the affected vocal fold on repetitive phonatory tasks [19]. Roy et al. demonstrated the deviation of the petiole of the epiglottis to the side of weakness during glissando maneuver in 60% of cases that were electromyographically confirmed [20, 21].

Tsai et al. demonstrated the common features of unilateral SLN paresis and paralysis and concluded that upon phonation there was ipsilateral vocal fold bowing and shortening, vocal process height asymmetry with the ipsilateral vocal process overriding the normal and the ipsilateral hyperadduction of the false vocal fold [22]. Many variables to the nerve injury such as a direct injury to cricothyroid muscle, cricothyroid joint dysmotility, coexistence with RLN palsy, compensatory action of other intrinsic muscles, and surgical scarring can coexist.

Laryngeal EMG is the most reliable method to diagnose EBSLN palsy [23, 24]. However, in routine practice, it is limited in its application because of the invasive nature and technical difficulty.

On acoustic analysis, there will be lowered fundamental frequency.

12.9 Treatment

Voice therapy is the mainstay of treatment. Voice therapy mainly aims at singing exercises, vocal hygiene tips, and stretching exercises [25].

Bilateral SLN palsy is also common and overlooked condition. This is again a worry some condition in a voice professional.

Here again, the primary step is voice therapy. And in those not responding well to voice therapy a Type 4 thyroplasty or cricothyroid approximation [26, 27] is the choice. A recent trial of non-selective reinnervation to the cricothyroid muscle using muscle-nerve-muscle with ansa-hypoglossi is promising [28].

References

1. Martensson H, Terins J. Recurrent laryngeal nerve palsy in thyroid gland surgery related to operations and nerves at risk. Arch Surg. 1985;120(4):475–7.
2. Lo C-Y, Kwok KF, Yuen PW. A prospective evaluation of recurrent laryngeal nerve paralysis during thyroidectomy. Arch Surg. 2000;135(2):204–7.
3. Chiang F-Y, et al. Recurrent laryngeal nerve palsy after thyroidectomy with routine identification of the recurrent laryngel nerve. J Surg. 2004;137(3):342–7.
4. Gauger PG, Delbridge LW, Thompson NW, Crummer P, Reeve TS. Incidence and importance of the tubercle of Zuckerkandl in thyroid surgery. Eur J Surg. 2001;167:249–54.
5. Zealer DL, Billante CR. Synkinesis and dysfunctional reinnervation of the larynx. In: Sulica L, Blitzer A, editors. Vocal fold paralysis. Berlin, Heidelberg: Springer; 2006.
6. Pavithran J, Menon JR. Unilateral vocal fold palsy: an etiopathological study. Int J Phonosurg Laryngol. 2011;1(1):5–10.
7. Benninger MS, Gillen JB, Altman JS. Changing etiology of vocal fold immobility. Laryngoscope. 1998;108:1346–50.
8. Cernea CR, Ferraz AR, Nishio S, et al. Surgical anatomy of the external branch of the superior laryngeal nerve. Head Neck. 1992;14:380–3.
9. Cernia CR, Nishio S, Hojaij FC. Identification of the external branch of the superior laryngeal nerve in large goiters. Am J Otolaryngol. 1995;16:307–11.
10. Kierner AC, Aigner M, Burian M. The external branch of superior laryngeal nerve: its topographical anatomy as related to surgery of the neck. Arch Otolaryngol Head Neck Sur g. 1998;124:301–3.
11. Friedman M, LoSavio P, Ibrahim H. Superior laryngeal nerve identification and preservation in thyroidectomy. Arch Otolaryngol Head Neck Surg. 2002;128:296–303.
12. Mu L, Sandres I. The human cricothyroid muscle: three muscle bellies and their innervations patterns. J Voice. 2009;23:21–8.
13. Wu BL, Sanders I, Mu L, et al. The human communicating nerve. An extension of the external superior laryngeal nerve that innervates the vocal cord. Arch Otolaryngol Head Neck Surg. 1994;120:1321–8.
14. Nasri S, Beizai P, Ye M, et al. Cross innervations of the thyroarytenoid muscle by a branch from the external division of the superior laryngeal nerve. Ann Otol Rhinol Laryngol. 1997;106:594–8.

15. Monfared A, Gorti G, Kim D. Microsurgical anatomy of the laryngeal nerves as related to thyroid surgery. Laryngoscope. 2002;112:386–92.
16. Barczynski M, Konturek A, Stopa M, et al. Randomized controlled trial of visualization versus neuromonitoring of the external branch of the superior laryngeal nerve during thyroidectomy. World J Surg. 2012;36:1340–7.
17. Yalcin B, Develi S, Tubbs RS, et al. Blood supply of the terminal part of the external branch of the superior laryngeal nerve. Surg Today. 2015;45:1160–6.
18. Tanaka S, Hirano M, Umemo H. Laryngeal behavior in unilateral superior laryngeal nerve paralysis. Ann Otol Rhinol Laryngol. 1994;103:93–7.
19. Rubin AD, Praneetvatakul V, Heman-Ackah Y, et al. Repetitive phonatory tasks for identifying vocal fold paralysis. J Voice. 2005;19:679–86.
20. Roy N, Barton ME, Smith ME, et al. An in vivo model of external superior laryngeal nerve paralysis. Laryngoscope. 2009;119:1017–32.
21. Roy N, Smith ME, Houtz DR, et al. Laryngeal features of external superior laryngeal nerve denervation: revisiting a century—old controversy. Ann Otol Rhinol Laryngol. 2011;120:1–8.
22. Tsai V, Celmer A, Berke G, et al. Videostroboscopic findings in unilateral superior laryngeal nerve paralysis and paresis. Otolaryngol Head Neck Surg. 2007;136:660–2.
23. Bevan K, Griffiths MV, Morgan MH. Cricothyroid muscle paralysis: its recognition and diagnosis. J Laryngol Otol. 1989;103:191–5.
24. Sulica L. The superior laryngeal nerve: function and dysfunction. Otolaryngol Clin N Am. 2004;37:183–201.
25. Dursun G, Sataloff RT, Spiegel JR, et al. Superior laryngeal nerve paresis and paralysis. J Voice. 1996;10:206–11.
26. Orestes MI, Chhetri DK. Superior laryngeal nerve injury: effects, clinical findings, prognosis and management options. Curr Opin Otolaryngol Head Neck Surg. 2014;22:439–43.
27. Shaw GY, Searl JP, Hoover LA. Diagnosis and treatment of unilateral cricothyroid muscle paralysis with a modified Isshiki type 4 thyroplasty. Otolaryngol Head Neck Surg. 1995;113:679–88.
28. El-Hashlan HK, Carroll WR, Hogikyan ND, et al. Selective cricothyroid muscle reinnervation by muscle-nerve-muscle neurotization. Arch Otolaryngol Head Neck Surg. 2001;127:1211–5.

13 Post-operative Care and Post-operative Management of Benign Thyroid Disease

Sabaretnam Mayilvaganan, Sai Krishna Vittal, Sai Vishnupriya Vittal, and Sarrah Idrees

13.1 Introduction

Advances in radiology and increased use of chest and neck imaging have led to increased discovery of asymptomatic incidentally detected thyroid nodules. The prevalence of nodules varies with different modalities, 2–6% with palpation, 60–65% with ultrasonography, 15% with computed tomography and 1–2% by position emission tomography. Although thyroid nodules are common as mentioned, the clinical significance is assessed by the presence of malignancy, compressive symptoms and thyroid dysfunction.

Benign thyroid diseases include euthyroid thyroid disease and hyperthyroid diseases (Fig. 13.1).

Euthyroid Diseases Include

1. Solitary thyroid nodule
2. Multinodular goitre
3. Dyshormonogenetic goitre

Hyperthyroid Diseases Include

1. Graves' disease
2. Toxic nodular goitre
3. AFTN (autonomously functioning thyroid nodule)

S. Mayilvaganan (✉)
Department of Endocrine Surgery, SGPGI, Lucknow, Uttar Pradesh, India

S. K. Vittal · S. V. Vittal
Endocrine and Breast Surgeon, Appolo Hospital, Chennai, Tamil Nadu, India

S. Idrees
Department of Endocrine Surgery, SGPGI, Lucknnow, India

C. G. Nair, S. J. Abraham (eds.), *Surgical Management of Thyroid and Parathyroid Diseases*, https://doi.org/10.1007/978-981-97-3774-1_13

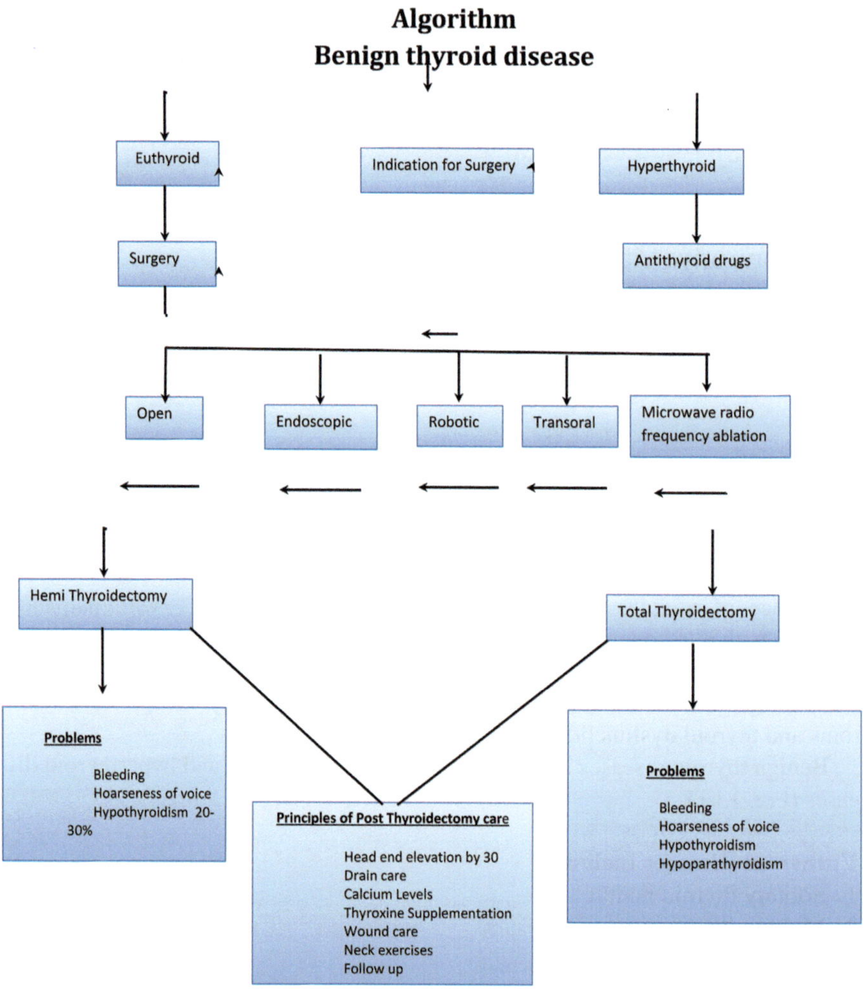

Fig. 13.1 Algorithm for treatment of benign thyroid disease

13.2 Post-operative Care

13.2.1 Immediate Post-operative Management

In the immediate post-operative period, the patient should be managed in a post-operative intensive care unit or High Dependency Unit (HDU) for at least 6 h and the most important parameters to be monitored include pulse rate, blood pressure measurement, PO_2 and drain monitoring. Bleeding and signs of airway compromise should be looked for in the immediate postoperative period. Post thyroidectomy bleed is most common during first 6 h and usually occurs within first 24 h. Patient has to be kept head end elevated by 30°. Oxygen supplementation is given for couple of hours and in malignant and long standing large goitres, one has to actively watch out for tracheomalacia and stridor following bilateral vocal card palsy. In such cases, it is preferable to have tracheostomy set and tracheostomy tube by the bed side. Thyrotoxic storm is rare nowadays but still can rarely occur when thyroidectomy is undertaken in uncontrolled hyperthyroid patients.

13.2.2 Post-operative Management

Once the patient is stable and has been shifted from post-operative intensive care unit, the patient can be nursed in sitting position. Depending on the type of anaesthesia and the patient's conscious level, oral fluids can be commenced after 6 h and semisolid food can be given if tolerated. The patient and their caregivers should be counselled regarding the operative procedure and also be instructed to immediately inform the nursing staff on duty and/or a member of the operative team in case of sudden increase in drain output. The most common complaint by the patient is the irritation of the throat (sensation of sore throat feeling) due to the endotracheal tube placement which can be taken care by steam inhalation. Vomiting in the post-operative period can be usually managed without difficulty but severe retching associated with vomiting can precipitate post-operative haemorrhage and can lead to disastrous effects.

In the post-operative period, the following issues have to be considered:

1. Head end elevation of patient
2. Active look out for post thyroidectomy bleed
3. Good post-operative pain relief
4. Hypocalcaemia management
5. Management of unilateral or bilateral recurrent laryngeal nerve injury
6. Scar management

It is advisable if logistics permit, to get a PTH (Parathyroid hormone) on the first postoperative day especially after total thyroidectomy. This is a good indicator of post thyroidectomy hypocalcaemia and aids in safe and early discharge of patients. Patients with low PTH levels (<5 pg/ml) may ideally be discharged with calcium and vitamin D supplementation.

In case of bilateral recurrent laryngeal nerve injury, tracheostomy is recommended. In case of unilateral nerve injury, patient should be counselled and also informed about neuropraxia of recurrent laryngeal nerve which is the most common cause and which usually resolves within few weeks. If it fails to get better within few weeks, the patient should be advised an otolaryngologist consultation for early vocal fold injection which can prevent aspiration and also improves the voice and quality of life of patients. Patients, who have undergone Total Thyroidectomy for benign thyroid disease, should be started on replacement doses of thyroxine.

13.2.3 Antibiotics

Usually, a single dose of third generation cephalosporin is given at the time of induction. However, in transoral thyroidectomy a minimum of three doses would be preferable (since it comes under the ambit of clean contaminated surgery).

13.2.4 Pain Relief

Analgesia is an important aspect of postoperative care of thyroidectomy patient. The patient is usually given intravenous infusion of paracetamol six hourly. Occasionally patient may in addition require either opioids or non-steroidal anti-inflammatory drugs (NSAIDS). Bilateral superficial cervical blocks with 0.25% Bupivacaine given at the time of induction has been shown to decrease postoperative pain and decrease analgesic requirements.

13.2.5 Ice Pack Dressing

In the immediate post operative period, ice pack dressing for the wound is advocated by some surgeons since it may reduce the degree of pain perception and may also reduce the analgesic requirement.

13.2.6 Head End Elevation

Studies have shown that 30° head end elevation of the bed resulted in maximum comfort for the patient. It also prevents venous congestion and facilitates maximum drainage of the wound. We advocate 30° elevation of the head.

13.2.7 Drain

Usually, a size 14 Romovac suction drain is used by most surgeons (Fig. 13.2). The drain is usually removed within 48 h. If the drainage is more than 20 ml/day, it is advisable to keep the drain till the output is <20 ml/day for a patient with Total Thyroidectomy. The nursing staff and the patient should be instructed to ensure that drain does not get entangled since the drain can be accidentally pulled out from the thyroidectomy wound. In case of sanguineous fluid (frank blood) in the drain, the wound is inspected to ensure that there is no active bleeding or obvious haematoma. In case of chylous drainage which is very unusual for thyroidectomy for benign disease, it is prudent to keep the drain in situ for longer duration.

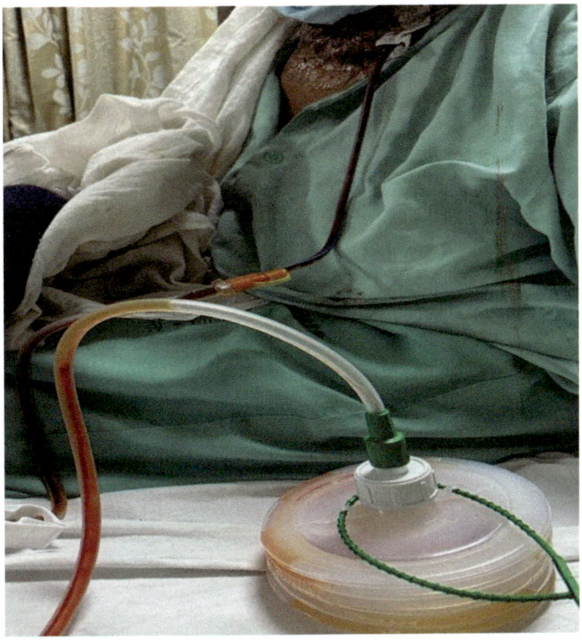

Fig. 13.2 Romovac drain in situ after total thyroidectomy

13.2.8 Hypocalcaemia

Clinical and biochemical monitoring for hypocalcaemia is advocated. Clinically, both symptoms and signs of hypocalcaemia should be assessed. The patient should be asked for numbness and paraesthesia around the mouth and the extremities. Signs of hypocalcaemia such as Trousseau's sign and Chvostek's sign should be sought (Figs. 13.3 and 13.4). "Trousseau's" sign monitoring every 6th hourly by nursing staff is a good way to elicit and clinically diagnose hypocalcaemia. In case of severe hypocalcaemia, calcium infusion is recommended (Fig. 13.5). Moderate hypocalcaemia is managed by intravenous bolus of calcium. Oral calcium and vitamin D supplementation is commenced simultaneously. Mild hypocalcaemia is managed with only oral supplementation of calcium and vitamin D.

13.2.9 Levothyroxine Dose

Thyroxine hormone replacement has to be titrated according to body weight of the individual. The usual replacement dose after total thyroidectomy is 1.6 mcg/kg for benign diseases of the thyroid [1]. Elderly patients (more than 60 years) may be administered with a lower dose (1.5 mcg/kg) to avoid cardiovascular problems. The absorption of an orally administered dose of Levothyroxine (T_4) is about 70–80% and peak concentration is reached 2–4 h after ingestion [2]. The half-life of T_4 is

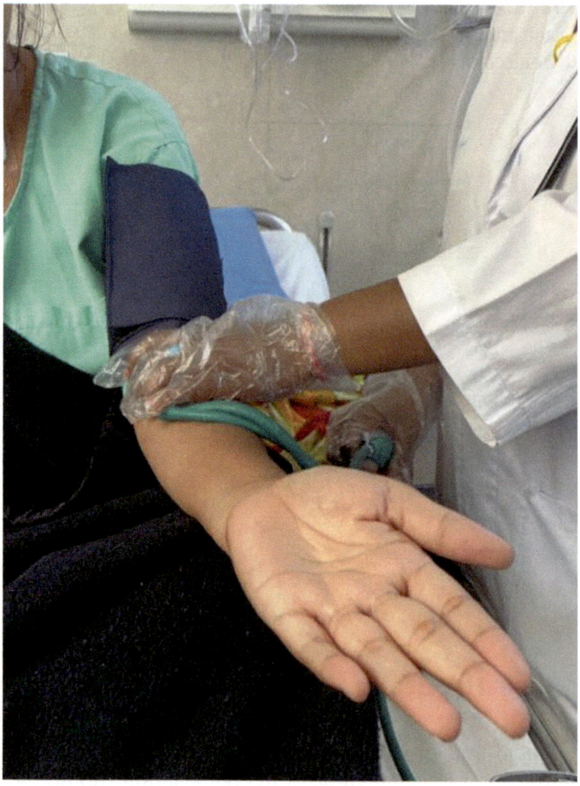

Fig. 13.3 Assessing Trousseau's sign in a patient

about 1 week. A fatty meal lowers absorption by 40% and coffee around 30%. Hence, patient is advised to take thyroxine tablet on an empty stomach as soon as the patient wakes up [3]. An interval of thirty minutes between the ingestion of tablet and consumption of beverage or breakfast is ideal.

13.2.10 Care of Scar Following Thyroidectomy

The skin is usually closed by either subcuticular technique or by the use of staples depending on the surgeon's preference. If skin is closed with subcuticular prolene or by skin staples it is usually removed on the sixth postoperative day. Adhesive strips (Steristrips) can then be applied for a week to reduce tension in the scar [4]. Subsequently, patient can be advised to apply moisturising cream or coconut oil to facilitate wound healing. Occasionally, silicone gel topical applications or silicone gel sheet is recommended to minimise unsightly scar [5].

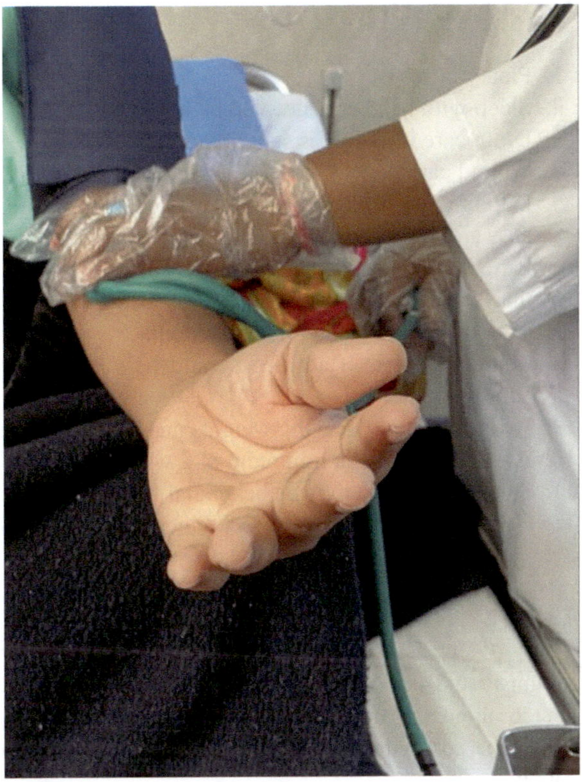

Fig. 13.4 Trousseau's sign positive in the same patient post-Total thyroidectomy

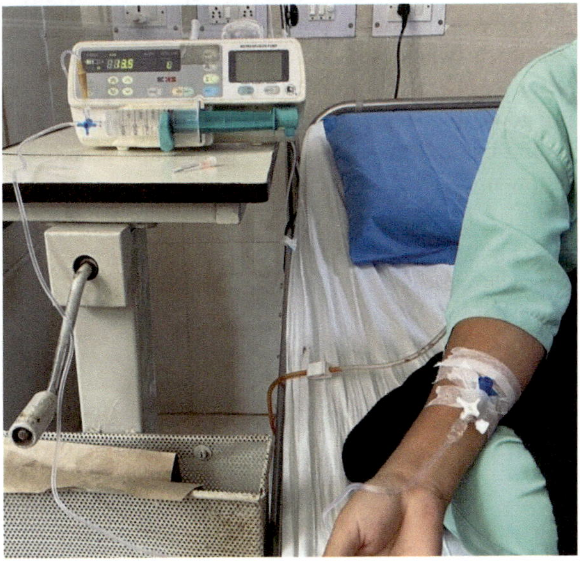

Fig. 13.5 Ongoing calcium infusion for a patient with hypocalcaemia after Total Thyroidectomy

13.2.11 Discharge Advice

1. Plenty of rest
2. Sleep with head end elevated at 30°
3. Drain care if patient is sent home with drain
4. Wound care
5. Patient is counselled regarding perioral and extremities paraesthesia and numbness and if present, advised to get in touch with the hospital
6. Patient is advised to get in touch with the hospital if he/she has fever of more than 101 °F
7. Take thyroxine tablet on empty stomach
8. Calcium tablets if advised should be taken a minimum of 3 h of thyroxine administration
9. Avoiding lifting heaving weights and strenuous exercises for 6 weeks
10. Avoid driving for a minimum of 2 weeks
11. Steam inhalation may be beneficial for a couple of weeks
12. Gentle neck exercises are recommended after the wound has healed (Fig. 13.6)

13.2.12 Follow-Up

Serum Free T4 and TSH levels are checked 6 weeks after hemithyroidectomy to decide about the need for thyroxine supplementation. Subsequently, the patient should be advised to come for follow up every 6 months for 2 years and yearly follow-up.

Women in reproductive age group should be counselled regarding titration of thyroxine doses during pregnancy.

Patients who have had Total Thyroidectomy are advised to come after 6 weeks to check their Free T4 and TSH levels to titrate the replacement dose of thyroxine. It is a good practice to measure the level of calcium as well. Subsequently, the patient should be advised to come for follow-up every 6 months for 2 years and yearly follow-up.

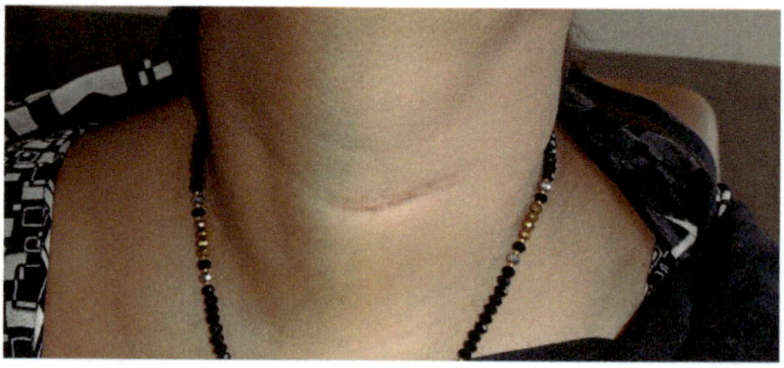

Fig. 13.6 Well healed scar of total thyroidectomy

References

1. Mistry D, Atkin S, Atkinson H, Gunasekaran S, Sylvester D, Rigby AS, England RJ. Predicting thyroxine requirements following total thyroidectomy. Clin Endocrinol. 2011;74(3):384–7.
2. Olubowale O, Chadwick DR. Optimization of thyroxine replacement therapy after total or near-total thyroidectomy for benign thyroid disease. J Br Surg. 2006;93(1):57–60.
3. Di Donna V, Santoro MG, De Waure C, Ricciato MP, Paragliola RM, Pontecorvi A, Corsello SM. A new strategy to estimate levothyroxine requirement after total thyroidectomy for benign thyroid disease. Thyroid. 2014;24(12):1759–64.
4. Mayilvaganan S, Bothra S, Bhargav PR, Chekavar A, Mattoo S, Rashid M, Agarwal A. Management of post-thyroidectomy scar? World J Endoc Surg. 2018;10(2)
5. Choi Y, Lee JH, Kim YH, Lee YS, Chang HS, Park CS, Roh MR. Impact of postthyroidectomy scar on the quality of life of thyroid cancer patients. Ann Dermatol. 2014;26(6):693–9.

14 Introduction to Surgical Treatments of Thyroid Cancers: Surgical Management of Advanced and Metastatic Thyroid Cancers

Riju Ramachandran and C. Gopalakrishnan Nair

14.1 Historical Perspective

Thyroid cancer was generally considered rare accounting for 1% of all human cancers. But many recent reports show a rapidly increasing trend in incidence in many countries. Analysis of incidences of thyroid cancer in representative countries in five continents for the period from 1973–1977 and 1998 to 2002 showed an increase in incidence among both males (48%) and females (66.7%) [1]. African countries showed the lowest incidences and the highest incidences and significant upward swing were in Israel and United States. The increase in incidence could also be due to frequent ultrasound and CT imaging, which revealed hitherto undiagnosed small lesions.

The histological features and the biological behavior differ grossly among the spectrum of thyroid cancers. One end of the spectrum is a slow-growing differentiated thyroid cancer of follicular cell origin with favorable outcome and the other end is fast-growing undifferentiated thyroid cancer with a very dismal outcome. It also encompasses cancers of follicular cell origin, parafollicular cell origin, and stromal cell origin.

Fourth edition WHO classification of endocrine cancers has introduced significant changes classification of thyroid cancer [2, 3]. Thyroid cancers of follicular cell origin account for 85–90% of all thyroid malignancies and include differentiated thyroid cancers (DTC), poorly differentiated thyroid cancers (PDTC), and anaplastic or undifferentiated thyroid cancer (UTC). The latter two constitute 2–5% and 1.7% of all thyroid cancers, respectively.

R. Ramachandran (✉)
Department of Surgery, AIMS, Kochi, Kerala, India

AG-1, Sterling Sarovar, Kosseri Lane, Edapally, Kochi, Kerala, India

C. G. Nair
Professor of Surgery, Endocrine Surgery Division, Amrita Institute of Medical Sciences and Research Centre, Kochi, Kerala, India

C. G. Nair, S. J. Abraham (eds.), *Surgical Management of Thyroid and Parathyroid Diseases*, https://doi.org/10.1007/978-981-97-3774-1_14

The DTC includes three different histologic subtypes, papillary thyroid cancer (PTC), follicular thyroid cancer (FTC), and Hürthle cell cancer which was previously considered a variant of FTC.

PTC is the commonest of DTC and includes 14 variants in addition to the conventional type and also has a new additional accepted entry; the Hobnail variant. Tall cell variant, columnar cell variant, Hobnail variant, and solid variant have higher chances of adverse outcome. The newer classification has recognized Encapsulated Conventional PTC (completely covered by the capsule) as a separate variant with a favorable outcome even when regional node metastases are present. Encapsulated follicular variant of papillary thyroid neoplasm was brought to limelight in 2016 [4] and the 4th edition of WHO classification downgraded it to the benign category and renamed it as non-invasive follicular thyroid neoplasm with papillary nuclear features (NIFTP) [3].

RET proto-oncogene rearrangement was the earliest known genetic background of PTC. Mutations of RAS genes that are involved in the regulation of MAPK and P13K/AKT pathways are generally found in follicular variants of PTC. BRAF gene is also signaling through MAPK pathways which control the proliferation of cells. A point mutation in BRAF accounts for the genetic basis of about 45% of PTC and the phenotype expressions show aggressive pathological features. Telomerase Reverse Transcriptase (TERT) promoter mutations are found in approximately 11.3% PTC and when coexisting with BRAF V600E exert a synergistic effect and induce aggressive histological features and adverse outcomes [4]. Follicular variants include two different histologic expressions; the encapsulated neoplasm and the infiltrative tumor and reveal RAS mutations or PAX8/PPAR gamma rearrangements.

Follicular Thyroid Cancer (FTC) is defined as a malignant tumor without the nuclear characteristics of PTC and exhibits no molecular genetic finding that can reliably predict the preoperative diagnosis in cytology or the outcome. Genetic signatures of FTC are predominantly RAS mutations or PAX/PPARy rearrangements [5] but show differences in prevalence based on the geography or ethnicity of the population [6]. Most FTC patients in Asian countries have a greater than 30% prevalence of RAS mutations (the most frequent being NRAS mutation) unlike western countries which exhibit low prevalence. On the contrary PAX8/PPARy rearrangement is a common genetic change of FTC in western countries [6].

Traditionally, FTC had two subtypes: the minimally invasive and widely invasive variants [7] but the 4th edition WHO classification has changed this classification, introducing a third category [2] considering the importance of vascular invasion.

1. The minimally invasive FTC (shows capsular invasion only).
2. Encapsulated angioinvasive FTC. This category includes a subtype of limited vascular invasion (invasion <4 vessels) and another with extensive vascular invasion (>4 vessels invasion).
3. Widely invasive FTC.

Hürthle cell carcinoma (HCC): WHO 2017 classification grouped Hürthle cell carcinoma as a separate entity with the three subtypes similar to FTC. Hürthle cells are large, polygonal cells with eosinophilic, granular cytoplasm and have abundant mitochondria. Oncocytic metaplasia occurs in many organs due to various forms of metabolic stress.

Hürthle cell tumors are composed of >75% oncocytic cells without characteristic nuclear features of PTC and are characterized by capsular and vascular invasions [7]. Like the other forms of DTC, HCC is more common in females and is frequently found to harbor concurrent regional and distant metastases at the time of diagnosis [8].

Genetic changes in Hürthle cell cancer RAS mutations or PAX8/PPAR gamma rearrangements but these mutations can also be found in benign Hürthle cell tumors. Mutations of the p53 or PI3kinase genes are seen in aggressive Hürthle cell cancer variants [9]. Molecular markers of HCC are very variable and in general include RAS, EIF1AX, PTEN, TP53, CAN, and mtDNA.

Histologically Hürthle cell cancers are classified as: (1) The minimally invasive HCC (capsular invasion only); (2) Encapsulated angioinvasive HCC. This category includes two subtypes: (a) limited vascular invasion (invasion <4 vessels) and (b) extensive vascular invasion (>4 vessels invasion); (3) Widely invasive HCC.

A preoperative confirmed diagnosis is not usual since the ultrasound has no characteristic features or cytology does not detect vascular or capsular invasion which are hallmark features in arriving at the diagnosis of HCC. Hürthle cells are seen in inflammatory diseases like Hashimoto's thyroiditis and are frequently encountered in aspiration cytology. Bethesda category III and IV cytology with predominant Hürthle cells are usually reported as Hürthle cell lesions and neoplasms.

Oncocytic cells are not concentrating iodine adequately and so are not effectively treated by I^{131}. A diagnostic hemithyroidectomy is an option for lesions <4 cm with no clinical or sonological evidence of extra-thyroid invasion. When the diagnosis of HCC is available the total thyroidectomy is the primary treatment and R0 resection is always attempted since there is no effective adjuvant [10]. Locoregional spread is common and so therapeutic level VI and lateral neck node dissection are concurrently performed when indicated.

Advanced age, large tumor size, male gender, extra-thyroid invasion, and recurrence following primary surgery indicate poor prognosis. The overall 5-year and 10-year survival rates of HCC with no loco-regional and distant metastases are 85.1% and 71.1%, respectively [11].

14.2 The Poorly Differentiated Thyroid Carcinoma (PDTC)

PDTC stays between DTC and Undifferentiated Thyroid Carcinoma (UTC) in clinical features and biological behavior. Both PDTC and UTC may occur de novo or as a progression from preexisting DTC.

PDTC was first recognized as a separate clinical and histologic entity in the WHO classification of tumors in 2004 [7]. The unique histologic feature of PDTC known as "cellular nests" or the insular groups was described earlier, but later non-insular forms with trabecular or solid patterns were also recognized as histological features of PDTC. Rarely, both histological features coexist in the same tumor [3].

The first major suggestion regarding a consensus in the correct histological definition of PDTC was from Memorial Sloan Kettering Cancer Center (MSKCC). The key features to predict PDTC were high mitotic rate and/or tumor necrosis [12]. The present-day consensus about diagnostic criterion (Turin-PDTC) included "solid/trabecular/insular pattern of growth, absence of conventional nuclear features of papillary carcinoma, and with at least one of the following features: convoluted nuclei, mitotic activity ≥3/10 high power microscopic fields (HPF), and tumor necrosis" [13]. But, the coexistence of a poorly differentiated area within PDTC will significantly affect patient prognosis [14] and so thorough scrutiny of all areas is necessary.

The genetic mutations behind PDTC are BRAF (27%) and RAS (24%) but TERT promoter mutations are the most common alterations in PDTC (40%). RET rearrangement seldom progresses to PDTC status [15].

Clinical features: PDTC is generally found in the 6th and 7th decades of life and is more frequently seen in females (sex ratio of 1:2). The majority of patients are diagnosed with advanced disease and the tumors usually have transgressed beyond the thyroid capsule with metastases in regional lymph nodes. There is a high chance of concurrent distant metastases at the time of initial diagnosis of PDTC [16]. Elderly patients with extensive locoregional disease disproportional to the duration of disease are clinically suspicious PDTC.

Preoperative diagnosis is not always easy due to the heterogeneous nature of cytological features and the rarity of the disease. Features suggestive of PDTC include severe crowding of cells with a high nuclear-cytoplasmic ratio and insular/solid/trabecular morphology (American Cancer Society 2009) [17]. A high index of suspicion helps the clinician and pathologist to arrive at a preoperative diagnosis.

Total thyroidectomy with complete clearance of locoregional disease preserving vital structures with the possible extent is the primary treatment with curative intent. Adequate workup to assess the extent of disease like CT and endoscopy of upper aero-digestive tracts will help to plan the disease and to comprehensively counsel the patient for adequate surgery. Details are discussed later in this chapter.

There is controversy regarding radioiodine avidity in PDTC patients but, I^{131} ablation is generally recommended [18]. External beam radiotherapy (EBRT) was found effective in controlling local disease and thereby improves overall survival also. So EBRT is advisable with all T4 lesions and shall follow I^{131} ablations even when thyroidectomy was adequate [18]. Majority of studies showed 5-year disease-free survival of PDTC is between DTC and ATC ranging from 62% to 70% [18].

14.3 Undifferentiated Thyroid Cancer (UTC)

UTC is a rare form of thyroid cancer from follicular cells with a dismal prognosis and median survival ranges between 3–5 months after the diagnosis. The majority of patients are above 70 years and present with a rapidly growing central neck mass.

Blood-born metastases in the lungs, bone, or brain are usually present at the time of initial presentation [19]. TP53 mutations are considered as a genetic hallmark of ATC but other mutations that are frequently seen in DTC such as BRAF and RAS may also occur in UTC.

UTC is generally considered stage 4 disease considering the unfavorable outcome, but the median survival was longer in thyroidectomized patients (6.6 months) compared to that for non-surgical patients (2.1 months) [20]. Diagnosis of UTC is arrived in more than 50% by routine cytology and immunohistochemistry studies on cell blocks. Core biopsy should be undertaken with extreme caution since resultant bleeding could be fatal. Genetic studies are recommended since targeted therapy is an option at a later stage. Concurrent occurrence of distant metastasis at the time of initial diagnosis is over 50% and so advanced imaging like FDG-PET CT is advisable before operation. Locoregional spread is better assessed by regional CT [1]. In view of relative gloomy outcome, patients and their relatives are counseled for treatment explaining the details.

Stage IVA lesions and Stage IV B are generally considered operable but T4 B lesions are generally inoperable since the carotid artery may be at risk. Generally, intrathyroidal ATC, or extrathyroidal extension without invasion of aerodigestive tract ideal candidate for surgery. Thyroidectomy, without microscopic margin positivity (R0) or with microscopic margin positivity (R1) is optimal surgery for UTI. In the case of nodal metastases, block dissection is performed with thyroidectomy [19, 20]. Despite all possible preoperative assessments and best surgical expertise, approximately <60% of surgical exploration is successful [21].

Postoperative external radiation is found to prolong overall survival. After the genomic characteristics are known targeted therapy is under trial and patients who have a BRAFV600E-positive ATC patients expressed encouraging results [19, 20].

Medullary thyroid cancer forms <2% of all thyroid cancers and originates from parafollicular cells. Medullary thyroid cancer is neuroendocrine cancer and includes a sporadic category (70%) and a less common familial variant. Familial MTC expresses mutation in RET gene and could be part of MEN-2 syndromes or familial MTC syndrome. Screening for genetic germline RET mutation helps to perform prophylactic thyroidectomy.

14.3.1 Risk Stratification

The 8th edition of the AJCC staging system has made changes in the risk stratification of thyroid cancers [22]. Details of the tumor, node metastases descriptions are noted in the following tables:

Tumor characteristics

T stage	Size
TX	Not assessed
T0	No evidence
T1a	<1 cm tumor confined to the thyroid
T1b	1–2 cm tumor confined to the thyroid

T stage	Size
T2	2–4 cm tumor confined to the thyroid
T3a	>4 cm tumor
T3b	Any size invading strap muscles, SCM, or surrounding soft tissue
T4a	Tumor with posterior invasion to larynx, trachea, or esophagus
T4b	Extension to pre-tracheal fascia, carotid sheath

Lymph node characteristics

Nodal status	
Nx	Not assessed
N0	No evidence of metastases
N0 a	No cytological or histological evidence
N0 b	No radiological evidence
N1 a	Lymph node metastasis of level VI station
N1 b	Lymph node metastases of deep cervical chain

Metastases

Metastases	
M0	No evidence of metastasis
M1	Confirmed metastasis

AJCC staging has included age as a criterion in stage grouping of DTC and noted 45 years as the cut-off. This cut-off level is increased to 55 years in the 8th manual and patients below this age are grouped to stage 1 or stage 2 only.

AJCC staging of DTC for patients below 55 years

Stage I	T-1 to T4; N0 to N-1; M0
Stage II	T-1 to T4; N0 to N-1; M1

AJCC staging of DTC for patients above 55 years

Stage I	T1; N0, M0 T2; N0, M0
Stage II	T1; N1, M0 T2; N1, M0 T3a; N1, M0 T3b; N1, M0
Stage III	T4a, any N, M0
Stage IVA	T4 b, any N, M0
Stage IVB	Any T, any N, M1

Stage grouping of DTC above 55 years stresses the significance of metastases in regional lymph nodes irrespective of the anatomical station. The extrathyroidal invasion has gained additional significance in TIII/TIV tumor stratification. The prognostic factors significantly influencing overall survival are lymph node metastases and extrathyroidal invasion. Anaplastic thyroid cancers have a poor prognosis and are always categorized as stage IV disease.

AJCC stage grouping of undifferentiated thyroid cancer

Stage IVA	Primary tumor without lymph node metastases and distant metastases
Stage IVB	Primary tumor with lymph node metastases but no distant metastasis
Stage IVC	Primary tumor with distant metastasis

Stage grouping of medullary thyroid cancer is not influenced by age and details are noted below.

AJCC stage grouping of medullary thyroid

Stage	Tumor	Node	Metastases
Stage I	T-1	N0	M0
Stage II	T2- T3	N0	M0
Stage III	T-1, T-2, T3	N1a	M0
Stage IVA	T-1; T-2; T3	N 1b	M0
	T4a	N0	M0
Stage IVB	T4	N0	M0
Stage IVC	Any T	Any N	M1

14.3.2 Risk Stratification of American Thyroid Association

The commonest form of thyroid cancers are DTCs which show slow progression and a high rate of overall survival. The treatment modifications are aimed to prevent recurrences and to increase disease-free survival. AJCC staging essentially estimates the overall survival and not the disease-free survival.

American Thyroid Association has proposed risk stratification in 2009 guidelines based on various histological characteristics [23]. This was modified in 2015 incorporating genetic features and postoperative serum thyroglobulin levels and categorized the DTC patients into low risk, intermediary risk, and high-risk categories [24]. The system is dynamic and patients shall move from one to another during the course of treatment enabling the clinician to modify management protocol.

The low-risk category includes PTC without extra-thyroid invasion, resected completely with no macroscopic residue, no aggressive variants, no vascular invasion, no radioiodine detected metastatic foci outside thyroid bed, maximal lymph node involvement <5 micro-metastases. Follicular thyroid carcinoma with capsular invasion or with minimal (<4 foci) vascular invasion is also included in the low-risk category.

Intermediary risk group: This group includes PTC with microscopic invasion of peri-thyroid tissue, histological aggressive variants, I^{131} detected foci in the neck, N1 involvement (>5) but the size <3 cm in largest dimension and multifocal PMC with extra-thyroid extension (with or without BRAF mutation).

High-risk group: Gross ETE with residual tumor, distant metastases, metastatic nodes larger than 3 cm, postoperative high serum thyroglobulin suggestive of metastases. FTC with extensive vascular invasion. This categorization is more predictive of the risk of recurrence since there is 78–91% 5-year disease-free survival in low-risk group compared to 14–31% in high-risk group (ATA Guidelines 2015).

14.4 Primary Treatment of DTC

The Bethesda III and IV categories are a diagnostic dilemma in Indian conditions, the scenario being further interfered by lack of widespread availability of molecular studies, presence of multiple nodules, and increased prevalence of thyroiditis. Diagnostic thyroidectomy is opted by many clinicians when preoperative evaluations fail to arrive at a proper diagnosis. Diagnostic lobectomy is the procedure of choice but many patients opt for total thyroidectomy for fear of a second operation and due to financial issues.

Thyroidectomy is the primary procedure of curative intent undertaken in a proven case of TC. So, the surgeon should attempt to remove the entire thyroid gland and suspected malignant tissues peri-thyroid area which may include soft tissue infiltration and metastatic lymph nodes. Adequate care is exerted to preserve vital structures so that quality of life is least affected. Remnants in the thyroglossal tract, at the sites of ligation of blood vessels, embedded in the ligament of Berry, nodules from tubercle of Zuckerkandl and the-thymic ligament rests are areas of concern for complete removal of the gland. When primary surgical treatment is incomplete the chance of disease-free survival diminishes. Adjuvants are not substituting and so adequate operation is of paramount significance.

The question of operating thyroid microcarcinoma is debated and active surveillance is advocated in selected categories. Microcarcinoma with a 2 mm rim of normal thyroid tissue all-around without regional nodal metastases are eligible candidates for surveillance [25, 26]. Selection criteria may be modified in the future with the addition of genetic mutations. Active surveillance is possible only when participants show good compliance with follow-up protocol and is difficult to achieve in rural India.

Traditionally total thyroidectomy was recommended when the diagnosis of thyroid cancer was available. But the well-established advantage of recurrence-free survival of total thyroidectomy over hemithyroidectomy was questioned in a large size study that included the size of the tumor, gender, age, and ethnicity were also included in the analysis [27]. There was no statistical difference in overall survival in patients undergoing total thyroidectomy versus hemithyroidectomy when tumor size was <4 cm. Apart from that advanced age, male gender, the low socioeconomic background also influenced the overall survival.

Total thyroidectomy is the procedure of choice when maximal tumor size exceeds 4 cm, extra-thyroid invasion is observed, cervical lymph node metastasis is present. Total thyroidectomy is advisable for patients who had prior external radiation or has a positive family history of thyroid cancer. British Thyroid Association recommends vascular invasion as an indication of total thyroidectomy. In other words, a hemithyroidectomy is an option in patients with tumors <4 cm size, confined to the thyroid, and with no local or distant metastasis. But the patient should be counselled for completion thyroidectomy if the final histology shows adverse features.

Hürthle cell carcinoma and medullary thyroid cancer are definite indications for total thyroidectomy. Thyroidectomy for these malignancies needs to be done with care and precision since adjuvants like radioiodine treatment and TSH suppression therapy have no role in the treatment. The tubercle of Zuckerkandl is the point of fusion of contribution from the ultimobranchial body and probably has a maximum accumulation of C cells. So meticulous dissection to remove all thyroid tissue, preserving laryngeal nerves and the superior parathyroid, is necessary.

Hemithyroidectomy is adequate only for minimally invasive FTC. Gray area persists regarding encapsulated vascular invasion type of FTC and aggressive variants of PTC. This selection criterion excludes a large number of DTC from total thyroidectomy. Active surveillance is essential for patients undergoing hemithyroidectomy and hence patient compliance is ensured.

14.5 Thyroidectomy with Advanced Locoregional Spread

Considering AJCC stage grouping, locally advanced thyroid cancers include stage III and IVA diseases. But many such patients are clinically asymptomatic or minimally symptomatic. Hoarseness of recent onset is a significant symptom that prompts many patients to seek medical aid. Recurrent episodes of dry cough and hemoptysis indicate tracheal invasion but difficulty in breathing is experienced when the tracheal lumen is reduced more than half by invasion. Dysphagia and regurgitation are characteristics of esophageal infiltration. But a large majority of patients are clinically free of symptoms. Thyroid nodule with relative fixity to surrounding tissues directs the clinician to perform imaging to exclude the possibility of infiltration.

Surgical procedure with a curative intent is also the treatment of option for locally advanced thyroid cancer. Macroscopic residual tissue is a crucial determinant factor predictive of poor prognosis. But thyroid gland is located close to the proximal aerodigestive tract and damage to it leads to poor quality of life. So, surgery is planned to remove complete thyroid tissue and the tumor with the least damage to the trachea, larynx, and esophagus.

Ultrasound scan is always done as the first line imaging in all thyroid nodules but should be repeated in the immediate preoperative setting and preferably done by the operating surgeon himself or in association with an experienced sonologist. Ultrasound is used to evaluate the primary tumor and neck nodes. The location and extension of tumor to neighboring structures and nodal stations are reassessed. When the invasion of the trachea, esophagus, and major blood vessels are suspected, higher imaging like contrast-enhanced CT or MRI is essential. CT and MRI have almost similar sensitivity and specificity in finding tracheal and esophageal infiltration. Authors prefer contrast-enhanced helical multiplanar CT harvesting axial, sagittal ad coronal images from skull base to the carina.

Upper GI endoscopy is performed when patients have dysphagia or imaging shows esophageal infiltration. Video laryngoscopy is routinely performed before thyroidectomy but detailed endoscopy of the larynx and trachea helps to plan the extend of resection and reconstruction. The authors perform on-table bronchoscopy before the surgical procedure.

Primary surgery with a curative intent requires the removal of all involved structures with no macroscopic residual tissue. Such R0 resection may lead to gross difficulty in phonation and swallowing and should be reserved for highly motivated candidates. Management of such patients is performed by team effort comprising endocrine/head and surgeon, cardiothoracic surgeon, trained anesthesiologist, physical medicine rehabilitation specialists.

14.6 Tracheal Infiltration

Two well-known classifications of tracheal infiltration are given below and are based on pathological studies of papillary thyroid cancers. There are minor differences in the classification proposed by Czaja and McCaffrey and those of Shin et al. [28, 29].

14.6.1 Grading of Tracheal Involvement

Stage	Czaja and McCaffrey [28]	Shin et al. [29]
I	Extra-thyroid invasion and adhesion to the trachea	Invasion of perichondrium
II	Invasion of the trachea, the mucosa is free	Invasion of cartilage
III	Gross invasion and excision require mucosal disruption, but mucosa is free	Invasion of submucosa
IV	Invasion of all layers of trachea	Invasion of mucosa

Dense adhesion to the trachea is not uncommon since thyroiditis is frequently associated with DTC. Type I and to some extend type II tracheal involvement are removed successfully by tangential tumor excision which involves sharp dissection to shave of the thyroid using number 16 surgical blade. The thick fibrous sheath between the cartilaginous rings could be eroded and hence should be closely inspected for residual tissue. The shaving maneuver is cautiously performed on the lateral aspect since blood supply is derived from the inferior thyroid artery as ascending branches. Second, RLN courses close to the trachea-esophageal grove. Frozen section studies to assess the completeness of excision are not always successful.

There is a strong argument that tumors can transgress the fibrous sheath and microscopic residual disease is possible in DTC with ETE [11]. But Kim et al. could not observe significant difference in incidence of recurrence with more radical excision [30].

But type III and IV invasion of the trachea requires full-thickness resection which could be a window resection or circumferential resection. The majority of patients are treated with window resection and primary closure with or without tracheostomy. The trachea is mobilized laterally taking care to preserve recurrent laryngeal nerve and arterial supply. The lateral extension of the window facilitates transverse closure. Primary closure of large size window is not always possible and require reconstruction with pedicled or free myo-cutaneous flaps [31].

More extensive tracheal infiltration requires circumferential resection and primary closure. An extend of 4–7 rings are resected circumferentially and primary anastomosis of ends is done using interrupted absorbable sutures. Adequate mobilization of the trachea is necessary to reduce tension at the anastomotic site by the release of the trachea at the carina and suprahyoid release of the larynx. Mobilization of the carina is done by open approach or endoscopically [31, 32]. The larynx slips down to about 2 cm by dividing attachments of mylohyoid, genioglossus, and geniohyoid muscles from the superior surface of the hyoid bone.

Absorbable suture materials (2-0 polyglactin) are used in the fashion "outside to inside and returning to outside" starting from the posterior midline. Interrupted sutures are tagged with small artery clamps and tied later starting from the lateral aspects. After closure air leak is tested by pouring saline over the trachea. The suture knots are always outside the lumen [32]. Chin is sutured to the anterior chest wall to minimize the tension at the suture line and is retained for a week. The patient is usually weaned off the ventilator in 4–5 days but the neck movement is restricted for a long time.

All patient requires training for swallowing food, phonation and may have to avoid extension of the neck for a long time. It is a formidable procedure requiring team effort from endocrine surgeons, cardiothoracic surgeons, trained nursing staff, and physiotherapy rehabilitation specialists.

Local invasion to the larynx is rare but synchronous with tracheal involvement is occasionally found. Hemi laryngectomy or total laryngectomy with reconstruction is the treatment of choice [33]. The quality of life after laryngectomy is generally not satisfactory considering factors such as reduced physical health, reduced ability to maintain social relations. They tend to be more depressed when compared to normative samples [34].

14.6.2 Recurrent Laryngeal Nerve (RLN)

Iatrogenic laryngeal nerve injury is a recognized complication of thyroidectomy and incidence is high when central compartment dissection is necessary. The course of the nerve is altered by enlarged lymph nodes and pre-laryngeal branching adds to the complexity of the situation. When transection is identified direct suturing or bridging with a cable graft is the routine practice. But the recovery of vocal cord function is not always satisfactory.

RLN involvement is found in up to 33% of thyroid cancers [35] and some of them are asymptomatic. Preoperative assessment is vocal cords is mandatory and palsy indicates extra-thyroid invasion and infiltration of upper aero-digestive tract also. The extent of anatomical infiltration is done during exploration and intraoperative nerve monitoring is recommended. When faced with unilateral cord palsy utmost care is given to preserve opposite RLN and such patients are better operated on in high volume thyroid center. Each situation merits an individualized approach but is generally based on the following recommendations [36].

1. RLN is resected when preoperative vocal cord paralysis is present and the nerve is found encased during the procedure.
2. Tumor is removed with sharp dissection preserving intact RLN when bilateral vocal cord function is preserved on preoperative assessment.
3. Tumor is shaved of RLN if the contralateral vocal cord paralysis is present.

Intraoperative nerve monitoring is recommended in these situations which enables the surgeon to identify and assess the functional status of the nerve.

Prophylactic central compartment lymph node dissection was routinely performed irrespective of tumor characteristics or age of the patient. It continues to be a matter of debate though ATA has revised its recommendation of prophylactic dissection. The incidence of complications in central compartment dissection is high even among experienced surgeons. The advantage of prophylactic central compartment dissection is yet to be convincingly proved. But for medullary thyroid carcinoma, prophylactic central compartment dissection is always recommended. The dissection is technically demanding since the lymph nodes are distributed close to laryngeal nerves and parathyroid glands.

14.6.3 Surgical Management of Recurrences and Metastases

Persistent local disease and recurrent disease after initial successful primary treatments are not uncommon in DTC. Approximately 10–30% of DTC patients with successful primary treatment with disease-free status develop local recurrence and/or distant metastasis [37]. Incidences of persistent or recurrent structural disease are as low as 3% in the low-risk category of DTC while it is as high as 68% in the high-risk category [38]. The risk of distant metastases in patients with DTC was found 10% for PTC, 22% for FTC, and 33% for Hürthle cell carcinoma [39].

14.6.4 Locoregional Recurrence

Locoregional recurrences develop in up to 20% of patients and surgical treatments are the preferred option for most of them [40]. Metastases in regional lymph nodes are frequent and account for more than 50% of recurrence. Clinically or sonologically evident lymph nodes of >10 mm size merit treatment. Since sonological features or elevated thyroglobulin levels are not convincing pieces of evidence of malignant deposits in lymph nodes, surgical treatment is better planned after cytological evidence becomes available. But in routine clinical practice, aspirations from lymph nodes are mostly inconclusive due to lack of cellularity. Thyroglobulin estimation of needle washout is a promising alternative but requires accurate validation. A pragmatic approach is advisable since the perinodal spread of malignancy may lead to debilitating morbidity.

Recurrences following block dissection are excised widely with all surrounding lymph nodes and soft tissue preserving the vital structures such as spinal accessory nerve, vagus nerve, recurrent laryngeal nerve, and phrenic nerve. Nodal recurrences in unexplored stations are treated with segment-oriented block dissection.

Locoregional soft tissue recurrences are occasionally seen in the thyroid bed. Surgical extirpation of such lesions is technically demanding since the laryngeal nerves and parathyroids are at risk. But successful excision prevents spread to the trachea and larynx. Soft tissue deposits in strap muscle and subcutaneous tissue are rarely encountered and excision is the treatment of choice.

14.6.5 Metastases Outside the Neck

Synchronous distant metastases are seen in up to 10% of patient at the time of diagnosis. Distant metastases develop in another 10% as part of disease progression [40]. Generally, distant metastases signify advanced disease stage but in majority of patients with DTC, adequate treatment results in favorable outcome. Despite the routine treatment with surgery and I^{131} a few may progress and require targeted therapies [41].

The most common locations of distant metastases are lung and bone though rare sites such as liver, adrenals, and skeletal muscles are recorded. The commonest location of distant metastases in PTC is lung parenchyma and is characterized by diffuse parenchymal infiltration and is effectively treated with I^{131} therapies. But Lin JD et al. found a significantly low 20-year survival rate for those with metastases (51.2% vs. 75.0%) compared to those with no metastases [42]. It is interesting to note that he did not appreciate any significant differences in 10-year survival.

Skeletal metastases occur in about 4% of DTC patients and are more frequently associated with FTC (12%) compared to PTC (2%) [43]. A subset of FTC presents with osseous metastases with a relatively hidden primary in the thyroid. Very often, the bone metastases could be occult and detected as hot spots in I^{131} whole body scan (WBS). Symptomatic bone lesions present as deformities or pathological fractures. Almost all lesions are osteolytic lesions and well projected serial x-rays or CT scans confirm the diagnosis.

Treatment of bone metastases: The principles of management shall be aimed at disease control, inhibition of bone resorption, and palliation. Isolated metastasis of long bone is excised and reconstructed and supplemented with external radiotherapy [44]. Surgical approaches are necessary to stabilize structural instability, fixation of the fracture, and relieve pain refractory other modalities of treatment. Prophylactic fixation is done when impending fracture is suspected especially in areas like vertebrae.

Agents such as bisphosphonates and denosumab are effective bone resorption inhibitors and effectively prevent bone-related events. But the optimal duration of treatment for DTC is not established but in osseous secondaries from the breast and prostate, the minimal recommended duration is 2 years. External radiation alone is useful in local pain relief and as an adjuvant following resection of isolated focus. Effective surgical resection and external radiation prolong progression-free interval.

Radioiodine is effective in iodine avid metastases but a much high dose is required to reduce the large tumor burden. But all those iodine avid metastases may not achieve complete remission even with high dose of I^{131}. Young age, small tumor, and iodine avidity are indicators of complete response. Pulmonary metastases from DTC are diffuse parenchymal infiltration and respond well to radioiodine treatment.

14.7 Conclusion

Surgical procedures are the primary treatment of DTC and on the majority of occasions they are aimed to cure the disease. The thyroid gland is placed in close vicinity to many vital structures and so procedures are planned in such a way to reduce functional disability considering the slow progression of the disease.

References

1. Kilfoy BA, Zheng T, Holford TR, et al. International patterns and trends in thyroid cancer incidence, 1973–2002. Cancer Causes Control. 2009;20(5):525–31. https://doi.org/10.1007/s10552-008-9260-4.
2. Lloyd RV, Osamura RY, Kloppel G, et al. WHO classification of tumours: pathology and genetics of tumours of endocrine organs. 4th ed. Lyon, France: IARC; 2017.
3. Lam A K-y. Pathology of endocrine tumors update. AJSP: Rev Rep. 2017;22(4):209–16. https://doi.org/10.1097/PCR.0000000000000183.
4. Nikiforov YE, Seethala RR, Tallini G, Baloch ZW, Basolo F, Thompson LD, Barletta JA, Wenig BM, Al Ghuzlan A, Kakudo K, Giordano TJ, Alves VA, Khanafshar E, Asa SL, El-Naggar AK, Gooding WE, Hodak SP, Lloyd RV, Maytal G, Mete O, Nikiforova MN, Nosé V, Papotti M, Poller DN, Sadow PM, Tischler AS, Tuttle RM, Wall KB, LiVolsi VA, Randolph GW, Ghossein RA. Nomenclature revision for encapsulated follicular variant of papillary thyroid carcinoma: a paradigm shift to reduce overtreatment of indolent tumors. JAMA Oncol. 2016;2(8):1023–9. https://doi.org/10.1001/jamaoncol.2016.0386.
5. Liu R, Xing M. TERT promoter mutations in thyroid cancer. Endocr Relat Cancer. 2016;3:R143–55. https://doi.org/10.1530/ERC-15-0533.
6. Odate T, Oshi N, Vuong HG, Mochizuki K, Kondo T. Genetic differences in follicular thyroid carcinoma between Asian and Western countries, a systematic review. Gland Surg. 2020;9(5):1813–26.
7. DeLellis RA, Lloyd RV, Heitz PU, Eng C, editors. World Health Organization Classification of tumours: pathology and genetics of tumours of endocrine organs. Lyon, France: IARC Press; 2004.
8. Zhou X, Zheng Z, Chen C, et al. Clinical characteristics and prognostic factors of Hurthle cell carcinoma: a population-based study. BMC Cancer. 2020;20:407. https://doi.org/10.1186/s12885-020-06915-0.
9. Ahmadi S, Stang M, Jiang XS, Sosa JA. Hürthle cell carcinoma: current perspectives. Onco Targets Ther. 2016;9:6873–84. https://doi.org/10.2147/OTT.S119980.
10. Ganly I, Rocarte F, et al. Genomic dissection of Hurthle cell carcinoma reveals a unique class of thyroid malignancy. J Clin Endocrinol Metab. 2013;98(5):E962–72.

11. Bhattacharyya N. Survival and prognosis in Hurthle cell carcinoma of the thyroid gland. Arch Otolaryngol Head Neck Surg. 2003;129(2):207–10.
12. Hiltzik D, Carlson DL, Tuttle RM, Chuai S, Ishill N, Shaha A, Shah JP, Singh B, Ghossein RA. Poorly differentiated thyroid carcinomas defined on the basis of mitosis and necrosis: a clinicopathologic study of 58 patients. Cancer. 2006;106(6):1286–95.
13. Volante M, Collini P, Nikiforov YE, Sakamoto A, Kakudo K, Katoh R, Lloyd RV, LiVolsi VA, Papotti M, Sobrinho-Simoes M, Bussolati G, Rosai J. Poorly differentiated thyroid carcinoma: the Turin proposal for the use of uniform diagnostic criteria and an algorithmic diagnostic approach. Am J Surg Pathol. 2007;31(8):1256–64.
14. Dettmer MS, Schmitt A, Komminoth P, et al. Poorly differentiated thyroid carcinoma. Pathologe. 2020;41:1–8. https://doi.org/10.1007/s00292-019-0600-9.
15. Cancer Genome Atlas Research Network. Integrated genomic characterization of papillary thyroid carcinoma. Cell. 2014;159:676–90.
16. Hannallah J, Rose J, Guerrero MA. Comprehensive literature review: recent advances in diagnosing and managing patients with poorly differentiated thyroid carcinoma. Int J Endocrinol. 2013;2013:317487., 7 p. https://doi.org/10.1155/2013/317487.
17. Bongiovanni M, Bloom L, Krane JF, Baloch ZW, Powers CN, Hintermann S, Pache J-C, Faquin WC. Cytomorphologic features of poorly differentiated thyroid carcinoma. Cancer Cytopathol. 2009;117:185–94. https://doi.org/10.1002/cncy.20023.
18. Sanders EM Jr, LiVolsi VA, Brierley J, Shin J, Randolph GW. An evidence-based review of poorly differentiated thyroid cancer. World J Surg. 2007;31(5):934–45.
19. Bible KC, Kebebew E, Brierley J, Brito JP, Cabanillas ME, Clark Jr TJ, Di Cristofano A, Foote R, Giordano T, Kasperbauer J, Newbold K, Nikiforov YE, Gregory Randolph M, Rosenthal S, Sawka AM, Shah M, Shaha A, Smallridge R, Wong-Clark CK. American Thyroid Association guidelines for management of patients with anaplastic thyroid cancer. Thyroid. 2021;31(3):337–86.
20. Tiedje V, Stuschke M, Weber F, Dralle H, Moss L, Führer D. Anaplastic thyroid carcinoma: review of treatment protocols. Endocr Relat Cancer. 2018;25(3):R153–61. https://doi.org/10.1530/ERC-17-0435.
21. Goffredo P, Thomas SM, Adam MA, Sosa JA, Roman SA. Impact of timeliness of resection and thyroidectomy margin status on survival for patients with anaplastic thyroid cancer: an analysis of 335 cases. Ann Surg Oncol. 2015;22:4166–74. https://doi.org/10.1245/s10434-015-4742-6.
22. Amin MB, Edge S, Greene F, Byrd DR, Brookland RK, American Joint Committee on Cancer, et al., editors. Thyroid—differentiated and anaplastic carcinoma. AJCC cancer staging manual. 8th ed. New York, NY: Springer; 2017. p. 881–98.
23. Cooper DS, Doherty GM, Haugen BR, Kloos RT, Lee SL, Mandel SJ, Mazzaferri EL, McIver B, Sherman SI, Tuttle RM, The American Thyroid Association Guidelines Taskforce. Management guidelines for patients with thyroid nodules and differentiated thyroid cancer. Thyroid. 2006;16:109–42.
24. Haugen BR, Alexander EK, Bible KC, Doherty GM, Mandel SJ, Nikiforov YE, Pacini F, Randolph GW, Sawka AM, Schlumberger M, Schuff KG, Sherman SI, Sosa JA, Steward DL, Tuttle RM, Wartofsky L. 2015 American Thyroid Association management guidelines for adult patients with thyroid nodules and differentiated thyroid cancer: the American Thyroid Association guidelines task force on thyroid nodules and differentiated thyroid cancer. Thyroid. 2016;26(1):1–133.
25. Ito Y, Miyauchi A, Inoue H, Fukushima M, Kihara M, Higashiyama T, Tomoda C, Takamura Y, Kobayashi K, Miya A. An observational trial for papillary thyroid microcarcinoma in Japanese patients. World J Surg. 2010;34(1):28–35.
26. Sugitani I, Toda K, Yamada K, Yamamoto N, Ikenaga M, Fujimoto Y. Three distinctly different kinds of papillary thyroid microcarcinoma should be recognized: our treatment strategies and outcomes. World J Surg. 2010;34(6):1222–31.
27. Adam MA, Pura J, Gu L, Dinan MA, Tyler DS, Reed SD, Scheri R, Roman SA, Sosa JA. Extent of surgery for papillary thyroid cancer is not associated with survival: an analysis

of 61,775 patients. Ann Surg. 2014;260(4):601–5.; discussion 605-7. https://doi.org/10.1097/SLA.0000000000000925.
28. Czaja JM, McCaffrey TV. The surgical management of laryngotracheal invasion by well-differentiated papillary thyroid carcinoma. Arch Otolaryngol Head Neck Surg. 1997;123:484–90.
29. Shin DH, Mark EJ, Suen HC, Grillo HC. Pathologic staging of papillary carcinoma of the thyroid with airway invasion based on the anatomic manner of extension to the trachea: a clinicopathologic study based on 22 patients who underwent thyroidectomy and airway resection. Hum Pathol. 1993;24:866–70.
30. Kim H, Jung HJ, Lee SY, Kwon TK, Kim KH, Sung MW, Hun HJ. Prognostic factors of locally invasive well-differentiated thyroid carcinoma involving the trachea. Eur Arch Otorhinolaryngol. 2016;273(7):1919–26.
31. Matsumoto F, Ikeda K. Surgical management of tracheal invasion by well-differentiated thyroid cancer. Cancers (Basel). 2021;13(4):797. https://doi.org/10.3390/cancers13040797.
32. Piazza C, Del Bon F, Barbieri D, Grazioli P, Paderno A, Perotti P, Lombardi D, Peretti G, Nicolai P. Tracheal and crico-tracheal resection and anastomosis for malignancies involving the thyroid gland and the airway. Ann Otol Rhinol Laryngol. 2016;125(2):97–104.
33. Chala AI, Vélez S, Sanabria A. The role of laryngectomy in locally advanced thyroid carcinoma. Review of 16 cases. Acta Otorhinolaryngol Ital. 2018;38(2):109–14.
34. Perry A, Casey E, Cotton S. Quality of life after total laryngectomy: functioning, psychological well-being and self-efficacy. Int J Lang Commun Disord. 2015;50(4):467–75. https://doi.org/10.1111/1460-6984.12148.
35. McCaffrey TV, Bergstralh EJ, Hay ID. Locally invasive papillary thyroid carcinoma: 1940–1990. Head Neck. 1994;16:165–72.
36. Stack BC, Twining C, Rastatter J, Angelos P, Baloch Z, Diercks G, Faquin W, Kazahaya K, Rivkees S, Sheyn T, Shin JJ, Smith J, Thompson G, Viswanathan P, Wassner A, Brooks J, Randolph G. Consensus statement by the American Association of Clinical Endocrinologists and the American Head and Neck Society Endocrine Surgery Section on pediatric benign and malignant thyroid surgery. Head Neck. 2021;27(3):174–84. https://doi.org/10.1002/hed.26586.
37. Mazzaferri EL, Jhiang SM. Long-term impact of initial surgical and medical therapy on papillary and follicular thyroid cancer. Am J Med. 1994;97(5):418–28.
38. Shaha AR, Shah JP, Loree TR. Patterns of nodal and distant metastasis based on histologic varieties in differentiated carcinoma of the thyroid. Am J Surg. 1996;172:692–4.
39. Tuttle RM, Tala H, Shah J, Leboeuf R, Ghossein R, Gonen M, Brokhin M, Omry G, Fagin JA, Shaha A. Estimating risk of recurrence in differentiated thyroid cancer after total thyroidectomy and radioactive iodine remnant ablation: using response to therapy variables to modify the initial risk estimates predicted by the new American Thyroid Association staging system. Thyroid. 2010;20(12):1341–9. https://doi.org/10.1089/thy.2010.0178.
40. Gallardo E, Medina J, Sánchez JC, et al. SEOM clinical guideline thyroid cancer (2019). Clin Transl Oncol. 2020;22:223–35. https://doi.org/10.1007/s12094-019-02284-8.
41. Hirsch D, Levy S, Tsvetov G, Gorshtein A, Slutzky-Shraga I, Akirov A, Robenshtok E, Shimon I, Benbassat CA. Long-term outcomes and prognostic factors in patients with differentiated thyroid cancer and distant metastases. Endocr Pract. 2017;23(10):1193–200.
42. Lin JD, Chao TC, Chou SC, Hsueh C. Papillary thyroid carcinomas with lung metastases. Thyroid. 2004;14:1091–6.
43. Iñiguez-Ariza NM, Bible KC, Clarke BL. Bone metastases in thyroid cancer. J Bone Oncol. 2020;21:100282. https://doi.org/10.1016/j.jbo.2020.100282.
44. Drost L, Ganesh V, Wan BA, et al. Efficacy of postoperative radiation treatment for bone metastases in the extremities. Radiother Oncol. 2017;124:45–8.

An Update on Medullary Thyroid Carcinoma

15

Supriya Sen and Deepak Thomas Abraham

15.1 Introduction

Medullary thyroid carcinoma (MTC) was first described by Jacquet close to a hundred years ago as thyroid carcinoma with amyloid. However, Hazard et al. narrated the conclusive histology as a solid non-follicular carcinoma with amyloid and named it the medullary thyroid carcinoma. However, the cellular origin of MTC was discovered by Williams, who established that cancer originated from the neural crest-derived parafollicular cells or the C-cells. These cells secrete calcitonin and occasionally other hormones like ACTH, which can present with ectopic Cushing's syndrome [1, 2].

MTC is a rare thyroid cancer and forms 2% of all malignancies. Generally, cancer presents in the 4th and 5th decade of life and has equal distribution among males and females. Approximately 75% of all MTCs are sporadic, but 25% have a hereditary background. MTC with hereditary background may be associated with multiple endocrine neoplasia (MEN) 2A or MEN 2B and the rest are known as familial medullary thyroid carcinoma (FMTC) [3].

15.2 Aetiology

The C-cells are primarily concentrated in the posteromedial aspect of each lobe almost at the junction of the upper one-third and lower two-thirds. The C-cells are derived from the neural crest cells and are found attached to basement membrane between the follicular cells but not reaching to the lumen [1, 4]. The unique genetic alteration characterised by the C-cell in MTC the gain function in RET gene

S. Sen (✉) · D. T. Abraham
Department of Endocrine Surgery, Christian Medical College, Vellore, Tamil Nadu, India
e-mail: deepakabraham@cmcvellore.ac.in

C. G. Nair, S. J. Abraham (eds.), *Surgical Management of Thyroid and Parathyroid Diseases*, https://doi.org/10.1007/978-981-97-3774-1_15

(*Re*arranged during *T*ransfection) on chromosome 10q11.2, and is seen in both in the hereditary as well as in certain sporadic forms. But the hereditary thyroid cancer in MEN 2A and 2B and FMTC have the germline mutations. Approximately 50% of sporadic forms have the somatic mutations [1, 3].

RET germline mutations in hereditary MTC induce increase in C-cell population popularly known as C-cell hyperplasia (Fig. 15.1) and regarded as the precursor of MTC.

15.2.1 Clinical Spectrum of MTC

MTC is usually a cytological or histological surprise except in hereditary forms when patients are diagnosed by genetic screening. MTC presents as nondescript thyroid nodule and does not have any characteristic clinical or sonological features. Patients with hereditary MTC (HMTC) presents at a much younger age, but vary based on type of mutations. However, the sporadic MTC (SMTC), the presents fourth to a sixth decade as a solitary nodule and up to 70% such patients have regional lymph node metastases, and 10–15% of them have distant metastasis.

The C-cells synthesise and secrete hormones or biological amines such as calcitonin, CEA, ACTH, corticotrophin-releasing hormone, chromogranin, serotonin, neurotensinase, histaminases and so on. Excess secretion of these chemicals brings about paraneoplastic symptoms such as flushing, diarrhoea, ectopic Cushing's syndrome, and cholestasis. Serum levels of calcitonin and Carcinoembryonic Antigen (CEA) are generally used for diagnosis, prognostication and follow-up of MTC. The serum concentrations of these markers are directly related to the C-cell mass [3].

Syndromic association of neoplasms of endocrine glands are rare disease combination. Medullary thyroid cancer is the most common neoplasm of multiple endocrine neoplasia type 2A and 2B and shows autosomal dominant inheriting pattern. When a genetic mutation of RET gene is identified, the chance of phenotype expression occurring is close to 100% and a prophylactic thyroidectomy shall prevent the cancer.

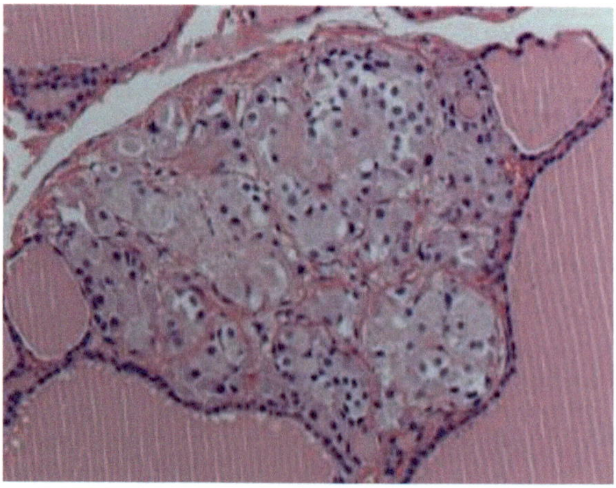

Fig. 15.1 C-cell hyperplasia with atypical C-cells

John H Sipple described the association of pheochromocytoma to thyroid cancer in 1961, and this syndrome is named after him. MEN 2A is now known to include parathyroid tumours, and is also the commonest MEN syndrome including 95% of them. There are four clinical variants of MEN 2A:

- Classical MEN 2A
- MEN 2A with cutaneous lichen amyloidosis (CLA)
- MEN 2A with Hirschsprung's disease (HD)
- Familial MTC (FMTC)
- Classical MEN 2A:
  RET germline mutations occur in codons 609, 611, 618, 620 or 634 in classic MEN 2A, and MTC is the presenting symptom of 95% patients; however, fewer patients present with pheochromocytoma and hyperparathyroidism. The incidence of latter two components of MEN 2A varies and depends on the type of RET-specific mutation. Majority of the pheochromocytomas are bilateral, benign and multicentric, and confined adrenal gland [3, 5]. PHPT is generally asymptomatic and due to hyperplasia of 1–4 parathyroids.
- MEN 2A with Cutaneous Lichen Amyloidosis (CLA):
  Approximately 10% of MEN 2A indicative skin lesions characterised by subepidermal deposition of keratin in the scapular region of the back, about T1–T6 dermatomal distribution. Patient used to have excessive pruritus which often exacerbates with stress and relieved with exposure to sunlight. The area becomes hyper pigmented and may present as 'notalgiaparasthetica' which is a sensory neuropathy involving dorsal spinal nerves (Fig. 15.2). The skin lesions may present at an younger age and herald the onset of clinically evident MTC later [3, 6, 7].
- MEN 2A with HD:
  Hirschsprung's disease occurs in about 7% of MEN 2A patients since RET mutation are also one of the genetic causes of the colonic lesion. As in HD, it is due to the failure of neural crest cells to form submucosal and myenteric cells [3, 8, 9]. The association of the colonic lesion with MTC is rare.

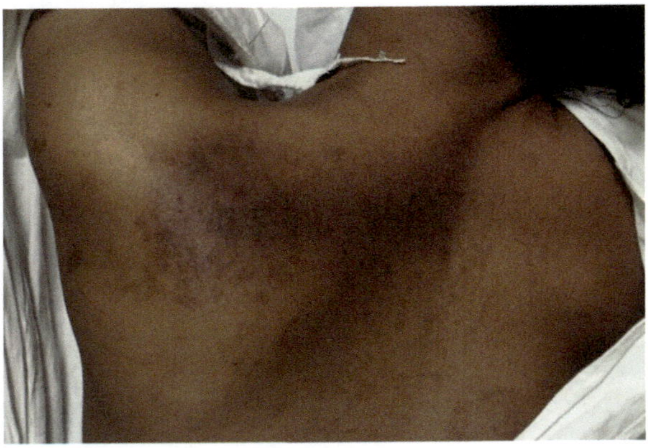

Fig. 15.2 Patient with MEN 2A and cutaneous lichen amyloidosis involving the scapular region

- Familial MTC:
  Familial medullary thyroid cancer (FMTC) is characterised by hereditary MTC but with no associated other endocrine tumours such as pheochromocytoma and PHPT. Presently, this is also recognised as a variant of MEN 2A [3, 9, 10].

15.3 MEN 2B

Unlike the MTC associated with the other hereditary syndromes, the biological behaviour of thyroid cancer in MEN 2B is different, developing early in life and is very aggressive. Mostly, MTC associated with MEN 2B presents with loco-regional and develop distant metastases. The commonest genetic changes MEN 2B is the *RET* mutation involving the 918 codons, while about 5% have *RET* mutation in the 883 codons. Pheochromocytoma is found in about 50% of patients but parathyroids are always spared.

These patients have a characteristic appearance which involves a distinctive faces with thick lips, fascial skeletal involvement with high arched palate, marfanoid body habitus (pectus excavatum, pes cavus, slipped femoral epiphyses) and ophthalmic abnormalities (inability to make tears, eyelid eversion, ptosis). About 50% have mucosal ganglioneuromas involving the lip, buccal mucosa, gingival, intestinal submucosal, and myenteric plexus [3, 7, 8] as depicted in Fig. 15.3.

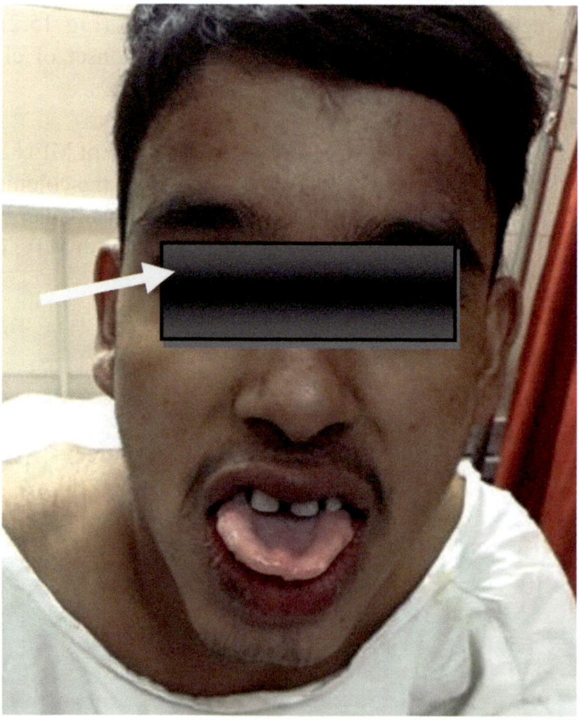

Fig. 15.3 Patient of MEN 2B with oral mucosal ganglioneuromas and eyelid eversion (arrow)

15.3.1 *RET* Proto-Oncogene

About one-fourth of MTC have germline mutation of *RET* proto-oncogene and this percentage is steadily increasing after mutational study has become routine for MTC patients. The genetic background decides the clinical behaviour of the tumour and the introduction of mutation analysis markedly revolutionised the treatment of MTC, especially in familial MTC. The surgical treatment and surveillance pattern can be tailor-made based upon the type of RET mutation to obtain a better clinical outcome. *RET* screening has opened to a new possibility of prophylactic surgery and help to choose the ideal time to do it. MTC is the first example of primary prevention of cancer in humans.

- Structure of *RET* proto-oncogene:

Genetic changes near the centromere of chromosome 10 which contains 21 exons was first describes by Takahashi in 1985. This transforming mutation controls the tyrosine kinase enzyme which mediates through plasma-membrane bound receptor and controls various functions of cells including proliferation. RET receptors seen in cells originating from neural crest like C-cells. RET receptor protein 1114 amino acid chain with extracellular, transmembrane and intracellular domain. The extracellular domain has four cadherin-like domain (CLD) and cysteine rich regions (Fig. 15.4). The cadherin-like domain is connected to transmembrane region and thereby to intracellular domain. The genetic activation leads on to genesis of a dimeric complex encompasses 2 RET proteins, 2 ligand molecules and 2 GFR α (GDNF co-receptors α). This change signals cell proliferation, differentiation and survival pathways.

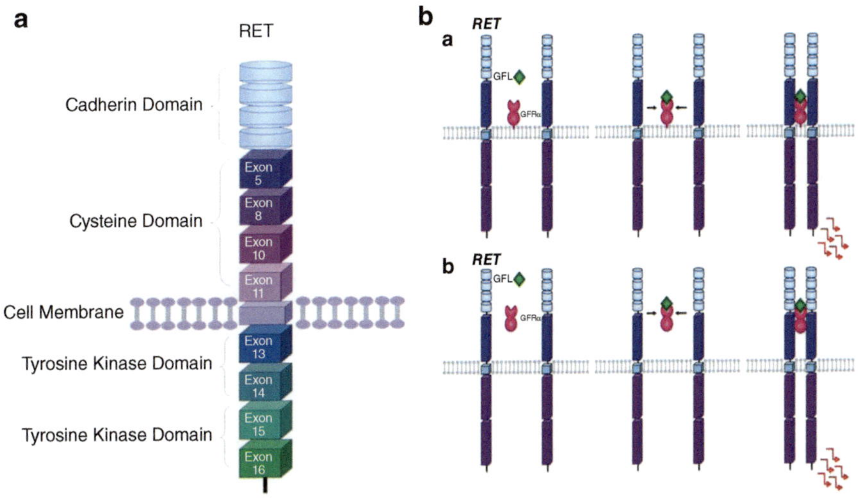

Fig. 15.4 (**a**) A diagram showing structure of RET receptor protein with exons involved, (**b**) A picture of a *cis* and *trans* model of RET activation with ligand binding [10, 11].

15.4 Evaluation and Diagnosis

The primary evaluation encompasses complete clinical examination, radiological assessment, and cytological/histological studies—generally known as 'triple assessment'.

- There is no distinctly different clinical presentation for MTC patients and most patients report with an incidentally detected thyroid nodule. However, a positive family history of thyroid cancer is a significant pointer. Patient who has functioning pheochromocytoma has episodes of panic with sweating and headache. Patients with palpable thyroid MTC nodule frequently have metastatic lymph nodes involvement [1, 3, 8]. Routine estimation of serum calcitonin and CEA is not a part of thyroid nodule evaluation but are always done when MTC is suspected. Implications of serum markers are discussed elsewhere.
- Radiological assessment:
  The first imaging is always ultrasound of the neck is a non-invasive and maps the regional nodal involvement. Chances of distant metastases are high when the serum calcitonin more than 500 pg/ml and higher imaging are strongly indicated [2, 3, 10]. The usual locations of distant metastases are lungs and liver and a Computed Tomography (CT) scan and Magnetic Resonance Imaging (MRI) are sensitive to pick up these metastases. Routine use of FDG-PET CT or F DOPA PET CT is not advisable in the preoperative set up. An isotope bone scan may be done skeletal metastases are suspected.
- Pathological evaluation:
  Fine needle aspiration cytology (FNAC) typically shows moderate to marked cellularity and multiple patterns like plasmacytoid or epithelioid types of cells. These cells are characterised by eccentric nuclei and have the characteristic neuroendocrine type of 'salt and pepper' chromatin with inconspicuous nucleoli (Fig. 15.5). The presence of amyloid is confirmed by Congo-red staining.

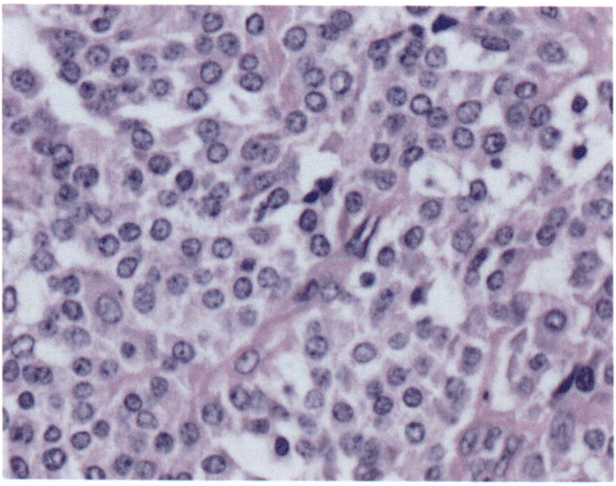

Fig. 15.5 HE stains 400×, showing polygonal cells with round nuclei, salt, and pepper (granular) cytoplasm

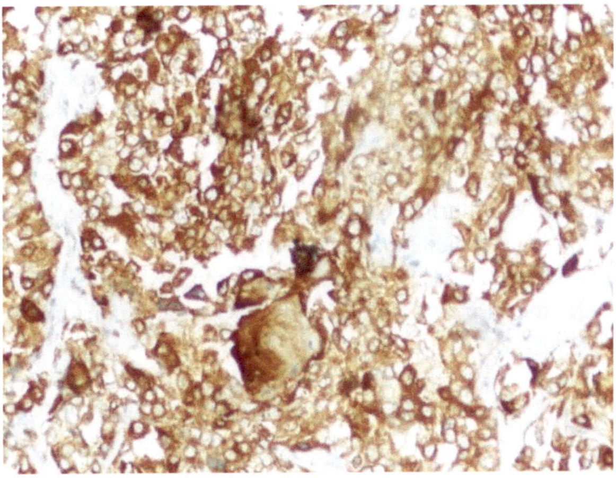

Fig. 15.6 Calcitonin immunohistochemistry showing positivity

After harvesting a cell block immunohistochemistry studies for calcitonin (Fig. 15.6), CEA, and chromogranin can be done on the FNAC sample as well [3, 11–13]

15.4.1 Tumour Markers

The two tumour markers of importance in diagnosis and prognostication in MTC are serum calcitonin and serum CEA. Serum level of calcitonin is highly sensitive and specific marker in diagnosing MTC. Serum levels above 100 pg/ml is highly predictive of MTC. The serum levels of these markers are also important in formulating the management protocol. A raised level of serum calcitonin may help in the early detection of locoregional metastasis and/ or metastasis. However, a raised CEA (though a more non-specific biomarker), as compared to serum calcitonin, indicates more aggressive disease or poorly differentiated MTC [3, 14, 15].

15.4.2 *Re*arranged During *T*ransfection (*RET*) Testing

Genetic studies are now routine recommendation for all MTC after proper counselling. When the index patient is found *RET* positive, all the first-degree relatives are also counselled for *RET* mutation analysis. The *RET* testing can be targeted based on the phenotype presentation. If a patient is suspected to have MEN 2A, then codon specific testing targeting exon 10 (for codons 609, 611, 618, 620) and exon 11 (for codons 630, 634) followed by exons 8,13,14,15 and 16, if no mutation is detected in exon 10 and 11. Likewise, when MEN 2B is suspected, then exon 16 (for codon 918) and exon 15 (for 883) are tested. However, there may be instances where no discernible phenotype is evident, in which case the whole *RET* sequence analysis is advisable [3, 16, 17].

When the RET mutation is found positive screening for the other components of the MEN syndrome like pheochromocytoma and primary hyperparathyroidism (PHPT) is necessary. The screening begins based on the type of mutation since the phenotype expression appears at different time span in life. In the highest risk category, screening for both pheochromocytoma and PHPT may start as early as 11 years and in the high and moderate risk categories, it may start at 16 years. The biochemical screening includes 24-h urinary metanephrine and normetanephrine for pheochromocytoma and serum PTH, calcium, albumin, and phosphorus for PHPT. If pheochromocytoma coexists with MTC at the time of initial diagnosis the surgery for pheochromocytoma precedes the surgery for MTC. PHPT is operated concurrently with MTC [3]

15.4.3 Risk Stratification of MTC-ATA 2015 Guidelines

Risk stratification aims at the timing of prophylactic thyroidectomy for the carriers of germline mutations. The prophylactic removal of organ shall be planned when it is expendable or the secretions or function can be effectively supplemented or supported.

Based on genetic testing, American Thyroid Association (ATA) 2015 guidelines stratified MTC into three groups (Table 15.1). The risk stratification helps to prognosticate the disease and indicates the appropriate time for prophylactic thyroidectomy on mutation carriers.

- Highest risk (MEN 2B and codon 918 mutations)
- High risk (MEN 2A and codon 634, 883 mutations)
- Moderate risk (Hereditary MTC with codon mutations other than mutations in 918, 883, and 634) (Table 15.1)

15.4.4 Surgical Management

The primary treatment for MTC is well planned surgery which should ensure a complete removal of tumour tissue since there are no effective adjuvant in the management of MTC. An effective primary surgery is the only hope for the cure. Also

Table 15.1 American Thyroid Association MTC risk stratification [3]

ATA risk category	Highest risk	High risk	Moderate risk
Codon specific mutation	918	634, 883	321 532, 533 609, 611, 618, 620, 630, 631, 635, 649, 666, 768, 790, 791, 804, 844, 891, 912
Aggressiveness	Very	Yes	Moderate
Age at onset	Within 1st year of life	<5 years	Adult-onset
Timing of prophylactic surgery	In 1st year of life	5 years or earlier depending on serum calcitonin levels	When serum calcitonin levels start increasing or earlier

for the mutation carriers are recommended appropriately timed thyroidectomy to avert chance cancer.

A central compartment lymph node dissection is always done in N0 patients along with total thyroidectomy since a re-entry in this location is technically more demanding and chances of complications are high. There are proponents of prophylactic ipsilateral lateral neck dissection in N0 patients with calcitonin >20 pg/ml. The probability of bilateral deep cervical lymph node metastases is high when the serum calcitonin is >100 pg/ml and so there are suggestions to modify the extent of surgery based on calcitonin levels. Contralateral prophylactic lateral neck dissection is also recommended when basal serum calcitonin is more than 200 pg/ml. Lymphadenectomy from the superior mediastinum is recommended when metastases in mediastinal lymph nodes are suspected. Incidental detection of MTC of diagnostic hemithyroidectomy is occasionally encountered, however a completion thyroidectomy is not always mandatory when a hereditary cancer is excluded and serum calcitonin is normalised [1, 3, 8, 18, 19].

15.4.5 Postoperative Management

Following total thyroidectomy, patients are placed on replacement dose of levothyroxine. Hypocalcaemia is frequently seen following a central compartment dissection and given oral calcium and vitamin D supplementation. Majority of patients recover by short-term course of supplementation.

The first review is usually done after 3 months of surgery for estimation of the serum calcitonin and due to long half-life of the tumour marker. During the first review, thyroid hormone levels, serum calcitonin, serum CEA, and serum total/albumin corrected calcium are estimated after a complete clinical examination.

- Replacement dose of levothyroxine
- Clinical examination
- Measurements of serum calcitonin and CEA. Serum calcitonin level within normal reference level suggests a favourable outcome. Serum calcitonin elevated but below 150 pg/ml is indicative of locoregional residual disease. Generally, neck ultrasound is done to assess the regional lymph nodes and residual disease in the thyroid bed.

  A serum calcitonin level >150 pg/ml is highly suggestive of distant metastases and suitable imaging such as CECT of the neck, thorax, and abdomen and symptom-directed high-resolution cross-sectional imaging are indicated.

The patient has to be instructed to comply with the follow-up protocol of 6–12 months visit thereafter. The tumour markers are measured at every visit and suitable imaging schedule is charted out based on these values. The formula to calculate the calcitonin doubling time is now available and is a valuable prognostic indicator. A doubling time of more than 25 months has a good prognosis while a doubling time of less than 6 months has a worse prognosis [1, 3, 20, 21].

15.4.6 Management of Locoregional Recurrence and Metastasis

The common locoregional residual disease is the persistence or recurrence in regional lymph nodes. There may be soft tissue residual disease usually following gross extrathyroidal disease at the time of primary surgery.

Lymph nodal recurrence in a previously unexplored lymph node stations is managed with well-planned compartment-oriented lymph node dissection. Majority of occasions, the metastatic lymph nodes in lateral neck are operable but for encasing the carotid vessels. The vital structures such as the laryngeal nerves, parathyroid glands trachea and oesophagus are carefully preserved intact to avoid severe morbidity. In high volume centres, a well-planned surgery ensures adequate clearance and a normal calcitonin level is achieved.

A curative surgery with preservation of vital structures is not always possible due to infiltration. External Beam Radiation Therapy (EBRT) is the next option when residual lesion is present. Generally, EBRT is indicated in patients with gross residue, high risk of local recurrence, risk of airway obstruction, and extensive extrathyroidal infiltration and massive lymph node metastases [3, 22, 23].

The treatment protocol for patients with distant metastases are designed on case-to-case basis but in principle based on balance between benefit of treatment and treatment related morbidity. The cancer per se progresses slowly and treatment related morbidity should be minimalised and should not compromise quality of life. Hence, to summarise, the main aim is to provide maximum locoregional control where feasible, palliate symptomatic metastasis, or treat life-threatening metastasis. There is no role of systemic therapy in low-volume metastatic disease or stable disease.

Cytotoxic chemotherapy has a very limited role in metastatic MTC and 5-flurouracil, doxorubicin, and dacarbazine were tried with no significant response.

Majority of MTC express somatostatin receptors and so radiolabelled molecules such as Yttrium^{90} DOTA, Lutetium^{177} DOTA Indium^{111} are used for diagnosis and treatment. This novel mode of delivering targeted radiation is evolving for MTC. Peptide receptor radionuclide therapy (PRRT) using ^{177}Lu-DOTATATE and ^{90}Y-DOTATOC are used for treatment purpose with promising results in halting disease progression. Octreotide, or Iodine^{131} MIBG have shown partial response and in the setting of metastatic disease [3, 24].

15.4.7 Targeted Therapy in MTC

Targeted therapy in cancer aims at the proteins that stimulate the excess cell proliferation and is useful when the genetic changes are well established. These are agents that block the steps of the over expression pathways and controls the growth and spread of cancer.

Tyrosine Kinase Inhibitors (TKIs) are now used to treat advanced MTC and molecules like Vandetanib (ZETA trial) and Cabozantinib (EXAM trial) are approved by FDA. Selective RET inhibitors such as Selpercatinib- LOXO-292 (Libretto-001 trial) and Pralsetinib (ARROW trial) are also now approved for treatment of advanced MTC. These drugs are reported to have improved safety and potency in MTC treatment. More molecules TKI family such as Sorafinib, Lenvitanib, Sunitanib and Pazopanib are in various stages of trials. Pembrolizumab, an immunomodulator, is also in under phase 2 trial and published data of the results are not yet available. Nelfinavir, a HIV protease inhibitor, was found to target heat shock protein 90 chaperone (HSP90) which is required for folding and stability of *RET* mutants [24–26].

15.5 Conclusion

MTC is a differentiated thyroid cancer of neuroendocrine origin arising from parafollicular C-cells. It is mainly of two types, familial and sporadic type, the familial type being positive for *RET* mutation. As it is not radioiodine avid, surgery is the mainstay of treatment. In the familial type, genetic counselling can be done and screening of families where indicated can be done. Hence, depending on the type of RET mutation and risk stratification, prophylactic surgery may be offered to mutation-positive patients before onset of clinical disease.

References

1. Al-Rawi M, Wheeler MH. Medullary thyroid carcinoma—update and present management controversies. Ann R Coll Surg Engl. 2006;88(5):433–8.
2. Stamatakos M, Paraskeva P, Stefanaki C, Katsaronis P, Lazarus A, Safioleas K, et al. Medullary thyroid carcinoma: the third most common thyroid cancer reviewed. Oncol Lett. 2011;2(1):49–53.
3. Wells SA, Asa SL, Dralle H, Elisei R, Evans DB, Gagel RF, et al. Revised American Thyroid Association Guidelines for the management of medullary thyroid carcinoma. Thyroid. 2015;25(6):567–610.
4. Mehrotra PK, Mishra A, Mishra SK, Agarwal G, Agarwal A, Verma AK. Medullary thyroid cancer: clinico-pathological profile and outcome in a tertiary care center in North India. World J Surg. 2011;35(6):1273–80.
5. Yip L. Multiple endocrine neoplasia type 2: evaluation of the genotype-phenotype relationship. Arch Surg. 2003;138(4):409.
6. Gagel RF, Levy ML, Donovan DT, Alford BR, Wheeler T, Tschen JA. Multiple endocrine neoplasia type 2a associated with cutaneous lichen amyloidosis. Ann Intern Med. 1989;111(10):802–6.
7. Stratakis CA. Hereditary syndromes predisposing to endocrine tumors and their skin manifestations. Rev Endocr Metab Disord. 2016;17(3):381–8.
8. Raam M, Cherian AJ, Paul MJ, Abraham DT. An update on medullary carcinoma thyroid. J Head Neck Physicians Surg. 2019;7(2):45.
9. Wells SA, Pacini F, Robinson BG, Santoro M. Multiple endocrine neoplasia type 2 and familial medullary thyroid carcinoma: an update. J Clin Endocrinol Metab. 2013;98(8):3149–64.
10. Machens A, Dralle H. Biomarker-based risk stratification for previously untreated medullary thyroid cancer. J Clin Endocrinol Metab. 2010;95(6):2655–63.
11. Desai SS, Sarkar S, Borges AM. A study of histopathological features of medullary carcinoma of the thyroid: cases from a single institute in India. Indian J Cancer. 2005;42(1):25.

12. Khurana R, Agarwal A, Bajpai VK, Verma N, Sharma AK, Gupta RP, et al. Unraveling the amyloid associated with human medullary thyroid carcinoma. Endocrinology. 2004;145(12):5465–70.
13. Mehdi G, Maheshwari V, Ansari HA, Sadaf L, Khan MA. FNAC diagnosis of medullary carcinoma thyroid: a report of three cases with review of literature. J Cytol Indian Acad Cytol. 2010;27(2):66–8.
14. Mendelsohn G, Wells SA, Baylin SB. Relationship of tissue carcinoembryonic antigen and calcitonin to tumor virulence in medullary thyroid carcinoma. An immunohistochemical study in early, localized, and virulent disseminated stages of disease. Cancer. 1984;54(4):657–62.
15. Thomas CM, Asa SL, Ezzat S, Sawka AM, Goldstein D. Diagnosis and pathologic characteristics of medullary thyroid carcinoma—review of current guidelines. Curr Oncol. 2019;26(5):338–44.
16. Elisei R, Romei C, Cosci B, Agate L, Bottici V, Molinaro E, et al. RET genetic screening in patients with medullary thyroid cancer and their relatives: experience with 807 individuals at one center. J Clin Endocrinol Metab. 2007;92(12):4725–9.
17. Mulligan LM, Kwok JBJ, Healey CS, Elsdon MJ, Eng C, Gardner E, et al. Germ-line mutations of the RET proto-oncogene in multiple endocrine neoplasia type 2A. Nature. 1993;363(6428):458–60.
18. Moley JF, DeBenedetti MK. Patterns of nodal metastases in palpable medullary thyroid carcinoma. Ann Surg. 1999;229(6):880.
19. Ito Y, Miyauchi A, Kihara M, Higashiiyama T, Fukushima M, Miya A. Static prognostic factors and appropriate surgical designs for patients with medullary thyroid carcinoma: the second report from a single-institution study in Japan. World J Surg. 2018;42(12):3954–66.
20. Meijer JAA, Cessie SL, Hout WBVD, Kievit J, Schoones JW, Romijn JA, et al. Calcitonin and carcinoembryonic antigen doubling times as prognostic factors in medullary thyroid carcinoma: a structured meta-analysis. Clin Endocrinol (Oxf). 2010;72(4):534–42.
21. Pellegriti G, Leboulleux S, Baudin E, Bellon N, Scollo C, Travagli JP, et al. Long-term outcome of medullary thyroid carcinoma in patients with normal postoperative medical imaging. Br J Cancer. 2003;88(10):1537–42.
22. Jin M, Megwalu UC, Noel JE. External beam radiotherapy for medullary thyroid cancer following total or near-total thyroidectomy. Otolaryngol Neck Surg. 2021;164(1):97–103.
23. Schwartz DL, Rana V, Shaw S, Yazbeck C, Ang K-K, Morrison WH, et al. Postoperative radiotherapy for advanced medullary thyroid cancer—local disease control in the modern era. Head Neck. 2008;30(7):883–8.
24. Maxwell JE, Sherman SK, O'Dorisio TM, Howe JR. Medical management of metastatic medullary thyroid cancer. Cancer. 2014;120(21):3287–301.
25. Priya SR, Dravid CS, Digumarti R, Dandekar M. Targeted therapy for medullary thyroid cancer: a review. Front Oncol. 2017;7:238. Oct 6 [cited 2021 Mar 18];7. Available from: https://www.ncbi.nlm.nih.gov/pmc/articles/PMC5635342/
26. Matrone A, Gambale C, Prete A, Cappagli V, Lorusso L, Bottici V, et al. Systemic treatment of advanced, metastatic, medullary thyroid carcinoma. J Cancer Metastasis Treat. 2021;7(23):10-20517. Available from: https://jcmtjournal.com/article/view/4013

16 The Role of Nuclear Medicine in Thyroid and Parathyroid Disorders

P. Shanmuga Sundaram, Padma Subramanyam, and G. Dinesh Kumar

16.1 Introduction to Nuclear Medicine

Nuclear medicine is a subspecialty that uses medically approved radioisotopes for imaging and therapy. Gamma-emitting radioisotopes are ideal agents for imaging, while beta and alpha emitters are used for radionuclide therapy. Nuclear imaging is commonly called "Scintigraphy" derived from the Latin word "scintilla," meaning "spark." The gamma camera is the instrument used to acquire images (gamma emissions or scintillations), while PETCT/MR captures the gamma-ray emissions from positron-emitting radioisotopes injected into the patient. All patients receive a radioactive isotope/radiopharmaceutical injection, and once it gets localized into the organ of interest, images are acquired. Processed high-quality images are visually and quantitatively analyzed to identify early organ physiological derangements.

16.1.1 Radiopharmaceuticals [1]

Radiopharmaceuticals are also known as radiotracers, comprising a radionuclide and a pharmaceutical that helps in the localization of radionuclide in the organ of interest. Generally, physiological changes precede anatomical derangements. Therefore, nuclear imaging identifies diseases much earlier than any other anatomical imaging counterpart. In the case of the thyroid, ^{131}I (radioiodine) gets maximally concentrated by thyroid follicular cells because of its affinity to iodine. The

P. S. Sundaram (✉)
Department of Nuclear Medicine and PET CT, Amrita Institute of Medical Sciences, Cochin, Kerala, India

P. Subramanyam · G. Dinesh Kumar
Department of Nuclear Medicine & Molecular Imaging, Amrita Institute of Medical Sciences, Cochin, Kerala, India

C. G. Nair, S. J. Abraham (eds.), *Surgical Management of Thyroid and Parathyroid Diseases*, https://doi.org/10.1007/978-981-97-3774-1_16

elemental form of iodine (stable iodine, i.e., ^{127}I) is essential for producing thyroid hormones. Hence, radioiodine is the universal radiotracer for imaging and treating thyroid diseases.

Iodides or iodide analogues (^{131}I or ^{123}I) are actively transported into the thyroid gland based on the sodium iodide symporter mechanism (NIS). The NIS is an intrinsic membrane protein implicated in iodide uptake into thyroid follicular cells. As the pure gamma-emitting ^{123}I is not available in India, gamma and beta-emitting ^{131}I is predominantly used.

The main use of ^{131}I is to determine the functional status of the thyroid gland in hyperthyroidism and also in the postoperative imaging and therapy of differentiated thyroid carcinoma (DTC).

Tumor seeking radiopharmaceuticals such as ^{99m}Tc - SestaMIBI (^{99m}Tc-MIBI) and PET tracer like ^{18}F- Fluorodeoxyglucose (^{18}F FDG) may be useful in discriminating benign from malignant thyroid nodule/s.

16.1.2 Radiopharmaceuticals Used in Thyroid Imaging

1. Radioactive iodine or iodine analogues ($^{123}I/^{131}I$)
2. ^{99m}Technetium ($^{99m}TcO4$ -Technetium pertechnetate)
3. Others: $^{124}I/^{18}F$ FDG (PET Tracers)

^{131}I and $^{99m}TcO4$ are the most commonly used radiotracers for routine thyroid imaging. Table 16.1 lists the various radiopharmaceuticals used in thyroid imaging.

16.1.3 Radioactive Iodine ($^{123}I/^{131}I$)

Both ^{123}I and ^{131}I are gamma-emitting radioisotopes used for imaging of thyroid gland. ^{124}I is a new PET counterpart that is still under investigation. ^{123}I, a pure gamma emitter, provides better quality images with lesser radiation exposure than ^{131}I due to its optimum gamma energy component (159 keV, kilo electron volt).

Both ^{131}I and ^{123}I can be used for radioactive iodine uptake (RAIU) studies.

^{131}I is an iodine radioisotope with gamma and beta emissions (energy of 364 and 606 keV respectively). The advantages of ^{131}I are manifold:

(a) dual potential as an imaging and therapy tool
(b) selective targeting and cytotoxic effect on thyroid tissue (i.e., magic bullet—due to selective uptake and organification in thyroid follicular cells)
(c) low cost, ready availability, oral route of administration
(d) no major side effects like hematologic, neurologic, or hepatotoxicity.

The major disadvantages of ^{131}I are its long physical half-life (8.02 days) and high beta emission, leading to a relatively higher radiation dose delivery to the thyroid and longer stay in the systemic circulation.

Table 16.1 Radiopharmaceuticals commonly used in thyroid imaging

Radiotracer	Type of emission	Availability	Use as a diagnostic agent	Therapy agent	Pt preparation	Time to complete imaging	Physical half-life
^{131}I	Both gamma (364 keV) & beta (606 Kev)	Nuclear reactor	Yes	Yes	Iodine restricted diet for 3–4 weeks (off Thyroxin, iodized salt, seafood, etc.)	48–72 h after oral administration	8.02 days
^{123}I	Gamma only (159 Kev)	Cyclotron produced (not available in India)	Best imaging tool	No	Iodine restricted diet	6–8 h after IV injection	13 h
$^{99m}TcO4$	Only gamma (140 Kev)	24 × 7 available in NM dept (daily elution)	Yes	No	No preparation needed	20 min after IV injection	6 h
^{124}I	Positron emitter	Not available	PET tracer	No data	No preparation	No waiting time	4.02 days
^{18}F FDG	Positron emitter	Available	PET tracer	No data	Euglycemic status mandatory before FDG inj.	Post injection waiting time of 1 h	110 min

Radioiodine (^{131}I), also popularly referred to as "atomic cocktail," is an ideal therapeutic agent to treat both benign thyroid conditions such as hyperthyroidism and malignancies like differentiated thyroid cancers.

16.1.3.1 ^{99m}Technetium

Technetium is prepared in-house on a daily basis using a ^{99}Mo–^{99m}Tc generator (commonly called Moly generator) by a process called elution. The final product eluted is Technetium pertechnetate ($^{99m}TcO4^-$). ^{99m}TcO4 is administered intravenously for thyroid imaging in the evaluation of hyperthyroidism/thyroid nodule evaluation.

^{99m}TcO4 uptake reflects only the thyroid trapping ability and not its organification ability but can nevertheless reliably differentiate various forms of hyperthyroidism.

16.1.3.2 ^{124}I Iodine

^{124}I is a cyclotron-produced PET radiopharmaceutical with a 4.2 days half-life. It offers superior imaging characteristics with enhanced spatial resolution and higher sensitivity. ^{124}I also has a favorable half-life permitting in-vivo iodine kinetics studies. It is an ideal agent for dosimetry studies in DTC. Numerous reports attest to the potential clinical benefits of ^{124}I imaging in patients with DTC. However, a uniform clinical protocol for imaging, image analysis and quantitation has not yet been firmly established.

16.1.3.3 ^{18}F Fluorodeoxyglucose, FDG

The most widely used positron-emitting radiopharmaceutical is the glucose analog FDG (Fluorodeoxyglucose) labeled with ^{18}F (Fluorine) having a relatively longer half-life of 110 min when compared to the rest of the shorter-lived PET agents (such as ^{11}Carbon, ^{13}Oxygen, etc.). Although it is not recommended as an agent for routine evaluation of thyroid, incidentalomas can be identified during oncologic evaluation for other malignancies. Imaging is performed using a PET camera with either a CT or MR counterpart (PET CT/MR hybrid imaging). FDG, being a glucose analogue, undergoes metabolism via the TCA cycle. FDG uptake in hyperplastic or malignant cells is governed by the enhanced GLUT (glucose transporter protein) expression. Overexpression of GLUT1 in malignant cells results in increased ^{18}F-FDG uptake. FDG is subsequently phosphorylated by hexokinase to FDG 6 phosphate. However, further metabolism of FDG 6 phosphate is curtailed and leads to its enhanced intracellular FDG accumulation. Aggressive tumors have higher FDG accumulation. This process of "metabolic trapping" of FDG in the cell constitutes the basis for imaging. It is possible to image the entire whole body in a single session, increasing the opportunity for finding unsuspected disease sites. Apart from oncology imaging ^{18}FDG radiopharmaceutical is increasingly used in the identification of sites of occult infection in patients with pyrexia of unknown origin, prosthetic infection, epilepsy, myocarditis, myocardial viability evaluation, etc.

16.1.4 Biokinetics of Radioiodine ^{131}I

Radioiodine ^{131}I is available in the form of sodium iodide (NaI). It is available both as an oral solution and capsules in different denominations that can be ordered as per requirement. Fasting for 4–6 h is recommended to enhance gastric absorption of ^{131}I. Once ingested, gastrointestinal absorption is rapidly at 5% per min and 50% with 30 min. The absorption is nearly complete within 12 h. Iodine is cleared from the plasma primarily by the thyroid gland but also by the salivary glands, gastric mucosal cells, mammary glands (NIS mechanism), and kidneys. Maximum excretion occurs within the first 24 h, and the remainder is localized in the thyroid gland. Renal excretion is through glomerular filtration at a mean rate of 34 mL (milliliter) per min. Around 73% of filtered iodide is reabsorbed by the renal tubules. Iodine is not bound in the kidneys. Renal clearance of iodine is constant over a wide range of plasma concentrations. Although radioiodine is not nephrotoxic, patients with renal failure should ideally undergo dialysis before the scheduled date of high-dose ^{131}I administration (therapy) to reduce whole-body radiation exposure (due to poor tracer clearance from systemic circulation).

16.1.5 Placental Transport and Lactation

Evidence supports placental transport of radioiodine, and high fetal-to-maternal thyroid ratios are achieved in the near term [2]. The fetal thyroid gland concentrates ^{131}I by the 12th–14th week of gestation [3]; hence, the use of radioiodine is strictly contraindicated in pregnancy. Iodine accumulates in lactating breasts and is secreted in breast milk [4]. Therefore, breastfeeding is temporarily discontinued following the ^{123}I scan but may be restarted after 2–3 days. However, breastfeeding is discontinued for a minimum period of 8 weeks for those undergoing high-dose ^{131}I therapy for DTC (to achieve an infant effective dose <1 mSv and an infant thyroid dose <10 mSv) [5].

16.1.6 Radiation Dose to Organs

The thyroid gland is the primary organ concentrating iodine, but the effective radiation is decided by the total uptake of the administered dose. The thyroid gland gets the maximum dose per millicurie, approximately 1300 rad per, compared to the rest of the body, which gets 0.71 rad.

16.1.7 Biokinetics of $^{99m}Tc04$ (Technetium Pertechnetate)

Technetium pertechnetate behaves like iodine in the thyroid gland and is trapped in the follicular cells within 20–30 min of intravenous administration. Normally, 1–4% administered dose is concentrated in the thyroid. The maximal concentration of the radiopharmaceutical is seen in the gastrointestinal tract, especially in gastric mucosa, approximately 5–10 times higher than the normal thyroid gland. The salivary glands also accumulate technetium more than seen in the thyroid gland. Routinely, four mCi of ^{99}Tc pertechnetate is administered intravenously. The follicular cells tend to concentrate Group VII anions of the periodic table that includes iodide and its analogues, i.e., pertechnetate, thiocyanate, and perchlorate. This forms the basis of the perchlorate discharge test in congenital hypothyroidism.

The distribution and blood clearance of ^{99m}Tc04 closely mimic those of iodides. Though the preferred route of administration is intravenous, oral administration is advisable in mentally challenged and non-cooperative patients. The material washes off the body in 24 h through urine but is excreted in breast milk in lactating ladies. The radiopharmaceutical crosses the placental barrier and so is generally contraindicated in pregnancy.

16.1.8 Radiation Dose to Other Organs

The amount of radiation to the thyroid gland following the routine dose of ^{123}I and ^{99} Tc is similar, but the whole-body dose is marginally high ^{99}Tc. However, both tracers emit considerably low radiation doses to the thyroid and the body. But I^{131}, on the contrary, emits about 100 times higher radiation in the therapeutic dosages than I^{123}. For the administered activity. (≈1 rad/μCi [10 mGy/0.037 MBq] versus 1 rad/100 μCi [10 mGy/3.7 MBq]). ^{99}Tc pertechnetate delivers 1 rad/5000 mCi (10 mGy/185 MBq) (μCi = micro curies, mGy = milligray, MBq = mega Becquerel).

16.1.9 Placental Transport and Lactation

^{99m}Tc crosses the placental barrier like iodine isotopes and should be avoided in pregnant ladies since the isotopes concentrate on the fetal thyroid at about the 12th week of gestation onwards. ^{99}Tc is concentrated in lactating breasts and contaminates breast milk.

16.2 Thyroid Scintigraphy

Technetium thyroid scintigraphy gives valuable information regarding the anatomy and function of the thyroid gland and is widely used in categorizing the nature of the hyperthyroid gland.

Thyroid scintigraphy is a simple investigation to evaluate the functional status of the thyroid gland. (Based on the selective uptake of various radionuclides by thyroid tissue). Scintigraphy also provides valuable information on thyroid anatomy and physiology and can play an integral role in diagnosing and managing hyperthyroidism. Technetium is a widely used radiopharmaceutical because of its ease of use and reliable information. But ^{131}I is selectively used in specific situations like congenital hypothyroidism along with the perchlorate discharge test.

16.2.1 Indications for ^{99m}TcO4 Thyroid Scintigraphy

1. Hyperthyroidism: The scintigraphy gives adequate information to differentiate a hyperfunctioning gland from viral thyroiditis with the release of hormones due to cellular damage. The qualitative and quantitative uptake of the tracer gives characteristic appearances in common hyperthyroid conditions such as Graves' disease and toxic nodular goiter toxic adenoma (Fig. 16.1).
2. Scintigraphy is rarely used as part of the evaluation of the thyroid.
3. When a primary mediastinal goiter is suspected, scintigraphy helps to differentiate it from other mediastinal mass lesions.
4. Scintigraphy is indicated in locating the source of hormone synthesis when extrathyroid locations are suspected, like lingual thyroid, thyroglossal tract and struma ovarii.
5. Congenital hypothyroidism—as a part of the assessment of native thyroid and developmental anomalies (Fig. 16.2).
6. Suspected amiodarone thyrotoxicosis (additional imaging with ^{99m}Tc MIBI recommended).
7. As preliminary workup for Graves' disease to assess the feasibility of low-dose ^{131}I therapy and for ^{131}I dose calculation.

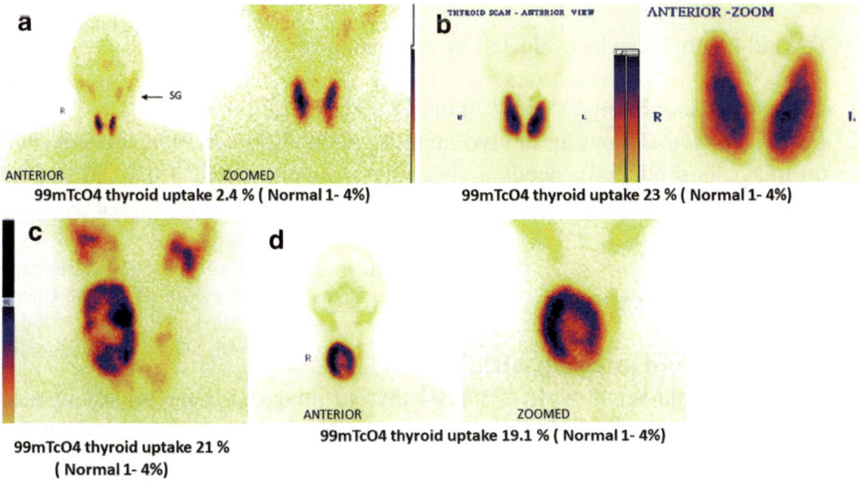

Fig. 16.1 ^{99m}TcO4 thyroid scan showing scintigraphic uptake patterns in benign thyroid diseases. (**a**) Normal functioning thyroid gland (SG denotes salivary gland); (**b**) Classical Graves' disease with high trapping function; (**c**) Toxic multinodular goiter; (**d**) Autonomous toxic nodule of right lobe suppressing left lobe of thyroid gland

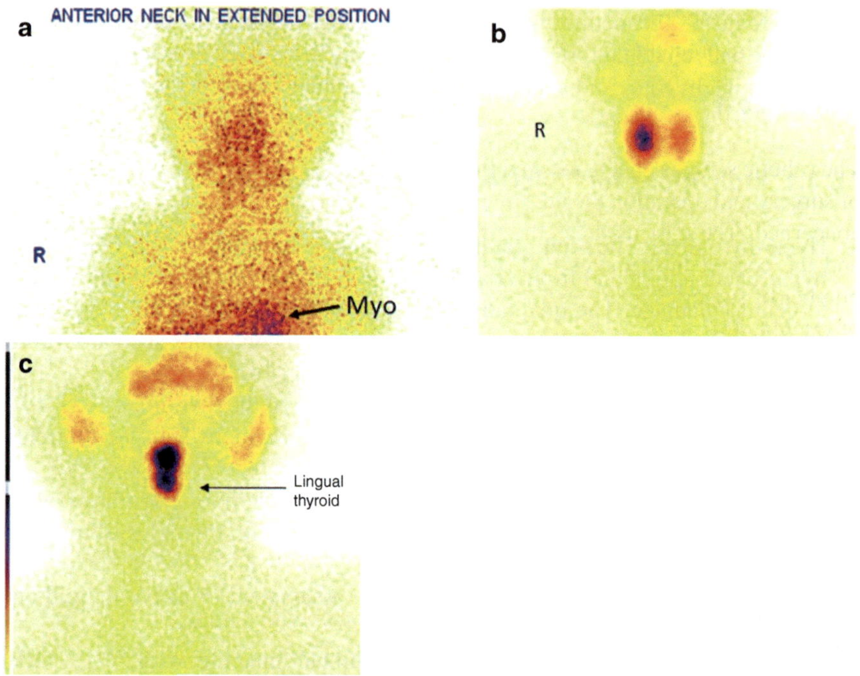

Fig. 16.2 (**a**) ^{99m}Tc MIBI thyroid scan in a hypothyroid child while on tab thyroxine. Scan shows absence of thyroid gland in its normal anatomical or ectopic positions (Myo denotes myocardium); (**b**) ^{99m}TcO4 thyroid scan in a hypothyroid child showing presence of native thyroid gland; (**c**) ^{99m}TcO4 thyroid scan showing lingual thyroid

Thyroid function can be evaluated by two methods:

1. *Camera method*: Using ^{99m}Tc04/^{131}I or ^{123}I.
2. *Uptake method*: This is an in vivo imaging method (radioactive count-based, no image is produced); needs a special instrument called a thyroid uptake probe. This procedure is popularly known as radioactive iodine uptake (RAIU).

16.2.2 Camera Method

16.2.2.1 Thyroid Imaging with ^{99m}Tc04

Technetium thyroid scintigraphy (^{99m}Tc04 thyroid imaging) is the widely chosen primary imaging for hyperthyroidism but a select Nuclear Medicine centers prefer Radioiodine Uptake Studies (RAIU) before planning low dose ^{131}I ablation. For planning treatment for differentiated thyroid carcinoma ^{131}I imaging is mandatory and ^{99}Tc04 is not used to perform whole-body imaging in DTC. A ^{99m}Tc04 thyroid imaging does not accurately reflect the disease burden.

Advantages of ^{99m}Tc04 imaging in hyperthyroidism:

1. There is no specific preparation essential for the procedure.
2. The entire procedure is short and completed in about 30 min.
3. There is no restriction to any age group, but it is avoided in pregnant and breast-feeding ladies (relative contraindication).
4. The diagnosis is almost straightforward since characteristic thyroid uptake patterns exist.
5. The procedure is cost-effective.
6. There are no significant side effects.

When the imaging is planned for patients receiving antithyroid medications, drugs are temporarily discontinued for 5–7 days for better imaging and to accurately estimate the thyroid gland's trapping function. Compared to ^{131}I imaging, TcO4 imaging provides all the necessary information to plan low-dose ^{131}I therapy.

16.2.2.2 Procedure

The radiopharmaceutical is injected (2–4 millicurie (mCi) of ^{99m}Tc04) intravenously, and static images of the extended anterior neck are collected about (or 300–500 kilo counts) about 15–20 min later. An initial image is also acquired using a Gamma camera to calculate the trapping function. Fast cine images (2 s/frame for 1 min) are used to study the characteristics of nodules when present. ^{99m}Tc04 accumulates in the thyroid gland because of the anionic transport mechanism and NIS (sodium iodide symporter). Salivary glands and oral mucosa are also normally visualized. The evaluation of the nodules is precisely done with pinhole collimators, which are expensive additional hardware that has to be attached to the gamma camera. Single-photon emission computed tomography (SPECT) is very helpful in assessing nodules.

16.2.2.3 Interpretation

The features of a structurally and functionally normal thyroid has a characteristic pattern and various pathological conditions thyroid alter these regular features that are almost typical of that pathology [6]. The features regularly looked at in the scintigram include size, presence of a pyramidal lobe, percentage distribution of tracer in each lobe, the total thyroid uptake, and the characteristics of nodules.

These abnormal tracer uptake patterns include diffuse or focal, homogeneous or heterogeneous, increased or decreased when compared to parotid glands. The pyramidal lobe is identified as a tongue-shaped projection arising from one of the thyroid lobes or isthmus. Based on uptake by the nodule they are called as hot/cold. Iodine uptake shall occur in outside the location of native thyroid in the esophagus, ectopic thyroid sites like thyroglossal duct cyst, and substernal/retrosternal goiter. Since the radiotracer is excreted in saliva and inadvertent swallowing can result in extrathyroidal uptake in the esophagus. This is confirmed by a second image after a gulp of water which washes off the activity. Thyroid scintigraphy confirms the pathology of congenital hypothyroidism of neonates. Scintigraphy is useful in locating functioning thyroid tissue in ectopic locations and perchlorate discharge test is useful to identify peroxidase deficiency conditions.

Table 16.2 Scintigraphic uptake patterns in thyrotoxicosis (findings are similar in Tc04 and iodine scans)

Etiology	Uptake patterns
Graves' disease	Diffuse increased uptake (homogenous) in both lobes of thyroid gland. SG not seen/faintly seen. Uptake >4%
Toxic hot nodule	(Autonomous) Increased uptake in nodule with suppression of surrounding normal thyroid tissue. SG may be normally seen/reduced uptake
Toxic MNG (Multinodular Goiter)	Multiple hot nodules are interspersed with cold nodules. SG faintly seen
Nontoxic MNG	Enlarged thyroid gland with multiple cold nodules, usually nonfunctioning. In patients with additional thyroiditis (biochemically detected hyperthyroidism), the gland may be faintly seen, and, nodules if any, may not be apparent. In this situation, apart from the ^{99}mTc04 scan, the I^{131} scan is also suggested to look for functioning nodules. Cold nodule in Tc scan and hot nodule in iodine scan, suspicious for malignancy
Subacute thyroiditis	The thyroid will be faintly visualized (depending on the duration/ severity of disease, i.e. nearly recovered case will show uptake close to 1% (normal 1–4%). Mimicking picture seen also with exogenous iodine interference like homeopathic drugs/oral or iodinated contrast agents
Chronic thyroiditis (Hashitoxicosis)	Only serial imaging can differentiate this disease entity along with biochemical evidence of waxing/waning thyrotoxicosis. Usually, Hashimoto's thyroiditis in hypothyroid status will show reduced uptake, but in toxicosis, the phase will simulate the GD uptake pattern
Factitious thyrotoxicosis (excess thyroxine intake)	Generalized reduced uptake (needs to be correlated with a history of drug intake)
Secondary causes (elevated TSG, HCG)	Increased uptake localized to ectopic sites such as pelvis/adnexa, etc.
Ectopic thyroid tissue	Focal tracer uptake in lingual/sub hyoid or other ectopic sites. Uptake may be focal normal or reduced. SG will be normally seen (Fig. 16.2)

SG Salivary gland

Scintigraphy characterizes the exact cause of hyperthyroidism and helps to formulate the treatment protocol. There are however no differences in thyroid scintigraphic findings in children compared to adults.

16.2.2.4 Scintigraphic Patterns in Different Thyroid Pathologies (Table 16.2)

1. **Normal pattern**: Structurally and functionally normal thyroid is known to have as butterfly-shaped appearance in the anterior neck at midline with homogenous tracer distribution. Parotid uptake of ^{99m}Tco4 is at par with par with thyroid uptake. Isthmus is a normally a thin bridging tissue and is occasionally visualized in scintigram. The pyramidal lobe may be seen in around 10% of patients. The uptake is estimated by comparison with the degree of thyroid trapping with

that in the salivary gland (NIS glycoprotein expression). The normal trapping function is 1%–4%. Suprasternal notch is marked and the retrosternal extension also is clearly seen in the imaging.

2. **Vascular status**: Generally thyroid gland is highly vascular and is evidenced by the increased trapping of the radiotracer. The vascularity of the gland in Graves' disease is high and adds on to increased trapping ability of the follicular cells. Thyrotoxicosis due to thyroiditis (viral or autoimmune) will display generalized reduced vascularity.
3. **Graves' disease (GD)**: The scintigraphic appearance in Graves' disease is characterized by intense homogenous concentration in comparison with parotid glands. The intensity of concentration is proportional to severity of the disease and the trapping ability will always be more than 4%. The pyramidal lobe and the isthmus are visible in the scintigram and if thyroid tissues are present in thyroglossal tract shall be lighted in the scintigram. The trapping function is blunted when patient is on antithyroid drugs and exogenous iodine supplementation.

   Graves' disease with Marine Lenhart syndrome: The fundamental reason of hyperactivity of follicular cells in Graves' disease is stimulating TSH receptor antibodies and the cells are no longer under control of TSH. The occasional presence of TSH-sensitive nodules coexisting in Graves' disease was first described 1911 by Marine and Lenhart. This rare association is known as Marine, Lenhart syndrome and has a 2.7–4% incidence. Thyroid scan shows poorly functioning or cold nodules scattered in bright areas. Presence of such nodules indicates uncertain outcomes following radioiodine ablation. They are found to be slightly resistant to ^{131}I therapy and need a larger dosage of ^{131}I when compared to GD patients without nodules.
4. **Toxic multinodular goiter** (Plummer disease): The disease is characterized by autonomous areas and nodules developing in nodular goiters. Thyroid scan shows heterogeneous appearance with patchy areas of intense concentration and cold areas. The overall trapping ability shall exceed 4%.
5. **Nontoxic multinodular goiter**: The scan shows a large volume goiter with scattered areas with cold/warm nodules. Usually, thyroid uptake will not be more than 4%.
6. **Subacute (viral) thyroiditis**: The initial clinical presentation is usually due to symptoms related to thyrotoxicosis with local pain at the region of thyroid gland. Dynamic imaging shows decreased vascularity and static image reveals poor trapping of radiotracer. The pattern of tracer uptake depends upon the phase of viral activity; acute phase, resolving phase, or chronic phase. Proper clinical assessment is absolutely essential for arriving at a correct diagnosis. Patients on iodine-containing drugs may also reveal a similar scintigraphic picture.
7. **Autonomous toxic nodule**: The situation is characterized as autonomously functioning thyroid nodule, the rest of the gland being suppressed. Unless the nodule is large to cause cosmetic disfigurement radioiodine ablation gives very good response.

16.2.3 Amiodarone Induced Thyrotoxicosis (AIT)

Amiodarone is a popular drug in the management patients with cardiac rhythm abnormalities. The drug is rich in iodine and is influence thyroid function due to the high iodine load. Amiodarone contains about 37% iodine by weight and the patient on daily dose of 200 mg, ingests 75 mg of organic iodine. The drug undergoes further metabolic degradation releasing more iodine, approximately 6 mg of free circulating iodine. Ultimately the patient receives about 20–40 times higher amount iodine daily. The clinical effects of amiodarone on thyroid function are both related hypothyroidism and hyperthyroidism. Amiodarone-induced hypothyroidism (AIH) is common in iodine sufficient population [7]. AIT is prevalent in iodine-deficient populations; with Hashimoto's thyroiditis being an important risk factor. Type I AIT usually occurs in patient with subclinical autoimmune thyroid disorder leading on to developing overt Graves' disease. Type II is toxic manifestations due to cellular destruction induced by excess iodine and later becomes hypothyroid. A combination of both types is not unusual (Type I, II, mixed). Amiodarone induced thyrotoxicosis has to be properly diagnosed and treated since thyrotoxicosis itself induces cardiac arrhythmias. Discontinuation of amiodarone with proper substitute is ideal and scintigram is good tool to make proper diagnosis (Table 16.3).

16.2.4 RAIU Studies: Radioactive Iodine Uptake Studies Using Thyroid Uptake Probe

A miniature gamma camera with sodium iodide crystal otherwise called the Gamma Probe is now in wide use. This cylindrical or conical shaped collimator mounted on a stand and helps to calculate the uptake quantitatively and helps to calculate the thyroid uptake in counts with the help of a computer software. There is no image and the report is interpreted as counts per minute (cpm).

Indications:

1. This probe helps to iodine uptake in low dose ^{131}I therapy in Graves' disease.
2. Probe helps to calculate the dosage of ^{131}I for low dose therapy.
3. As a diagnostic tool to differentiate Graves' disease from subacute thyroiditis or factitious hyperthyroidism.

   The probe is also useful in management of differentiated thyroid carcinoma (DTC) for assessment of regional and distant metastases. But the procedure can be done with radioiodine (^{125}/^{131}I) given as 10-25 micro-Ci capsules, and no images are obtained. The number of counts emanating from the thyroid gland is recorded over a specified period of time to determine the functional status of the gland. The uptake study gives information of percentage of administered ^{131}I thyroid over a specified period.

Table 16.3 Differentiation of Type I and II amiodarone induced thyrotoxicosis

	Type I	Type II
Underlying thyroid disease	Yes	No
Incidence	68%	Remaining
Time after starting amiodarone	Short (median 3 months)	Long (median 30 months)
Cause	Sudden iodine load increases thyroid hormone production in autonomous thyroid tissue—Jod-Basedow phenomenon	Direct toxic effect of the drug on normal thyroid gland resulting in subacute and destructive thyroiditis
24-h iodine uptake	Low–normal (may be high in iodine-deficient regions)	Low to suppressed
USG findings	Diffuse may be nodular	A normal or small gland
^{99m}TcO4 Scintigraphic findings–vascularity	Increased	Absent
^{99m}TcO4 Scintigraphic findings—static images	The pattern of GD/MNG	Gland not visualized
T4/T3 ratio	Usually, <4	Usually, >4
Tg AB/TPO ab/TSI	Maybe present	Usually, absent
Circulating interleukin 6	Normal to high	Frequently markedly elevated
^{99m}Tc MIBI findings	Goiter noted (GD/MNG)	Outlines of gland delineated (even though no uptake on ^{99m}Tco4 scan)
Therapy	Treat primary disease with low dose I^{131} therapy	Self-limiting on stoppage of drug, supportive with NSAID, steroids

16.2.5 Special Indications for ^{131}I Scintigraphy in Benign Thyroid Diseases

1. The diagnosis of mediastinal primary goiter.
2. Diagnosis of functioning thyroid tissue in ectopic locations like Struma ovarai.
3. Congenital defect in organification of iodine.

16.2.6 Congenital organification Defect Evaluation—Perchlorate Discharge Test

Defect in organification of iodine is cause of congenital hypothyroidism and is evidenced in the first few months of life. Laboratory studies show markedly elevated TSH. The diagnosis is confirmed with perchlorate discharge test

Table 16.4 Camera method—scintigraphic uptake patterns in congenital organification defects

Diagnosis	Initial scan findings	Delayed scan findings
Normal patient	Homogenous tracer uptake	Retained homogenous tracer uptake
Partial organification defect	Homogenous/inhomogeneous tracer uptake will be seen	>10% clearance of tracer
Complete organification defect	Homogenous/inhomogeneous tracer uptake will be seen	>90% clearance of tracer (washout)
Agenesis	No uptake	No uptake

(Table 16.4). Tracers used for this test are ^{131}I or ^{99m}Tc04, but the recommended tracer is ^{131}I.

Perchlorate is a NIS inhibitor and in normally functioning thyrocytes, oral perchlorate (500–1000 mg) challenge does not induce the clearance of tracer from the thyroid gland. The is delayed when challenged with oral perchlorate a reduction in thyroid tracer uptake by more than 10% in imaging after 1–2 h indicates a partial defect. Iodine discharge more than 90% is considered total defect of iodide organification.

But when ^{99m}TcO4 is used as tracer, an initial image at 10 min and a delayed image at 60 min are gathered. An additional ^{99m}Tc MIBI thyroid scan may highlight the presence or absence of native thyroid, uninfluenced by thyroxine/presence of stable iodine in the body pool.

16.3 Thyroid Nodule Evaluation

The challenge in routine evaluation of thyroid nodule is detection of neoplastic nodules. Thyroid scintigrams differentiate a functioning and nonfunctioning nodules and functioning nodules are usually benign. Based on the scintigram the nodules which fail to concentrate tracer are called cold and hot when the concentration is more than surrounding area. A nodule with concentration of tracer similar to that of surrounding area is called warm nodule. Nonfunction nodule has a higher risk for malignancy and aspiration cytology is routinely done. Coupling with high resolution ultrasound imaging enhances the accuracy and also helps to take targeted cytology.

16.3.1 Tracers Used for Thyroid Nodule Evaluation

1. ^{99m}Tc O4/^{131}I/^{123}I
2. ^{99m}Tc MIBI (SestaMIBI)/^{99m}Tc Tetrofosmin/^{201}Thallium
3. ^{99m}Tc DMSA (V)—dimercaptosuccinic acid
4. ^{18}F FDG PET—18 Fluorodeoxyglucose
5. ^{68}Ga DOTA PET—somatostatin (SST) analogs labeled with ^{68}Gallium
6. ^{18}F DOPA—Fluorine-18-dihydroxyphenylalanine

   Generally, ^{99m}Tc4 thyroid scan is recommended when the TSH is low however, it is used in US and European Union countries for evaluation of nodules > 10 mm size regardless of TSH levels [8].

Indications for MIBI thyroid scan in evaluation of thyroid nodule:

MIBI concentrates in metabolically active tissues and is used as a tracer in neoplastic thyroid nodules.

1. Thyroid scan with MIBI as tracer helps to locate the nodule in the hyper functioning thyroid gland.
2. MIBI scan also helps to characterize USG detected suspicious thyroid nodule.
3. MIBI scan is also used as part of evaluation of amiodarone-induced thyrotoxicosis (AIT).
4. MIBI scan could be part of evaluation of nodules with indeterminate cytology (Bethesda III and IV).

16.3.2 FDG PETCT Imaging

A fluorodeoxyglucose positron emission tomography (FDG-PET) is now popular imaging used in assessment of extent of many malignancies. Now the imaging has applications in many fields other than malignancy. The imaging is usually performed using vendor-specified simultaneous PET CT protocol. The image acquisition can be of the whole body or region specific according to the need. Ideally the patient is fasting overnight and the blood sugar level <150 mg/dL and serum creatinine levels within the normal ranges. The CT scan image acquisition precede the FDG administration.

FDG-PET imaging is not routinely performed as part of thyroid nodule evaluation. But a nodule incidentally detected on PET scan done for other indications stand higher chances of malignancy [9]. ^{18}F FDG accumulates in neoplastic cells which use glucose as the source of energy source in proliferative activity. But FDG PET-CT is able to identify and characterize with accuracy small thyroid nodules (i.e., ≥5 mm) and also cytological indeterminate thyroid nodule [10]. The ^{18}F FDG PETCT scores over the qualitative ^{99m}Tc MIBI scan in predicting risk of malignancy in thyroid nodules. FDG-PET is on regular use in assessment of advanced differentiated and poorly differentiated thyroid carcinoma especially when ^{131}I scan fail to localize regional or distant metastases in the presence of elevated thyroglobulin levels.

Introduction of newer tracers like ^{18}F-DOPA, ^{68}Ga DOTATATE has widened the scope of the evaluation of thyroid nodules with Medullary Thyroid Cancer (MTC). However, neuroendocrine tumors (NETs), including MTC, have low proliferation index and so has low ^{18}F-FDG uptake. ^{68}Ga-DOTATATE- PET-CT is useful in detection of recurrent MTC demonstrating a significantly higher number of lesions.

16.4 Low Dose ^{131}I Therapy for Hyperthyroidism

Conventionally, the term low dose ^{131}I therapy denotes oral administration of ^{131}I to treat hyperthyroidism (dosage range 5–15 mCi); while high dose ^{131}I therapy denotes the use of ^{131}I in dosage above 30 mCi for DTC management.

Ablation using ^{131}I established economical, safe, and effective treatment thyrotoxicosis management for over six decades. However, dosage of ^{131}I for effective treatment of Graves' disease remains controversial. Despite the wide spread usage of the radioactive iodine, few controversies exist in selection of patients, goal of therapy, and dosage schedule.

16.4.1 Indications

1. Graves' disease (GD)
2. Toxic multinodular goiter (TMNG)
3. Autonomously functioning toxic nodule/s (ATN)

16.4.2 The Goal of Low Dose ^{131}I Therapy [11]

The debate over the aim of treatment was whether to make the patient euthyroid or to make him hypothyroid. However, the wider accepted protocol is to administer a dose of ^{131}I dosage to render patients euthyroid is not always possible and a higher dose to render the patient hypothyroid is more practical.

Selection Criteria:

1. Ensuring iodine avidity by a ^{99m}TcO4 thyroid scan or RAIU study is recommended.
2. A ^{99m}TcO4 thyroid scan or RAIU study will exclude the other forms of thyrotoxicosis such as viral thyroiditis and factitious thyrotoxicosis.

The low dose ^{131}I therapy as a primary treatment for hyperthyroidism was found more effective than following anti-thyroid drugs (ATD) failure [12] probably due to drug induced impairment of follicular cell activity. Propyl Thiouracils have lasting effects on the cells even after the drug is withdrawn for 15 days. Generally, anti-thyroid drugs are withdrawn at least 5–7 days before low dose ^{131}I therapy (a minimum of 5 days for methimazole and 3 days for propylthiouracil) and are restarted 7 days after treatment.

The success rate of one sitting of low dose ^{131}I therapy for hyperthyroidism management is 93% in our institute with adequate patient preparation. A second or third doses are required in high volume glands and those with florid Graves' disease evidenced by a high technetium trapping function. Hypothyroidism sets in due course of time and levothyroxine supplementation is essential.

Patients with Graves' ophthalmopathy may worsen following ^{131}I therapy. It is not unusual to develop ophthalmic complications following ablation. When Graves' ophthalmopathy is suspected a proper ophthalmology assessment to assess the clinical activity score is mandatory before scheduling for low dose ^{131}I therapy. Systemic steroid in a dose 0.5 mg to 1 mg/kg body weight started a week before low dose ^{131}I therapy and is continued for 3–4 weeks as a prophylaxis against worsening of ophthalmopathy.

16.4.3 Methods of Calculating ^{131}I Dosage for Low Dose ^{131}I Therapy

1. The dosage of ^{131}I calculated based on volume of the thyroid gland calculated by ultrasound imaging.
2. Otherwise, a fixed empirical dosage is given.

There is no significance evidence supporting dosage calculated has superiority over the fixed dose regimen. Majority of patients receiving low dose therapy ends up in hypothyroid state irrespective of the dosage schedule adopted and a fixed dose regimen is convenient to follow.

Ideally, 150–300 Gy (gray) radiation absorbed dose needs to be delivered to thyroid tissue for exerting cytotoxicity and get successful ablation in hyperthyroidism. The ^{131}I dosage necessary to deliver the above radiation effect is ideally 160 microCi per gram of thyroid tissue. The routine practice followed in our Institution a fixed dosage protocol as noted below.

Graves' disease: 8–10 mCi
Toxic MNG: 10–15 mCi
Autonomous toxic nodule: 15–20 mCi

The radiopharmaceutical is administered orally as a single dose and the patient report after overnight fasting. The procedure is conducted as a day procedure and the patient is instructed to restart the antithyroid medications a week after. The patient is reviewed after 6–8 weeks later and thyroid function is estimated after discontinuing antithyroid medications for a week. In case of resolution of hyperthyroid state and hypothyroidism has set in levothyroxine supplementation is started.

Those who show persistence of hyperthyroid state need additional doses of ^{131}I which is planned 8–12 weeks after the initial ablation.

16.4.3.1 Factors Affecting the Clinical Outcome

Patient characteristics such as gender and age, disease severity, and goiter volume are the common features influencing the outcome. Discontinuing antithyroid medications for a minimum period one helps to achieve a better outcome.

16.4.3.2 Low Dose ^{131}I Therapy in Children

Generally, the use of ^{131}I is not recommended in children with Graves' disease but a few centers have treated children with radioiodine. The dosage of ^{131}I has to be adjusted with reference to the clinical and biochemical severity, and body mass index. Hypothyroidism almost always sets in with a dose more than 200–250 Gy (220–275 microCi/g). The size of the gland is a significant predictor of predictor with large volume goiter > 80 g. Usually, ^{131}I therapy is avoided in children <5 years.

16.4.4 Complications

The common side effects include nausea, neck pain.

Side effects of low dose ^{131}I therapy are transient and include nausea, mild neck pain lasting for few days. Thyroid storm had been reported rarely 1–14 days post-^{131}I administration in patients with severe, uncontrolled hyperthyroidism and very large goiters. Worsening of ophthalmopathy has been reported in a small percentage of adults who have received ^{131}I, but now is controlled by prophylactic steroid. Symptoms of hyperthyroidism if any in the interval period can be readily controlled with adequate use of beta-blockers (usually oral metoprolol 20–40 mg per day).

16.5 High Dose ^{131}I Therapy in Differentiated Thyroid Carcinoma

Since the DTC cells retain iodine avidity to certain extent ^{131}I is widely used adjuvant in the management of these cancers following thyroidectomy. The term high dose ^{131}I ablation aims to destroy locoregional residual tissues and high dose therapy aims at locoregional or distant metastases. Distant metastases from DTC at any location can be treated with high dose therapy.

16.5.1 ^{131}I Therapy Indications in DTC [13]

1. Gross extrathyroidal extension or incomplete tumor resection regardless the size of the tumor.
2. Primary tumor size more than 4 cm regardless of age.
3. When distant or regional metastases are confirmed irrespective of the size of primary tumor.
4. DTC patients of age more than 45 years.
5. Presence of tumor invasion to the perithyroidal soft tissues.
6. Papillary tumor with aggressive histological subtypes.
7. Patients with ^{131}I concentrating distant metastases (lymph node, pulmonary, and skeletal).
8. Patients who has inoperable residual disease.

16.5.2 Patient Preparation

The uptake of ^{131}I depends upon the TSH and low serum iodine where by the cells are starving. Serum TSH level $\geq 30\ \mu IU/L$ enhances uptake of iodine by an increased NIS expression and the hypothyroid state is achieved in 3 weeks after thyroidectomy. If the patient is on levothyroxine (LT4) discontinuation for 3–4 weeks is necessary. Such severe hypothyroid state is intolerable to a few patients and in those patients Recombinant human TSH (rhTSH) is an alternative. Recombinant human

TSH (rhTSH) is a genetically engineered molecule containing thyrotropin Alfa (0.9 mg in each vial) (Thyrogen by Genzyme). The molecule is commercially available vials containing 1.1 mg of thyrotropin Alfa, 36 mg Mannitol, 5.1 mg sodium phosphate, and 2.4 mg sodium chloride. The drug is reconstituted prior to use with 1.2 mL of sterile water to deliver 1 mL injection from each vial. One milliliter reconstituted medication is injected intramuscularly in the gluteal region followed by the second injection 24 h later. There are no major side effects except for an occasional mild flu-like symptoms. The common recommendations for rhTSH include for those patients intolerable to hypothyroid state, other major illness preventing a hypothyroid state and extensive functioning metastases [14].

16.5.3 Scan Procedure

Pre-therapy scan indicates a preliminary scan done after administering 2–3 mCi of ^{131}I which shall expose the disease status. Images of anterior neck and whole body are collected. The argument against this procedure is possibility of stunning of follicular cells when the actual ablative dose of ^{131}I is administered. Before the full dose of ablation is administered a smaller dose of 2–3 mCi of ^{131}I given and images are acquired and the procedure is called pre-therapy scan. The chances of stunning are minimalized by the dose not exceeding 5 mCi.

16.5.3.1 Interpretation

There is physiological uptake of iodine in extra-thyroidal regions such as salivary glands, breasts, and stomach. The absorbed iodine is metabolized in liver and excreted through the urinary system and so liver and urinary tract are also visible in whole body scan. When the pre-therapy scan shows no significant residual tissue and the serum thyroglobulin levels are also low the patient may there is no need for ablation. False-positive ^{131}I uptake may be encountered in inflammatory, cystic, non-thyroid neoplastic disorders, for example, pleural or pericardial effusions; aspergilloma, arthritis; ovarian or breast cysts (Figs. 16.3 and 16.4).

Usually, ^{131}I uptake in nasopharynx, esophagus, smearing of urine in pelvis/any other sites of contamination is recognized easily by experienced specialists, and repeat imaging after decontamination clarifies the case. Stunning is the phenomenon of low or absent uptake of ^{131}I by residual or metastatic tissue due to previous iodine loading.

16.5.4 Nuclear Medicine Algorithm of Management of DTC [15, 16]

Conventionally, 3–4 weeks after total thyroidectomy, a residual thyroid and whole body (low dose) ^{131}I scintigraphy is performed to assess the presence of residual thyroid tissue. In presence of residual thyroid tissue, as per risk stratification, patients undergo ^{131}I residual thyroid ablation/therapy. Two important determinants of the success of thyroid ablation are the mass of remnant thyroid tissue in the neck, and the initial radiation dose delivered to the tissue. Normal dosage of ^{131}I used to ablate residual thyroid tissue is in the range of 30–50 mCi (between 30,000 to 100,000 rads to be delivered to the remaining thyroid tissue). A post ^{131}I therapy whole-body scintigraphy, routinely performed 5–7 days after ^{131}I ablation/therapy, reconfirms a good concentration of ^{131}I in thyroid bed and also identifies unsuspected distant metastases (Figs. 16.5 and 16.6).

TSH suppression regimen of thyroxine started following the radioiodine treatment and most of the patents are scheduled for review after 6 months. Patients presenting with distant metastases need higher dosages of ^{131}I therapy and the dosage of ^{131}I may be calculated on a dosimetry calculation method or on empirical fixed dosage schedule (Figs. 16.5 and 16.6).

The fixed dosage schedule followed in the Institution is shown below.

1. Residual thyroid ablation—30–50 mCi
2. Lymph nodal metastases—around 100 mCi
3. Pulmonary metastases—120–150 mCi
4. Skeletal metastases—200 mCi

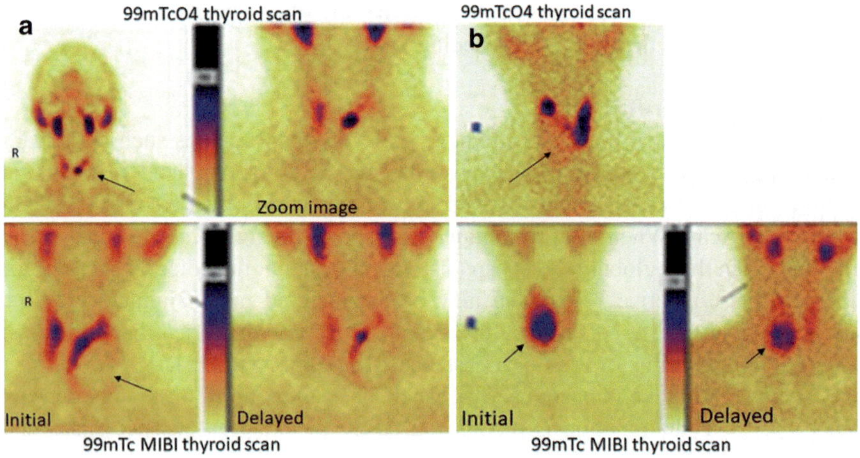

Fig. 16.3 (**a**) Upper row: $^{99m}TcO4$ thyroid scan showing normal tracer distribution in right lobe and left upper lobe of thyroid gland. Large cold nodule (denoted by arrow) occupies major part of left lobe with no filling up of tracer in MIBI thyroid scan (lower row images, both initial and 2 h delayed imaging). R denotes right lobe thyroid. Combined $^{99m}TcO4$ and MIBI thyroid scan suggests a benign nodule, probably a large colloid nodule involving major part of left lobe of thyroid gland; (**b**) Images of a 26-year-old lady with palpable right lobe thyroid nodule. Upper row image demonstrates a large cold nodule involving major part of right lobe (in $^{99m}TcO4$ thyroid scan) which shows filling up of MIBI in lower row images (arrow) in initial and delayed scan. Findings strongly favor a neoplastic thyroid nodule

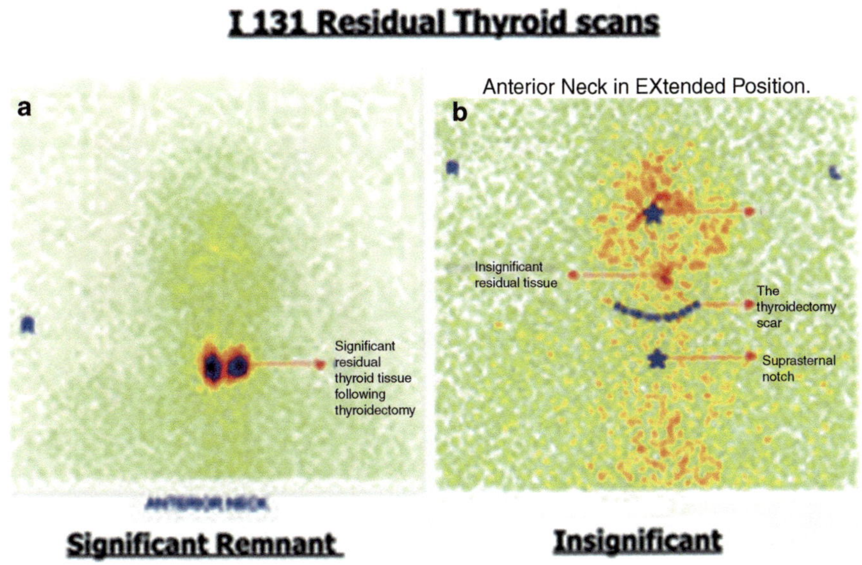

Fig. 16.4 Diagnostic ^{131}I thyroid images of two patients. (**a**) Shows significant ^{131}I uptake in residual thyroid tissue in both lobes; (**b**) Negligible residual thyroid tissue post thyroidectomy status (if stimulated thyroglobulin is <2 ng/mL with normal antiTg antibody, then radioiodine therapy is not necessary)

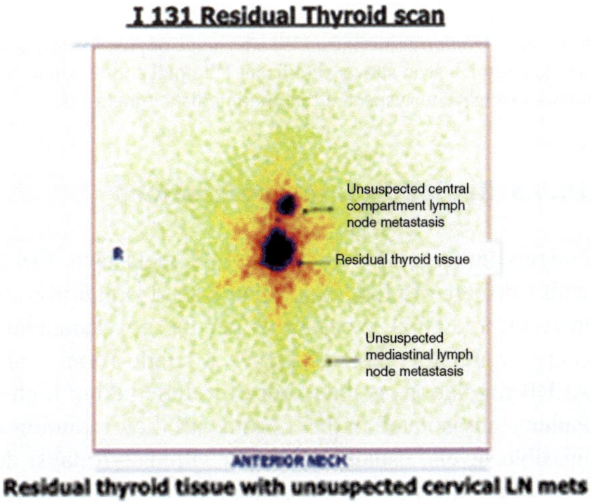

Fig. 16.5 Scan demonstrates risk stratification by performing whole body ^{131}I scan. Added advantage of residual thyroid scan in identifying residual thyroid tissue and unsuspected upper, lower cervical nodal metastases (i.e., functioning cervical nodal metastases) is to be noted in this 48-year-old papillary thyroid cancer patient

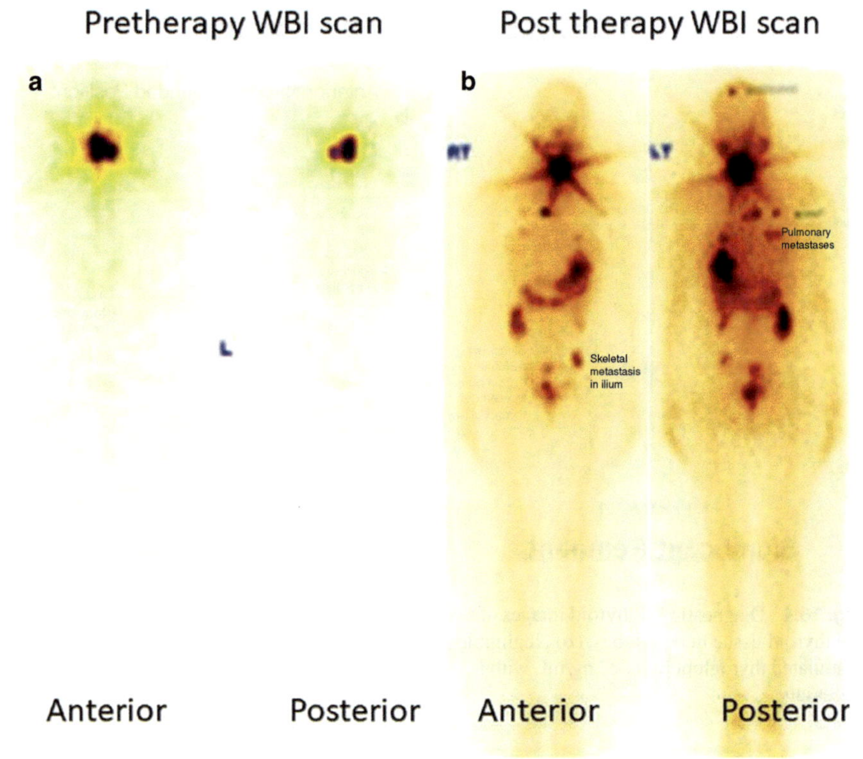

Fig. 16.6 Demonstrates the added advantage of post therapy whole body I^{131} scan. (**a**) Pre-therapy whole body I^{131} scan shows significant I^{131} uptake in residual thyroid tissue. (**b**) Post therapy scan shows extensive unsuspected functioning distant metastases

16.5.5 Radiation Safety Precautions

Patients undergoing a high dose ^{131}I metastasis therapy/residual thyroid ablation with a dosage of more than 30 mCi need isolation from the general public to minimize the radiation exposure as per the governmental regulations (AERB, atomic energy regulatory board). Hence a single room isolation designed according to AERB regulations is mandatory for undertaking high-dose ^{131}I therapy (Fig. 16.7). Patients are isolated in these rooms till their radiation exposure rate reduces to permissible levels, usually achieved within 2–3 days, depending on the ^{131}I dosage administered. They are advised home isolation for the next 5–7 days and can use a separate toilet for next 1 week.

The importance of post-therapy WBI scintigraphy is its high sensitivity in identifying and characterizing the extent of thyroid remnant, tumor, and detecting previously occult distant metastases. Incorporating SPECT/SPECT-CT enhances the efficacy in imaging the regions of interest (ROIs) as applicable and this imaging is generally performed 4–7 days after the day of ablation.

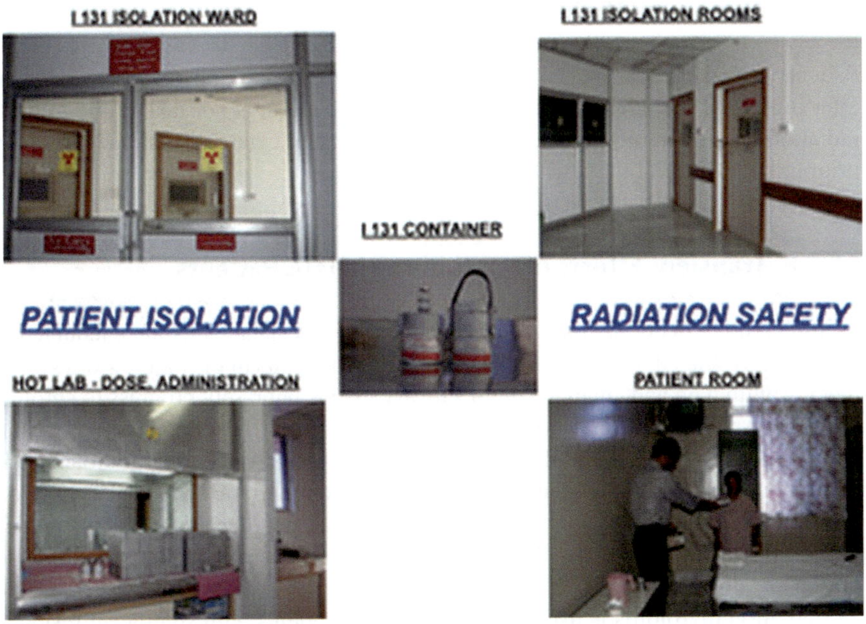

Fig. 16.7 Isolation room for high dose I^{131} therapy, package of I^{131} when received from supplier, radiation exposure readings of patient and 1 and 3 meter distance being monitored by radiation safety officer (RSO) in isolation room prior to discharge from room

16.5.6 Pediatric/Adolescent Patients with DTC [17]

Thyroid cancers are <1% among those <10 years but increases in incidences to 3.6% among the age group of 10–14 years and 7.8% among the age group of 15–19 years. Papillary thyroid cancers account for the 90% pediatric thyroid cancers. A good number of these patient have high tumor burden with frequent regional and not so infrequent pulmonary metastases. These cancers generally TSH dependent and iodine avid.

They also show excellent therapeutic response with ^{131}I and dosage is based on body weight, histology, stimulated thyroglobulin level, and scan findings. But generally, it is recommended to individualize ^{131}I dosage to a lower than the adult level in pediatric patients due to (a) children's longer life expectancy, and hence, vulnerability to undesired treatment effects, (b) the greater absorbed dose to bone marrow and extra-thyroidal tissue, given their smaller body sizes, and (c) the increased cross-radiation due to shorter distances between organs. Although skeletal metastases are rare (05%), lung metastases are present at diagnosis in approximately 10–20% of children with DTC. Children have a higher rate of local and distant recurrences than adults. Despite this, the prognosis in children with DTC is excellent, with a 10-year mortality of <10% and overall survival of 98%; at 20 years, overall survival is 95%.

16.5.7 Therapy in High-Risk Patients

Patients with vertebral metastases have the risk of developing cord compression after high dose ^{131}I administration if adequate precaution is not taken (due to post radiation inflammation), like decompressive surgery or external beam radiotherapy prior to high dose ^{131}I therapy.

16.5.8 Transient Effects of ^{131}I Therapy in DTC Patients

Transient side effects such as gastritis, neck pain, and sialadenitis are encountered occasionally. Sialadenitis occurs in about 10–20% patients in the first few days after therapy, with pain and enlargement of salivary glands but rarely progress to chronic xerostomia. Xerostomia is very rare following the first ^{131}I therapy but may occur following a very high dose of ^{131}I. In majority of occasion good hydration, sialagogues and perfect oral hygiene prevent development chronic sialadenitis. Loss of taste and discoloration of tongue due to destruction of papillae are less frequent and are reversible. Gastritis is usually self-limiting, H2 blockers following ^{131}I administration are suggested.

Radiation thyroiditis is usually encountered in patients with significant residual thyroid tissue and occurs immediate post ablation period. Patient experiences pain and swelling in the front of neck and systemic steroid and non-steroidal anti-inflammatory drugs are used to combat this issue.

Gonadal radiation exposure can be minimized by liberal hydration and frequent urination.

16.5.9 Long Term Effects of ^{131}I Therapy in DTC Patients

16.5.9.1 Infertility, Gonadal Failure, and Genetic Effects

Ablation using ^{131}I is contraindicated in pregnant and lactating ladies. Transient ovarian failure in ladies is not unusual and is common in older premenopausal group.

Administration of repeated radioiodine in men shall cause impairment of spermatogenesis, increased levels of the follicular-stimulating hormone, and decreased levels of inhibin B. Patients in reproductive age are advised not to plan pregnancy for the next 6 months following high dose ^{131}I ablation or therapy.

There is a general perception that radiation has mutagenic and affects germ cells resulting in genetic damage. There are no convincing evidences that exposure to ^{131}I alters preterm birth, low birth weight, stillbirth, congenital malformations, and death during the first year of life, thyroid disease, or non-thyroidal malignancies in offspring.

16.5.9.2 Carcinogenicity

The incidence of leukemia and solid organ malignancies in post ^{131}I treated cases versus the general public has been long debated. Follow-up studies of patients exposed to ^{131}I do not demonstrate any tumorigenic effect of ^{131}I on the thyroid gland in adults but demonstrate such an effect in children. The dose delivered to other tissues is relatively low, and a significant risk of cancer and leukemia has been found only in patients exposed to high cumulative activities of ^{131}I (>500–600 mCi). No genetic effect has been found in studies on the outcome of subsequent pregnancies in women treated with ^{131}I for thyroid carcinomas. When treated with ^{131}I, simple measures such as overall good hydration and the use of laxatives help in reducing overall tissue radiation doses.

Absolute Contraindications to ^{131}I Therapy for DTC

1. Pregnancy—it is recommended to delay conception for 6 months after high-dose ^{131}I therapy.
2. Breastfeeding.
3. Patients with elevated urine iodine levels (over 200 μg/L) either from IV iodinated contrast CT study or from dietary intake should have, ^{131}I therapy postponed until levels return to normal.

16.5.9.3 Iodine Refractory Thyroid Cancer [18]

The DTC cells retain NIS expression and ability to concentrate radioiodine and is used as therapeutic too in treating the cancer. But rarely the thyrocytes lose the ability to concentrate the iodine and becomes refractory to ^{131}I treatment. Approximately four cases in a million of the population are refractory to radioiodine therapy. The patients do not show iodine uptake in whole body scan immediately and few days after few days of administration of iodine in the setting of clinical or biochemical evidences persistent cancer. Occasionally a few of the metastases may be concentration iodine while some may not show uptake. Despite a suitable dose or the maximal dose of 600 mCi of radioiodine administered empirically, the recurrence or metastases progress. But before finally labeling the patient refractory possibility of improper patient preparation has to be excluded. The ten-year overall survival of these patients drops considerably to 10% from the time of detection of metastases. Tyrosine-kinase inhibitors therapy has been found to be promising in these clinical settings. But the treatment protocol of these patients is individualized based on tumor characteristics and patient features. Occasionally these patients have somatostatin receptor expression and show uptake on ^{68}Ga DOTANOC PETCT and ^{177}Lutetium DOTATATE therapy has been tried. Obviously refractory cases need a multimodality approach.

16.5.9.4 Martinique Principles

A new set of principles has been put forth in the past 3–4 years pioneered by the British medical fraternity to provide a framework for many important management issues in DTC. This aims to bridge the gap created by ATA guidelines 2015. The result of the first meeting is a set of nine principles called "The Martinique Principles" that (i) describe a commitment to proactive, purposeful, and inclusive interdisciplinary cooperation; (ii) define the goals of ^{131}I therapy as remnant ablation, adjuvant treatment, or treatment of known disease; (iii) describe the importance of evaluating postoperative disease status and multiple other factors beyond clinicopathologic staging in ^{131}I therapy decision making; (iv) recognize that the optimal administered activity of ^{131}I adjuvant treatment cannot be definitely determined from the published literature; and (v) acknowledge that current definitions of ^{131}I refractory disease are suboptimal and do not represent definitive criteria to mandate whether ^{131}I therapy should be recommended. The Martinique Work Group and its past 2018 and 2019 meetings have been extremely valuable in organizing collaborative efforts to establish common terminology, to help identify areas of differences of opinions, to better understand what our differences are, and to recognize that, overall, our intentions of improved diagnosis and management of patients with differentiated thyroid cancer are aligned.

In postoperative DTC patients, it is imperative to assess (in an immediate postoperative setting) and follow them with a whole-body ^{131}I diagnostic scan. High dose ^{131}I ablation/metastases therapy is indispensable for better long-term survival and prognosis. Serum thyroglobulin is a sensitive marker to look for recurrence/distant thyroid metastasis in post ^{131}I residual thyroid ablation setting. Recombinant TSH-assisted ^{131}I therapy is also available these days in patients who cannot stop T4 withdrawal or in patients with distant extensive metastases where TSH does not raise over 30 IU/mL.

Postoperative metastatic/recurrent medullary thyroid carcinoma can be imaged and treated with ^{131}I labeled MIBG (Meta Iodo Benzyl Guanidine) or ^{68}Gallium DOTATATE PETCT imaging followed by ^{177}Lutetium DOTATATE therapy.

16.6 Role of Nuclear Medicine in Parathyroid Imaging

16.6.1 Introduction

Primary hyperparathyroidism is one of the most common endocrine disorders, after diabetes and hyperthyroidism. It occurs with a frequency of about 0.5 per 1000, with approximately half of the persons being asymptomatic and detected by serum calcium screening. Surgical removal of the diseased parathyroid gland(s) stays as the definite treatment of hyperparathyroidism. Bilateral neck exploration was the accepted surgical procedure and had a good success rate in hands of experienced surgeons. But the procedure needed extensive dissections and usually ended in prolonged hypocalcemia due to exposure of all parathyroid glands for inspection. So minimally invasive parathyroidectomy is becoming the procedure of choice due to its lower complication rate. This shift towards minimally invasive procedures emphasizes the need for preoperative localization of the abnormal parathyroid glands.

Challenges of imaging parathyroid gland [19]:

1. The size and location of the normal parathyroid glands since 80–85% adenomas are located adjacent to the thyroid. Conventionally there are four parathyroid glands, but the number varies occasionally.

2. One or more of the glands may be placed in ectopic locations in 10–15% patients. Ectopic locations are mediastinum, carotid sheath, retro-esophageal, and intra-thyroidal.
3. Failure of identifying and removing adenomas in ectopic locations was found as high as 5%.

A single adenoma is the cause of hyperparathyroidism in 80–85% of occasions, but a multi-glandular pathology accounts for 12–15% patients. Parathyroid carcinoma is rare and forms <1% etiology of all hyperparathyroidism patients.

Preoperative localization techniques: The widely available non-invasive techniques include:

1. Ultrasound (US)
2. Computed tomography (CT)
3. Magnetic resonance imaging (MRI)
4. Nuclear medicine imaging studies with ^{99m}Tc-Methoxyisobutylisonitrile (MIBI) planar scintigraphy and single photon emission computed tomography (SPECT)/CT
5. Four-dimensional CT (4D-CT) scan
6. Positron Emission Tomography (PET) CT and PETMR

The latest imaging procedures such as 4DCT [20] and PET are showing promising results in localization of parathyroid lesion(s).

Invasive techniques include US or CT guided fine needle aspiration (FNA) with concomitant PTH assay, parathyroid angiography, and selective venous sampling (SVS) for the PTH gradient. Invasive techniques are reserved for re-operative and failed cases.

16.6.2 Functional Imaging with Nuclear Medicine

Various radiotracers were used in localization of parathyroid lesions. The popular tracers are Thallium chloride-201 (^{201}Tl-chloride), Technetium-99m SestaMIBI (^{99m}Tc-SestaMIBI), Iodine-123 (^{123}I-iodide), and ^{99m}Technetium pertechnetate (^{99m}TcO4).

Recently, PET tracers such as Fluorine-18 Fluoro-Deoxy Glucose (^{18}F-FDG) and Fluorine-18 fluorocholine (^{18}F-Fluorocholine) are employed in selectively.

16.6.3 History of Parathyroid Scintigraphy

The earliest scintigraphic technique was using ^{201}Tl–^{99m}TcO4^{-} or thallium/pertechnetate subtraction imaging which was introduced in early 1980s. The inorganic cationic analogue of potassium (^{201}Tl) is a cyclotron produced radionuclide with a half-life of 73 h and was used for myocardial perfusion imaging. This agent is actively transported into the cell via sodium-potassium ATPase pump, and gets integrated into the intracellular potassium pool. The ability of ^{201}Tl to accumulate in thyroid and parathyroid tissue was utilized in parathyroid scintigraphy. The uptake of ^{201}Tl in the thyroid was removed in order to visualize the parathyroid activity and is done by administering ^{99m}TcO4, which accumulates only in thyroid gland. By digitally subtracting the ^{99m}TcO4 image of thyroid from the combined image by the parathyroid localization is done. This method was called **"dual tracer subtraction technique."** However, this method has its own disadvantages. The photon energy is suboptimal for imaging using the current gamma camera and movements of patient during image acquisition affects the accuracy of the localization. The sensitivity greatly varies from 44% to 95% probably due to difference laboratory protocols.

Hexakis 2-methoxy isobutyl isonitrile (^{99m}Tc SestaMIBI) replaced ^{201}Tl and continues as the radiopharmaceutical of choice since Coakley et al. reported use of MIBI in parathyroid imaging in 1989.

16.6.3.1 Imaging Protocols

1. Single isotope, dual phase techniques (^{99m}Tc MIBI)
2. Dual isotope technique
3. Additional subtraction technique

Mechanism of uptake: MIBI is a lipophilic, monovalent cationic isonitrile compound that diffuses passively across the cell membrane. In the cells MIBI is sequestered in the mitochondria due to the negative transmembrane potential. The tracer preferentially accumulates oxyphil cells of the parathyroid lesion whereas normal cells do not concentrate it. The tracer is excreted partly through the kidneys (25%) and the rest through GI tract.

16.6.3.2 Patient Preparation

The imaging procedure is attempted only after confirmation of hyperparathyroidism by complete clinical examination and laboratory studies. The imaging is not a substitute for confirmation of the diagnosis. There is no specific preparation prior to imaging procedure.

The tracer has differential affinity to thyroid and abnormal parathyroid gland where material stays for longer time even after it is washed of thyroid. Taillefer et al. [21] proposed the concept of "Single isotope, dual phase techniques" based on the differential washout rates of SestaMIBI from thyroid and diseased parathyroid glands. This selective prolonged retention of SestaMIBI by abnormal parathyroid glands could be attributed to the presence of oxyphil cells rich in mitochondria, which are the sites of intracellular sequestration of SestaMIBI.

16.6.3.3 Procedure

Intravenous administration of 15 mCi of MIBI is done in adults and the initial dynamic images are acquired for 1 min. The early high-resolution images of anterior neck and anterior mediastinum are acquired at 10–20 min post injection. The delayed images are acquired at 1.5–2.5 h post injection. When facility is available

SPECTCT of neck is also obtained which helps for the anatomical localization of the lesion.

16.6.3.4 Interpretation

A persistent delayed focal uptake is the characteristic feature of a parathyroid lesion. Occasionally the presence of focal thyroid pathology may concentrate MIBI for longer time causing difficulty in localization. When thyroid pathology interferes with localization a subtraction protocol using a second tracer specific for thyroid (^{99m} TcO4 or ^{123}I) is done. The parathyroid lesion brightens when the thyroid image is digitally subtracted from the combined image acquired with MIBI.

The sequencing of tracer administration is first imaging of thyroid with 2–4 mCi of ^{99m}TcO4 followed by 10–20 mCi ^{99m}Tc SestaMIBI after acquiring thyroid imaging 20–30 min after injection of the former. The patient has to remain without movement during the total duration of imaging of both tracers. A computer software removes the thyroid image from the ^{99m}Tc SestaMIBI (thyroid + parathyroid) images. Combination of subtraction and dual phase techniques has been used in patients with thyroid abnormalities to avoid false positive results from ^{99m}Tc MIBI retention in thyroid lesions.

SPECT-CT enhances further the efficacy of dual phase scan technique and helps the surgeon to plan the procedure especially the lesions trachea-oesophageal grove. SPECTCT provides more precise anatomic localization of ectopic parathyroid adenomas and assist in directing the surgical approach, such as median sternotomy versus right or left thoracotomy. SPECTCT can be acquired immediately after the planar (static) imaging both in early and delayed imaging. An early SPECT-CT is great help in occasions of parathyroid lesions with which loses the affinity to MIBI early (Rapid washout) (Fig. 16.8). Studies have shown that addition of SPECT improved the sensitivity of parathyroid scintigraphy from 86% to 90.5% and also improved preoperative localization and characterization of parathyroid lesions.

16.6.4 Factors Affecting the Scan Findings

The selection of radiopharmaceutical and technique determine the sensitivity and detection efficiency of parathyroid scintigraphy. Currently ^{99m}Tc-SestaMIBI scintigraphy is the preferred method for parathyroid imaging, due to its superior sensitivity, availability, and cost effectiveness.

The reported sensitivity of ^{99m}Tc-SestaMIBI scintigraphy ranges from 80% to 100% probably due to the differences in the protocols. The timeframe in imaging affects the scan sensitivity. When dual phase technique is employed acquisition of early images in 10–30 min, and delayed images at 1.5–2.5 h with planar images and three-dimensional SPECT or SPECTCT gives good results in preoperative localization. Dual tracer subtraction technique using ^{99m}Tc SestaMIBI and ^{99m}Tc-pertechnetate can be used in special cases such as synchronous thyroid disease or in rapid washout parathyroid adenoma.

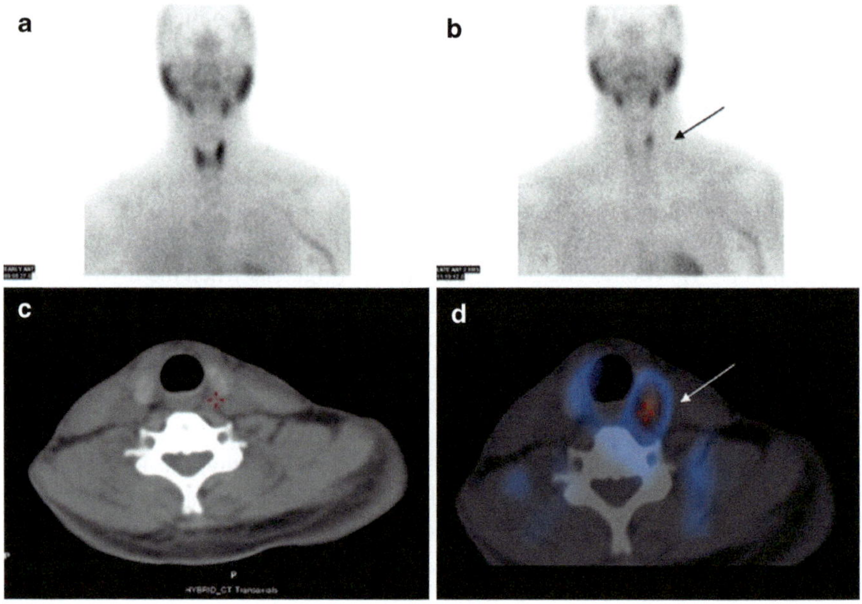

Fig. 16.8 36-year-old patient with S.PTH 126.8 pg/mL and hypercalcemia (S.Ca of 11.2 mg/dL). Dual phase ^{99m}Tc MIBI parathyroid scintigraphy demonstrating left superior parathyroid adenoma with image: (**a**) Initial image at 15 min post injection showing homogenous normal tracer uptake in both lobes of thyroid gland; (**b**) 2-h delayed image showing focal abnormal tracer uptake in left superior parathyroid region with washout from thyroid gland; (**c**) Non-contrast CT neck; (**d**) SPECTCT image shows accurate localization of parathyroid adenoma at left superior region

Certain specific features of the lesion such as size, cellularity, presence of P-glycoprotein, and phase of cell growth affect the scan sensitivity. Oxyphil cell rich parathyroid adenoma were found to show good ^{99m}Tc SestaMIBI and ^{201}Tl (Thallium) uptake. The sensitivity of ^{99m}Tc SestaMIBI scintigraphy was low in parathyroid adenomas with P-glycoprotein or Multi Drug Resistance (MDR) gene expression.

The sensitivity of ^{99m}Tc SestaMIBI is approximately 60% in detecting hyperplastic glands. The size, cellularity of the glands and the PTH level influence sensitivity and specificity of the scans. MIBI scans fail to localize the multi-glandular pathology.

16.6.5 False Positive and False Negative Results in ^{99m}Tc-SestaMIBI Scan

False positive images are frequently associated with thyroid pathologies such as nodules, thyroiditis, and with concurrent lymph node pathologies. Small size of parathyroid lesions and multi-glandular diseases, and p-glycoprotein expression are few conditions which lead on to negative scan results.

16.6.6 Recent Advances in Parathyroid Imaging Using PET

Many PET radiotracers have been used in parathyroid imaging. Methionine is a precursor amino acid in PH synthesis and PET imaging using ^{11}C-methionine showed promising results. Since the half-life of the tracer is 20 min, images are harvested immediately after the intravenous administration. The tracer is costly and is not widely available. The sensitivity of the scan in localizing multi-glandular lesions is low. ^{18}F-FDG PETCT has poor sensitivity in unselected patients with primary hyperparathyroidism but is useful in patients with parathyroid carcinoma in assessing spread of lesion.

Choline is the precursor for cell membrane synthesis, thus being concentrated by parathyroid adenoma and hyperplastic parathyroid glands. PET imaging with ^{11}C-Choline was found superior to ^{99m}Tc SestaMIBI/^{99m}Tc-pertechnetate scintigraphy. ^{11}C-Choline has been found to be superior in detecting parathyroid lesions. However, limited availability of radiopharmaceutical as well as the shorter half-life of the radiotracer remains the area of concern. ^{18}F-Fluorocholine is a PET tracer found promising results in parathyroid localization with a reported sensitivity and positive predictive value of 93–100% and 90–100%, respectively. The sensitivity of ^{18}F-Fluorocholine PETCT was greater than that of 4D-CT regarding the number of lesions (96% vs. 75%) and patients (85% vs. 63%), respectively. A meta-analysis of 11 studies on ^{18}F-Fluorocholine PETCT found the rate of detection of lesions to be 97% in patient-based analysis and 94% in lesion-based analysis. ^{18}F-Fluorocholine PETCT is being considered as one of the imaging methods for parathyroid gland visualization before invasive procedures in patients with persistent or recurrent disease with negative conventional imaging results.

Similarly, ^{18}F-Fluorocholine PETMR, where anatomical imaging with CT is being replaced by MRI, has also shown promising results. In 10 patients with negative or inconclusive results of ultrasonography and ^{99m}Tc-SestaMIBI imaging, ^{18}F-Fluorocholine PETMR was performed after intravenous administration of 3 MBq/kg of ^{18}F-Fluorocholine. The sensitivity and PPV of ^{18}F-Fluorocholine PETMR were found to be 90% and 100%, respectively.

16.6.7 Gamma Probe Guided Parathyroidectomy [22]

Minimally invasive parathyroidectomy became popular after successful preoperative imaging, particularly with SestaMIBI scintigraphy and ultrasound. Radioguided surgery using intraoperative gamma probe facilitates the surgical exploration, reduces the operative time, and confirms the correct excision of the pathological parathyroid tissue. The intraoperative gamma probe is a handheld radiation detector device which gives both auditory signals and digital counts to guide the surgeon to localize and dissect the radioactive target tissue. Minimally invasive radioguided surgery (MIRS) (Fig. 16.9) is recommended to reduce blood loss, intraoperative surgical time, faster postoperative recovery (reduced hospital stay days) and cosmetic purpose to have a small scar. Patients undergo a prior SestaMIBI scintigraphy

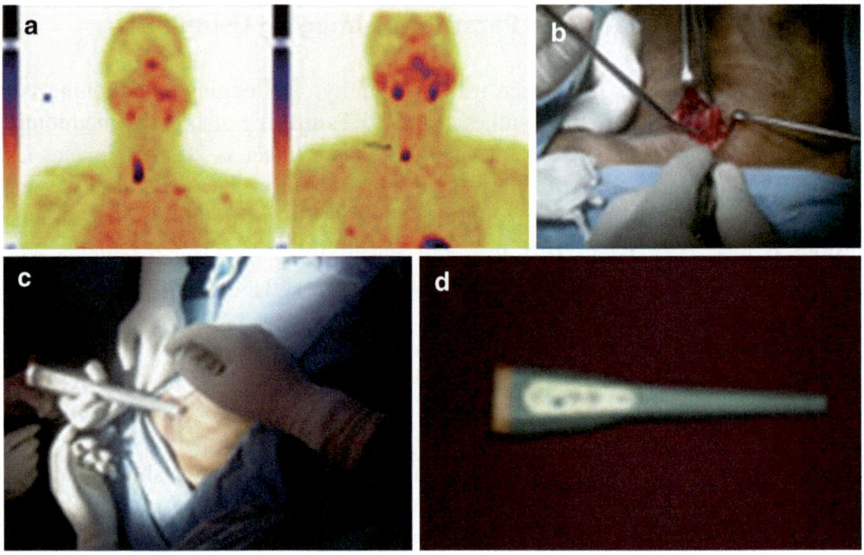

Fig. 16.9 MIRS, minimally invasive radioguided parathyroidectomy in a 56-year-old male with history of left hemithyroidectomy. (**a**) ^{99m}Tc MIBI concentrating right inferior parathyroid adenoma; (**b**) Exposing the adenoma intraoperatively based on Gamma probing; (**c**) Sterile draped battery-operated hand-held Gamma probe showing count statics in neck at the site of adenoma; (**d**) Collimated Gamma finder

to identify the site of solitary parathyroid adenoma. On the day of surgery 2–3 mCi of intravenous ^{99m}Tc SestaMIBI is injected 1 h prior to surgery. This is done to minimize radiation exposure to surgeon and theatre staff. Gamma probe localizes the site of adenoma by keeping the probe over the overlying skin surface, incision is made accordingly and neck is exposed. This helps in complete excision of adenoma, reducing the intraoperative time by 50% and also reduces the length of incision. Studies also show less evidence of post-surgical hypocalcemia, less pain, and complications.

16.7 Conclusion

Single isotope, dual phase ^{99m}Tc MIBI parathyroid scintigraphy is the procedure of choice for localizing parathyroid adenomas in patients with suspected hyperparathyroidism apart from high-definition US. MIRS has placed a greater emphasis on preoperative localization of abnormal parathyroid glands. However, there is no perfect imaging modality with widespread availability and high sensitivity, to localize all abnormal parathyroid glands. ^{18}F-Fluorocholine PETMR is another emerging technology especially used in those patients with negative MIBI findings for preoperative localization of abnormal parathyroid glands.

References

1. Mettler FA, Guiberteu MJ. Essentials of nuclear medicine imaging (Chapter 4). 6th ed. Philadelphia: Elsevier, Inc; 2012.
2. Burrow GN, Fisher DA, Larsen PR. Maternal and fetal thyroid function. N Engl J Med. 1994;331:1072–8.
3. Gorman CA. Radioiodine and pregnancy. Thyroid. 1999;9(7):721–6.
4. Rubow S, Klopper J, Wasserman H, et al. The excretion of radiopharmaceuticals in human breast milk: additional data and dosimetry. Eur J Nucl Med. 1994;21:144–53.
5. Robinson PS, Barker P, Campbell A, Henson P, Surveyor I, Young PR. Iodine-131 in breast milk following therapy for thyroid carcinoma. J Nucl Med. 1994;35(11):1797–801.
6. Intenzo CM, dePapp AE, Jabbour S, Miller JL, Kim SM, Capuzzi DM. Scintigraphic manifestations of thyrotoxicosis. Radiographics. 2003;23(4):857–69.
7. Martino E, Aghini-Lombardi F, Mariotti S, et al. Amiodarone iodine-induced hypothyroidism: risk factors and follow-up in 28 cases. Clin Endocrinol. 1987;26:227–37.
8. Haugen BR, Alexander EK, Bible KC, Doherty GM, Mandel SJ, Nikiforov YE, et al. 2015 American Thyroid Association Management guidelines for adult patients with thyroid nodules and differentiated thyroid cancer: The American Thyroid Association Guidelines Task Force on Thyroid Nodules and Differentiated Thyroid Cancer. Thyroid. 2016;26(1):1–133.
9. Shanmuga Sundaram P, Padma S. Diagnostic utility of PETCT in thyroid malignancies: an update. Ann Nucl Med. 2013;27:681–93. https://doi.org/10.1007/s12149-013-0740-6.
10. Hales NW, Krempl GA, Medina JE. Is there a role for FDG PET/CT in cytologically indeterminate thyroid nodules? Am J Otolaryngol. 2008;29:113–8.
11. Silberstein EB, Alavi A, Balon HR, Clarke SE, Divgi C, Gelfand MJ, Goldsmith SJ, Jadvar H, Marcus CS, Martin WH, Parker JA, Royal HD, Sarkar SD, Stabin M, Waxman AD. The SNMMI practice guideline for therapy of thyroid disease with 131I 3.0. J Nucl Med. 2012;53(10):1633–51.
12. Andrade VA, Gross JL, Maia AL. The effect of methimazole pretreatment on the efficacy of radioiodine therapy in Graves' hyperthyroidism: one-year follow-up of a prospective, randomized study. J Clin Endocrinol Metab. 2001;86:3488–93.
13. Luster M, Clarke SE, Dietlein M, Lassmann M, Lind P, Oyen WJ, Tennvall J, Bombardieri E, European Association of Nuclear Medicine (EANM). Guidelines for radioiodine therapy of differentiated thyroid cancer. Eur J Nucl Med Mol Imaging. 2008;35(10):1941–59.
14. Luster M, Lippi M, Jarzab B, Perros P, Lassmann M, Reiners C, Pacinie F. rhTSH-aided radioiodine ablation and treatment of differentiated thyroid carcinoma: a comprehensive review. Endocr Relat Cancer. 2005;12(1):49–64.
15. Tuttle RM, Sukhjeet A, Avram AM, et al. Controversies, consensus and collaboration in the use of I-131 therapy in differentiated thyroid cancer: a joint statement from the American Thyroid Association, the European Association of Nuclear Medicine, the Society of Nuclear Medicine and Molecular Imaging, and the European Thyroid Association. Thyroid. 2019;29:461–70.
16. Padma S, Sundaram PS. Radioiodine as an adjuvant therapy and its role in follow-up of differentiated thyroid cancer. J Can Res Ther. 2016;12:1109–13.
17. Samuel AM, Rajashekharrao B, Shah DH. Pulmonary metastases in children and adolescents with well-differentiated thyroid cancer. J Nucl Med. 1998;39(9):1531–6.
18. Fugazzola L, Elisei R, Fuhrer D, Jarzab B, Leboulleux S, Newbold K, Smit J. 2019 European Thyroid Association guidelines for the treatment and follow-up of advanced radioiodine-refractory thyroid cancer. Eur Thyroid J. 2019;8:227–45. https://doi.org/10.1159/000502229.
19. Smith JR, Oates ME. Radionuclide imaging of the parathyroid glands: patterns, pearls, and pitfalls. RadioGraphics. 2004;24:1101–15.
20. Kukar M, Platz TA, Schaffner TJ, et al. The use of modified four-dimensional computed tomography in patients with primary hyperparathyroidism: an argument for the abandonment

of routine sestamibi single-positron emission computed tomography (SPECT). Ann Surg Oncol. 2015;22(1):139–45.
21. Taillefer R, Boucher Y, Potvin C, Lambert R. Detection and localization of parathyroid adenomas in patients with hyperparathyroidism using a single radionuclide imaging procedure with Tc-99m sestamibi (double phase study). J Nucl Med. 1992;33:1801–7.
22. Norman JG. Minimally invasive radioguided parathyroidectomy: an endocrine surgeon's perspective. J Nucl Med. 1998;39:15N–24N.

Surgery of the Parathyroid Gland

17

Aravindan Nair and P. Balaji Viswanath

17.1 Introduction

From the time of Sir Richard Owen's first description in 1850 of the parathyroid glands, the clinical features, diagnosis, and treatment of parathyroid disorders have changed dramatically over the past 150 years. Advances in diagnosing, imaging, and improvements in surgical techniques have made parathyroid surgery safe and hardly with any morbidity. In many institutions, minimally invasive parathyroid surgery under local anesthesia has become a daycare procedure.

17.2 Anatomy of the Parathyroid

Generally, there are four parathyroid glands, one pair of glands on each side located on the posterior border of the thyroid. Approximately 3% have <4 glands and 5–10% have more than 4 glands. Sine thymus and inferior parathyroid gland share embryological origin from third pharyngeal pouches the inferior pair have close relation with thymus or even migrate along with it to the mediastinum. Frequently, islets of parathyroid tissue are seen in the thymus which gains significance in Multiple Endocrine Neoplasia and secondary hyperparathyroidism.

The superior parathyroids have a more or less fixed location on the posteromedial aspect of the superior pole of the thyroid lobe close to the trachea esophageal groove. The inferior parathyroids are less constant in location and are usually found on the posterolateral aspect of the thyroid lobe in a circular area with a radius of around 2.5 cm with the center being the point of contact between RLN and inferior thyroid artery. The average parathyroid measures 5 × 3 × 1 mm and weighs about 35–40 mg. The adult gland is yellow and almost pear-shaped with a tuft of

A. Nair (✉) · P. Balaji Viswanath
Department of General and Endocrine Surgery, Naruvi Hospitals, Vellore, Tamil Nadu, India

C. G. Nair, S. J. Abraham (eds.), *Surgical Management of Thyroid and Parathyroid Diseases*, https://doi.org/10.1007/978-981-97-3774-1_17

normal-looking fat attached to it. The symmetry in location and size of the superior parathyroid glands is around 80% whereas it is about 70% for the inferior glands [1].

The parathyroid glands have variable relations with the thyroid gland and may be placed within the fibrous capsule or outside it. The intracapsular parathyroid gland remains within the confines of the thyroid gland even after undergoing pathological changes. Extracapsular parathyroid tends to take a path of least resistance when the volume increases and can be seen from the cricothyroid junction up to the posterior mediastinum or in the superior mediastinum.

At operation, the parathyroid glands are globular or oblong and soft in consistency (unlike a thyroid nodule or a lymph node). They are quite vascular and usually have what is known as the floating sign, the gland is seen to move within its fatty capsule when gentle pressure is applied. The superior glands are usually seen posterior to the nerve and the inferior glands are seen anterior to the nerve.

Parathyroid glands receive the blood supply from the inferior thyroid on the majority of occasions but in about 15% superior thyroid artery directly or through the posterior anastomotic channel supply superior parathyroids.

17.2.1 Variations

Since parathyroid glands originate from 3rd and 4th pharyngeal pouches and migrate to their usual locations close to the thyroid gland and since they share an embryological origin with other organs, ectopia is not rare. Approximately 10–15% of parathyroid glands may be ectopically placed. Ectopic locations include the thyro- thymic tract, trachea-esophageal groove, retro esophageal space, intrathyroidal, carotid sheath, and mediastinum. Supernumerary glands are recorded in different frequencies varying from 2.5% to 22% of the population. Most often they are located in the thymus or relation to the thyro-thymic tract. Parathyroid glands are described from Type A to G based on embryological development and regional anatomy. Understanding this helps in localization during surgery.

Type A: Adherent to posterior thyroid capsule with normal superior gland location
Type B: Tracheo-esophageal groove
Type C: Tracheo-esophageal groove inferior to the lower end of the thyroid lobe
Type D: Directly over recurrent laryngeal nerve
Type E: Close to the inferior thyroid pole (normal inferior parathyroid location)
Type F: within the thyro-thymic ligament
Type G: Intra thyroidal

17.3 Calcium Metabolism

The adult human body contains approximately 1 kg of calcium of which only 1% is in the extracellular fluid. Total calcium in the blood is maintained within relatively narrow ranges of 8.5–10.2 mg/dl (2.2–2.5 mmol/L) and 55% of this is protein bound. Calcium remains in three forms in blood as ionic form, complexed form, and protein-bound form. About 90% of protein-bound calcium is attached to albumin and the rest to a variety of globulins. A portion of total calcium that forms ion couplets with anions such as bicarbonate and/or citrate is known as complexed calcium. Ionic form and complexed form constitute the diffusible portion of serum calcium which can pass through biological membranes. A diffusible portion of serum calcium is a determinant factor in calcium metabolism.

Traditionally serum total calcium is measured but this includes all three forms of circulating calcium. Calcium levels are corrected (serum corrected calcium) based on serum albumin, by the fraction of 0.8 mg/dl decrease for 1 mg/dl decrease in albumin, which represents the diffusible portion of calcium. Estimation of ionic calcium levels is not routine.

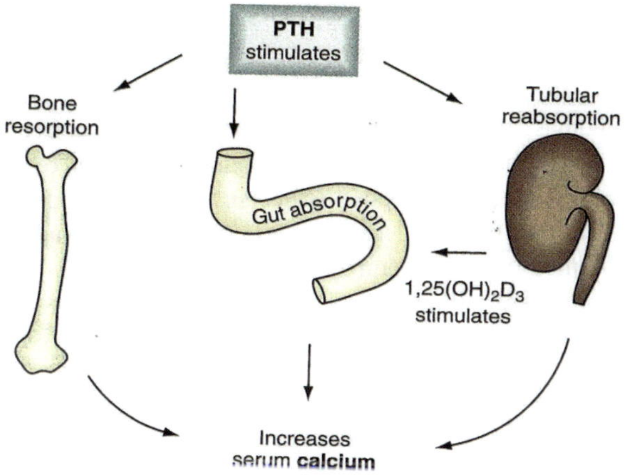

Cycle of calcium metabolism

The ionized calcium levels are inversely affected by the pH of the blood. One unit of rising in the PH of blood will decrease the calcium by 0.36 mmol/L.

Calcium homeostasis is regulated mostly by parathyroid hormone (PTH) synthesized and secreted from the parathyroid glands. Serum ionized calcium controls the release of PTH from parathyroid glands by a negative feedback mechanism. The setpoint at which the release of PTH is initiated is serum ionic calcium level of 1.1–1.2 mM (4.41–4.81 mg/dl) which corresponds to serum total calcium level of 2.2–2.4 mM (8.82–9.62 mg/dl). Calcium sensing receptors in the parathyroid gland detect the changing levels of calcium in the blood and modulate the synthesis and release of parathyroid hormone by the chief cells. PTH favors osteoclastic activity in bone, renal reabsorption of calcium, and synthesis of 1,25, OHD in kidneys. 1,25, OHD enhances gastrointestinal absorption of calcium. Serum levels of inorganic phosphate and vitamin D are other determinants of serum calcium level.

Calcitonin produced by the parafollicular cells of the thyroid in response to increased calcium levels inhibits bone resorption thus reducing calcium levels. Its role in humans is not very clear when calcitonin levels are constantly very high as in medullary carcinoma thyroid. Salmon calcitonin has been useful in reducing serum calcium in patients with high calcium levels.

17.4 Pathology of Hyperactive Parathyroid Glands

17.4.1 Adenoma

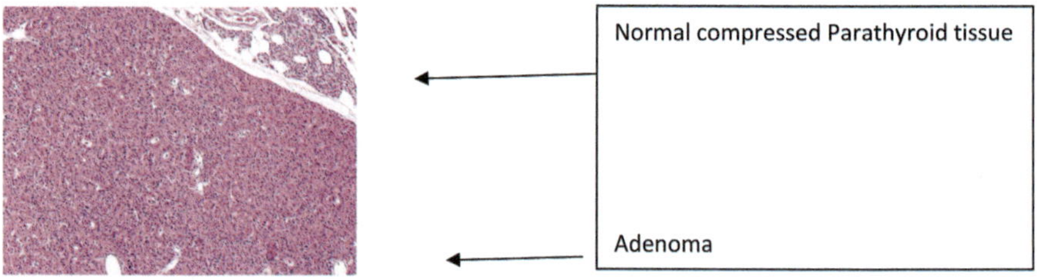

Parathyroid adenomas are usually single but multiglandular lesions are seen in about 10–15% of instances. The parathyroid gland contains chief cells, oxyphil (oncocytic) cells, fibrovascular stroma, and adipose tissue. An adenoma could be monoclonal or polyclonal and is an encapsulated lesion, tan brown with occasional hemorrhagic and cystic areas. The majority of adenomas are of chief cell origin but oncocytic and mixed cell adenomas are infrequently detected as a histological surprise. The adenoma differs from hyperplasia by the presence of a rim of compressed normal tissue.

17.4.2 Hyperplasia

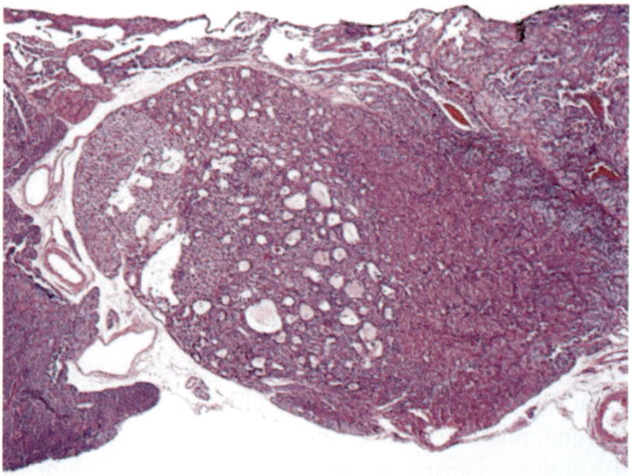

Hyperplasia is usually multiglandular and could be part of multiple endocrine neoplasias, and secondary hyperparathyroidism. Nodular hyperplasia is rarely found in sporadic primary hyperparathyroidism. Unlike adenoma, the lesion does not have a thick capsule or a rim of compressed normal tissue.

17.4.3 Carcinoma

Parathyroid carcinoma is rare and accounts for about 1% of primary hyperparathyroidism. Most of the parathyroid carcinomas are functional and used to express high disease severity indices such as serum PTH and calcium. Parathyroid carcinoma presents as an ill-defined mass, usually palpable with adherence to surrounding tissues. The cut surface is nodular, firm, and has a gray tan appearance. Microscopically it shows atypical neoplastic parenchymal cells with mitotic figures interspersed with thick fibrous bands showing capsular and vascular invasion.

17.5 Hyperparathyroidism

Customarily hyperparathyroidism is categorized into three different forms primarily based on the pathophysiology of the disease.

1. Primary
2. Secondary
3. Tertiary

17.5.1 Primary Hyperparathyroidism

Primary hyperparathyroidism is the third most common endocrine disorder after diabetes mellitus and thyroid disorders. One or more parathyroid glands undergo neoplastic changes or hyperplasia and continue to function autonomously. A single adenoma is the etiology of PHPT in approximately 85% of cases and 2–5% of cases are due to double adenomas. Hyperplasia and parathyroid carcinoma are the other causes.

Occasionally more than one member of the family suffers from PHPT but evidence of germline mutation is not available. Mutation in the *MEN1*, *CDC73*, or *CASR* gene was detected in some cases.

17.5.2 Secondary Hyperparathyroidism

Secondary hyperparathyroidism is seen in the setting of dialysis-dependent end-stage renal disease and is associated with persistent vitamin D deficiency. In Chronic Kidney Disease (CKD), there is renal wastage of calcium and accumulation of PO4 in blood and this hypocalcemia induces PTH secretion. In renal failure, there is hyperphosphatemia which induces Fibroblast Growth Factor 23 (FGF-23) from bone resulting in the decreased renal conversion of 25–hydroxy cholecalciferol to 1,25–dihydroxycholecalciferol. Thus, intestinal absorption of calcium is decreased. Both these effects cause hypocalcemia which stimulates PTH secretion and parathyroid hyperplasia sets in. Secondary hyperparathyroidism is a multiglandular disease and usually reverts to normality when CKD is corrected by transplant.

Secondary HPT is generally managed conservatively with vitamin D analogs and calcimimetic drugs. Occasionally few patients develop persistent hyperparathyroidism despite conservative management and this is referred to as refractory SHPT. There is no consensus of PTH level that defines refractory state but persistent serum levels of PTH above 800 pg/ml is considered significant.

Medical treatment: In all patients with CKD one should be on the lookout for concurrent parathyroid dysfunction. Medical management includes serum phosphate lowering agents, vitamin D analogs, and calcimimetic agents such as cinacalcet hydrochloride. Continuous regular monitoring of serum phosphate, calcium, and PTH is essential.

Parathyroidectomy: Parathyroidectomy is indicated in refractory secondary HPT and also in conditions such as calcium to phosphorous product >70 and persistent hypercalcaemia (corrected calcium > 11.5 mg/dl). Symptoms related to bone loss, generalized pruritis, calciphylaxis are other indications for parathyroidectomy in secondary HPT.

Surgical options include total parathyroidectomy with cervical thymectomy and auto-transplantation of relatively normal parathyroid tissue, subtotal parathyroidectomy with cervical thymectomy, and auto-transplantation of relatively normal parathyroid tissue, and total parathyroidectomy and cervical thymectomy without transplantation. Hungry bone syndrome in this setting requires very cautious management as a fluid overload may result from overzealous calcium infusion. These patients require life-long supplementation of calcium and vitamin D analogs.

17.5.3 Tertiary Hyperparathyroidism

Tertiary hyperparathyroidism is almost always preceded by a prolonged secondary form of hyperparathyroidism. Elevated PTH levels persist in more than 25% of patients after transplantation despite the presence of normal renal function. The parathyroid cell proliferation continues and nodular areas of monoclonal cell masses are formed. This condition is often referred to as tertiary or autonomous hyperparathyroidism. The term autonomous hyperparathyroidism refers to a specific functional state of the parathyroid being relatively unresponsive to negative feedback mechanisms. Renal graft function is also an important determinant of post-transplant serum PTH levels. Immunosuppressive drugs and steroids also contribute to tertiary HPT. Features include hypercalcemia, hypertension, deterioration of graft function, neuropsychiatric and gastrointestinal symptoms, hypophosphate bone loss. Medical management is seldom successful and subtotal with a marker in the remaining gland is an acceptable alternative.

17.5.3.1 Primary Hyperparathyroidism

Primary hyperparathyroidism (PHPT) encompasses asymptomatic endocrine disorder, symptomatic endocrine disease, and recently detected normocalcemic hyperparathyroidism. It is most likely that the community prevalence of PHPT always has the mixture of the three different forms but has different regional proportions. In

Asian countries, PHPT still remains an asymptomatic disease and many patients are properly diagnosed late in the course of the disease process probably due to relative rarity and differing clinical presentations.

Normocalcamic hyperparathyroidism is recognized recently and the actual prevalence course of the disease is yet not clear. The clinical expression is characterized by consistently normal ionized calcium levels and a raised PTH level. These patients may become symptomatic in due course. Most of these patients have normal serum levels of vitamin D and 24-h urinary calcium levels. Before confirming the diagnosis, it is mandatory to exclude secondary causes of elevated PTH such as vitamin D deficiency, reduced creatinine clearance, hypercalciuria, and gastrointestinal causes of calcium malabsorption. History of regular use of drugs like thiazide group of diuretics and lithium salts should be excluded. In patients with normocalcemic hyperparathyroidism, it is prudent to keep them on follow-up with routine calcium, PTH, and bone mineral density assessments. Surgery needs to be considered when there is evidence of early bone involvement or renal involvement.

17.5.3.2 Neonatal Hyperparathyroidism

Neonatal hyperparathyroidism is rare and presents in familial and sporadic forms. The clinical features usually become evident by the first week of life but sometimes are diagnosed at 3–4 months of age. The clinical features include failure to thrive, hypotonia, constipation with anorexia, vomiting, difficulty in feeding, and dehydration. Defects in the calcium Sensing Receptor (CaSR) gene have been found in these neonates. Medical management in these neonates includes hydration and cinacalcet. Total parathyroidectomy with autotransplantation of the parathyroid is the treatment of choice. These neonates have parathyroid hyperplasia on histopathology.

17.5.3.3 Familial Hypocalciuric Hypercalcemia

Benign familial hypocalciuric hypercalcemia is characterized by a mild increase in both serum calcium and intact PTH levels. The 24-h urine calcium in these cases is low. Here the calcium to creatinine clearance ratio is <0.01, unlike primary hyperparathyroidism where the ratio is more than 0.02. Benign familial hypocalciuric hypercalcemia is an autosomal dominant trait. Parathyroidectomy is not indicated in this condition.

17.5.4 Familial Hyperparathyroidism

Most of the familial cases occur in association with Multiple Endocrine Neoplasia (MEN). However, there is a distinct clinical entity known as non-MEN familial hyperparathyroidism. These patients present with profound hypercalcemia, and some present with hypercalcemic crises. These patients have multiple abnormal parathyroid glands, either synchronously or metachronously, and are at risk of persistent and recurrent hyperparathyroidism. They need to be treated aggressively.

Hyperparathyroid Jaw Tumor Syndrome: This rare syndrome has phenotype expression of hyperparathyroidism, jaw tumor, renal cysts, and uterine tumors. They are characterized by autosomal dominant germline transmission. There is an increased chance of parathyroid carcinoma among these patients.

Hypercalcemia unrelated to parathyroid lesions:

There are a variety of diseases expressing disturbances in calcium metabolism. Hypercalcemia is seen in malignancy, granulomatous disorders such as tuberculosis, sarcoidosis, other endocrine disorders such as Addison's disease, hyperthyroidism, phaeochromocytoma, and VIPoma. Long-term usage of medications such as thiazide diuretics, lithium, calcium supplements may induce hypercalcemia.

17.5.5 Hypercalcemia of Malignancy

Hypercalcemia is detected in about 10–30% of patients with advanced cancer and is associated with osseous secondaries in many patients. Hypercalcemia can occur with or without bone secondaries since the pathogenesis is either humoral or local osteolysis.

Malignancy associated hypercalcemia has three distinct entities:

(a) Humoral hypercalcemia occurs in patients with solid malignancies of organs such as the breast, kidney, and ovary. Parathyroid hormone-related protein (PTHrP) was identified in the 1990s and was found to share the same receptors of PTH and exhibit characters similar to PTH.
(b) Hypercalcemia associated with bone metastases. Local osteolytic hypercalcemia is caused by the presence of tumor cells in the bone microenvironment causing increased osteoclastic bone resorption and hypercalcemia. This mechanism is seen in breast carcinoma.
(c) Hypercalcemia is associated with hemotological malignancies like multiple myeloma.

The important differences in humoral hypercalcemia and primary hyperparathyroidism are the low plasma chloride, high bicarbonate, and urinary cyclic AMP in the former.

Hypercalcemia is a sign of advanced malignancy and one needs to treat the underlying malignancy. Life-threatening complications of high serum calcium require emergency management. Medication used to treat hypercalcemia of malignancy includes Bisphosphonates, Lenosumab, (a monoclonal antibody that inhibits the differentiation of osteoclasts), and Calcitonin.

17.5.6 Hypoparathyroidism

Hypoparathyroidism can result from decreased secretion of parathyroid hormone or target organ resistance to PTH and the majority of cases are iatrogenic. Surgical destruction or devascularization of the parathyroid gland during parathyroid surgery or head and neck surgery for cancer is known to happen. Thyroidectomy is one of the common surgical procedures which carries a reported risk of 1–3% for permanent hypoparathyroidism. Other causes include genetic defects and mutations in the PTH gene or the calcium sensor receptor gene, developmental anomalies as in DiGeorge's syndrome, poly organ autoimmune endocrinopathies, and infiltrative diseases such as haemochromatosis and Wilson's disease.

The serum calcium values are always subnormal and serum PTH values also remain low in hypocalcemia of parathyroid origin. Postoperative hypoparathyroidism manifests usually 24–48 h after surgery and estimation PTH is a good predictor of hypocalcemia. Postoperative hypoparathyroidism can be transient and calcium values return to normal in a week. The hungry bone syndrome is typically associated with symptomatic hypocalcemia and treatment for several weeks is required. Severe symptomatic hyperparathyroidism with marked bone mineral loss is a usual candidate to develop post-parathyroidectomy hypocalcemia requiring calcium and dihydroxy vitamin D supplementation for a long time.

The typical symptoms of hypocalcemia are neuromuscular irritability, perioral or acral paranesthesia, muscle spasm, carpopedal spasm, laryngeal spasm, bronchial spasm, and tetany. Simple bedside maneuvers like Chvostek's sign and Trousseau's sign can reveal the presence of neuromuscular irritability. Hypocalcemia can also manifest as prolongation of QT interval.

Calcium homeostasis is closely dependent on phosphate, magnesium, and vitamin D. For effective absorption of calcium from the G I tract dihydroxy vitamin D is required. Serum magnesium levels also require close monitoring and correction of hypomagnesemia is essential for maintaining normocalcemic levels as well as recovery of neuromuscular symptoms. Treatment of postoperative hypoparathyroidism is based on the degree of hypocalcemia and severity of symptoms. If the serum calcium level is below 7.5 mg/dl and the patient is symptomatic an infusion of 10 ampoules of calcium gluconate in 1 L of 5% dextrose is given at 100 ml/h with serum calcium estimations at regular intervals. For patients who are mildly symptomatic and whose calcium, values are between 7.5–8.5 mg/dl, 1000 mg of oral calcium can be given every 6 h with 0.25–2 μg/day of Calcitriol. It is also advisable to check vitamin D and magnesium levels and supplement them if the levels are below normal. Currently, two types of recombinant PTH are available as subcutaneous injections given twice daily.

17.5.7 Pseudohypoparathyroidism

Pseudohypoparathyroidism is a heterogeneous group of disorders characterized by PTH resistance. Patients present with hypocalcemia and hyperphosphatemia due to impaired target tissue response to PTH. This is an autosomal dominant disorder. In Type-Ib: pseudohypoparathyroidism there is resistance to multiple hormones PTH, TSH, Luteinizing hormone, and gonadotrophic releasing hormone. The defects include short stature, obesity, round facies, brachymetacarpia, subcutaneous ossification, and mental retardation (Albright's hereditary osteodystrophy). Patients with pseudohypoparathyroidism 1b lack the features of Albright's hereditary osteodystrophy and show renal resistance to PTH as the only manifestation. Pseudohypoparathyroidism 1b arises as a result of imprinting defects that affect the expression of GNAS I in the proximal renal tubules.

Laboratory evaluation of pseudohypoparathyroidism will show hypocalcemia, hyperphosphatemia, and a raised PTH. Confirmation of diagnosis of PHP can be done by estimating Urinary cyclic AMP and phosphate excretion following infusion of PTH. Treatment is with calcium and vitamin D. Hypocalcemic crisis requires intravenous calcium. Research is also going on for stem cell therapy in hypoparathyroidism.

17.5.8 Cryopreservation of Parathyroid Tissue

The bits from parathyroids for cryopreservation are sent for frozen, confirmed as parathyroids, and then prepared for preservation. Freezing media is an 80% Rosewell park memorial institute (RPMI) 1640 solution with the addition of DMSO in 10% of autologous serum. The vials are placed in a freezer where the temperature goes down gradually from –1 °C to –80 °C. The vials are relocated into a liquid nitrogen freezer at –170 °C. When required the vials are placed in a water bath at 37–42 °C. They are rinsed in an RPMI solution at 37 °C three times. The parathyroids are now ready for implantation.

17.6 Primary Hyperparathyroidism (PHPT)

17.6.1 Clinical Manifestations

Hyperparathyroidism results due to hyperfunctioning of one or more parathyroid glands. Three different clinical expressions of PHPT are now established. The mildly symptomatic or asymptomatic disease is the common presentation in North American and European countries. This wide prevalence of this condition had been recognized following the introduction of multichannel autoanalyzer in 1970 and predominantly affects females in the postmenopausal period. The asymptomatic disease affects approximately 1% general population and about 2% of postmenopausal women in North American and European countries. These patients do not present with overt symptoms and clinical signs related to target organ damage. The serum calcium level usually remains within 1 mg of the upper limit of reference level is seldom elevated more than 1.5–2 fold higher than the upper limit. Only a few of these categories progress to damage the target organs and symptomatic stage [2].

At times the distinction between asymptomatic and symptomatic is not very clear as many of them have subtle symptoms which come out on detailed history taking. Symptoms such as fatigue, exhaustion, weakness, polydipsia, polyuria, bone pain, joint pain, back pain, depression, memory loss, nausea, pruritis, and loss of appetite may be present. Most often many of the so-called asymptomatic patients on detailed evaluation have subtle symptoms and surgery leads to improvement in the quality of life.

Symptomatic PHPT is not an uncommon disease in Asian countries. The demographic profile differs with the higher frequency of male patients and the average age of patients is a decade younger than the western counterpart. The skeletal profile of osteitis fibrosa cystica, brown tumors of the long bones, and fragility fractures are still seen in patients in India. Other less severe symptoms related to the musculoskeletal system include diffuse bone pain, arthralgia, and myopathy affecting the pelvic girdle muscles. Because of bizarre clinical presentation diagnosis of PHPT is delayed considerably.

Renal stone disease is the common primary symptom of more than half of Asian patients. Urinary stones are common in the general public and <1% of patients have a causal association with hyperparathyroidism. But recurrent bilateral stone disease indicates the possibility of metabolic disorders like hyperparathyroidism. The majority of these patients show hypercalcemia and exaggerated 24-h urinary calcium excretion. Nephrocalcinosis is a relatively rare imaging finding and usually leads to deterioration of renal function. Occasional patients develop renal parenchymal damage with evidence in ultrasound imaging and are associated with progressive renal dysfunction.

Gastrointestinal manifestations are mostly related to hypercalcemia, stone diseases in the gall bladder and pancreas, or primary tumors seen in MEN 1 syndrome. Recurrent acute pancreatitis is a primary presentation of PHPT and is considered as complication hypercalcemia.

Metabolic disturbances such as loss of appetite, severe fatigue, dehydration, asthenia, and rarely hemodynamic instability are a rare presentation of PHPT.

Hypercalcemic crisis: Hypercalcemic crisis is a reversible but life-threatening condition occasionally found in PHPT and advanced stages of certain malignancies like lymphomas, multiple myeloma. Hypercalcemic crisis associated with primary HPT is also known as parathyroid storm, parathyrotoxicosis, or parathyroid intoxication.

A constellation of symptoms such as anorexia, nausea, vomiting, polyuria, dehydration, drowsiness, stupor, coma, renal insufficiency, cardiac dysrhythmias are seen. The diagnosis of hypercalcemic crisis can be made when the serum calcium is more than 14.5 mg/dl and acute symptoms and signs can be reversed by correcting the hypercalcemia. All these patients need emergency admission in high dependency wards, close monitoring, emergency measures to lower the serum calcium, and support to deteriorating general condition. Volume expansion by intravenous infusion of normal saline (150 ml/h) is initiated and maintained. For quick control of hypercalcemia calcitonin is used as a nasal spray or subcutaneous injection. This brings down but serum calcium levels revert to previous high in 48 h. So, calcitonin is generally used to prepare the patient for emergent parathyroidectomy. For sustained lowering of serum calcium bisphosphates, infusions are given.

The classical clinical manifestations such as osteitis fibrosis cystica, recurrent nephrolithiasis, nephrocalcinosis, peptic ulcer disease, pancreatitis are still common in Asian countries, unlike the western world. Majority of patients present with a combination of nonspecific complaints such as fatigue, bone pain, weight loss, pruritis, polydipsia, polyuria, nocturia, joint pain, cardiac dysfunction, constipation, depression, nausea, and hematuria. Because of vivid symptoms related to various systems the diagnosis is often delayed.

Musculoskeletal manifestations are a primary symptom of about 50% of cases in India. The symptoms include fragility fractures, brown tumors, skeletal deformities, diffuse bone, and joint pain, and proximal girdle myopathy. Urinary stones are the most common presentation of PHPT but are greatly being overlooked since the causal association is significant in <1% of urolithiasis. Nephrocalcinosis is almost always an incidental finding on imaging but has a serious impact on renal function.

Hypercalcemia-induced hyperacidity and peptic ulcer disease are common in PHPT patients but are under-evaluated. Hypercalcemia is one of the metabolic factors associated with acute pancreatitis.

Psychological symptoms, personality disorders, depression, and psychoses have been found in PHPT. Neuromuscular symptoms include fatigue, generalized weakness, and depression.

Glucose metabolism in PHPT:

There is a considerably higher risk of diabetes mellitus among patients with PHPT. Approximately 40% of patients with PHPT have impaired glucose metabolism. Diabetes mellitus associated with PHPT is difficult to control probably due to alterations in peripheral insulin sensitivity. Insulin resistance in hyperparathyroidism is probably due to raised intracellular free calcium concentration, which interferes with normal insulin-stimulated glucose transport. Parathyroidectomy corrects to some extent insulin hypersecretion and insulin resistance.

Lipoprotein metabolism: Studies show discrepant findings of lipoprotein metabolism in PHPT. Decreased serum level of high-density lipoproteins (HDL), increased serum level of triglycerides and VLDL are some evidences of lipid metabolism disturbances found in PHPT. Increased lipolysis is also noted with increasing PTH levels thus causing elevation in VLDL levels.

17.6.1.1 Metabolic Complications of Primary Hyperparathyroidism

Over the years the clinical picture of hyperparathyroidism has changed from a rare illness featuring bone disease with or without renal disease to a disease characterized by more subtle or nonspecific clinical manifestations. Occasionally, metabolic complications are seen in patients with a long history of primary hyperparathyroidism. Mortality in these patients is usually due to cardiovascular disease or renal failure. The severity of the metabolic complications cannot be predicted by the presence or absence of symptoms. Some symptoms and metabolic changes improve after parathyroidectomy.

17.6.1.2 Arterial Hypertension

Incidence of hypertension in PHPT patients is significantly high compared to general population but etiopathogenesis of this association is not completely understood. Parathyroid hormone is said to play a permissive role in the hypertensive action of hypercalcemia. Animal studies indicate that chronic hyperparathyroidism and paraneoplastic hypercalcemia can be instrumental in causing secretion of aldosterone ensuing relative hyperaldosteronism. Excess levels of PTH and aldosterone have a damaging effect vessel wall [3]. The resultant vasoconstrictor activity which enhances the renin-angiotensin aldosterone system and probably explains higher frequency of hypertension among patients with PHPT. The resultant vasoconstriction and extracellular fluid over load may be the pathogenesis of hypertension in PHPT.

17.6.1.3 Cardiovascular Disease

Patients with hyperparathyroidism are a higher risk of developing arrhythmias, left ventricular hypertrophy, cardiac failure, and calcific diseases of arteries. Studies showed that the LVH is independent of hypertension. Cardiac failure is found more frequent among patients with PHPT and could be due to direct action of PTH on cardiac myocytes. Calcium deposition was identified in valvular annuli and cusps, coronary arteries, individual myocardial fibers, and the interventricular septum. Mitral and aortic valves are more frequently disturbed.

Hyperuricemia—Data from a national representation survey in the U.S. indicates that serum PTH levels were independently associated with hyperuricemia. Increased frequencies of hyperuricemia and gout have been associated with PHPT.

17.6.1.4 Chondrocalcinosis and Pseudogout

This deposition of calcium salts in the articulate hyaline cartilage and fibrocartilage is seen specially in the knee joint. Chondrocalcinosis and pseudogout are frequently seen in patients with PHPT. Pseudogout can be an initial manifestation characterized by one or more joint involvements associated with calcium pyrophosphate dehydrate crystals in the synovial fluid. These attacks can be precipitated by transient or rapid changes in calcium concentration. Pseudogout rarely affects the foot. Parathyroid surgery prevents the progression of chondrocalcinosis and relieves symptoms. However active episodes of gout and pseudogout arthritis are known to occur post parathyroidectomy.

Clinical examination does not routinely show any pathognomonic signs. Evidences of target organ damages as deformities, brown tumors and arthritis are not confirmatory of PHPT. Neck swellings closely mimicking thyroid nodules are rare findings.

17.6.2 Evaluation of Primary Hyperparathyroidism

17.6.2.1 Biochemical

1. Serum total calcium is always advisable as screening test and repeated elevated levels merit detailed evaluation. Normocalcemic hyperparathyroidism is a rare clinical entity likely to be missed by estimating serum calcium alone. Serum calcium estimation is preferably done during morning hours on an empty stomach. It is preferable to avoid strenuous exercises before blood draw. Tourniquet is removed after venous puncture since the resultant stasis may alter the serum

calcium level. When facilities are not available to estimate ionic calcium, correction based on serum albumin level is preferred. Normal calcium level ranges from 8.5 to 10.2 mg/dl (2.2–2.5 mmol/L).

2. When hyperparathyroidism is suspected simultaneous estimation of serum calcium, serum intact PTH and serum PO4 is done. Since decay of PTH occurs in minutes blood sample should not be stored before loading to analyzer. The second generation immunoradiometric assays are advisable since the third-generation assays are not universally available. (The PTH assay—the normal levels are between 10–55 pg/ml.)
3. When PHPT is diagnosed biochemically further laboratory studies are aimed at target organ damages. Liver function studies and isolated elevation of serum alkaline phosphatase indicates bone damage.
4. Renal function studies include serum creatinine and 24-h urinary creatinine and calcium estimation. Creatinine clearance (eGFR) is calculated to assess the renal function. Creatinine clearance—the normal is more than 60 ml/min.
5. Serum 25 OHD estimation is necessary in further management of the PHPT.
6. Alterations of electrolytes especially those of serum chloride, serum magnesium are routinely estimated.
7. Bone mineral density studies indicate fracture risk of the patient and is an indicator of decision-making regarding parathyroidectomy. Severe bone loss predicts postoperative hungry bone syndrome.
8. Well prepared X-ray of kidney ureter and urinary bladder and abdominal ultrasound imaging are done to screen for stone disease, nephrocalcinosis, renal cortical damage and back pressure effects. Further imaging like intravenous urogram and contrast enhanced CT are planned based on creatinine values and initial imaging.
9. Assessment of cardiac function is done with ECG and echocardiogram as routine screening.

17.7 Overview of Management

The only curative treatment of PHPT is parathyroidectomy. Symptomatic disease is an absolute indication for parathyroidectomy whereas it is recommended in selected patients with asymptomatic disease. Parathyroidectomy is reserved to selected patients of asymptomatic hyperparathyroidism.

The guidelines [4] for surgery in asymptomatic patients are:

- Age <50 years
- Serum calcium > 1 mg/dl (0.25 mmol/L) above the normal limit
- T-Score of −2.5 or less or fragility fracture
- Vertebral fracture on imaging
- Creatinine clearance below 60 ml/min
- 24-h Urine calcium more than 400 mg/day (10 mmol/day)

- Presence of nephrolithiasis or nephrocalcinosis
- Patients unwilling to come for regular follow up

Asymptomatic patients who do not fall into the category requiring operation should be reviewed annually. Clinical judgment shall be supported by estimation serum calcium, phosphate, and creatinine. Biannually skeletal screening for bone loss is done by DXA of wrist with lower radius, lumbar spine and hips is suggested. Annual testing for kidney function by estimating 24-h excretion of calcium and creatinine and eGFR estimation. Patients who show disease progression are suggested parathyroidectomy.

Vitamin D deficiency whenever diagnosed should be corrected with appropriate supplementation. Calcium lowering agents are not generally required since such patients are always candidates for surgical treatment.

The surgical treatment of parathyroidectomy has considerably changed from laborious bilateral neck exploration to targeted removal of involved gland or glands. This change was facilitated by localization studies many of which are contribution from advances of Nuclear Medicine. A list of radiotracers such as Thallium-201, technetium-99 methoxyisobutylisonitrile (Tc99-MIBI), Tc-99m tetrofosmin and 18F-fluorocholine are used in different protocols.

The various surgical approaches to hyperparathyroidism include bilateral neck exploration, unilateral neck exploration and focused parathyroidectomy. Bilateral neck exploration is aimed to expose all four parathyroid glands and physically ensuring their normality and frozen section may be done whenever in doubt. Routine frozen section of all glands has high chance of permanent hypoparathyroidism and is not recommended. When a smaller number of glands are identified, ectopic foci are explored resulting long duration procedure, the recurrent laryngeal nerve and blood supply of parathyroid are at risk. During unilateral exploration the abnormal parathyroid lesion is excised and the other parathyroid on the same side is exposed and inspected.

Advances in imaging technology and estimation of PTH enables surgeon to perform minimal exploration. Focused parathyroidectomy involves the removal of the only the abnormal parathyroid gland identified by imaging. When the procedure is performed through a 2–4 cm incision, it is generally known as mini-parathyroidectomy. The completeness of removal of all hyperfunctioning glands is ensured by estimating PTH during the procedure. PTH has a short half-life (<5 min) and 10 min decline of serum PTH to <50% of pre-excision level ensures completeness of removal of all abnormal glands [5].

Remote access procedures are extended to parathyroid diseases also and parathyroidectomy can be performed by endoscopic exploration. With the introduction mini-Gamma probes hand held probes are used for intraoperative guidance (Radio guided parathyroidectomy).

Modifications in parathyroidectomy is effectively implemented as a result of advances in localization studies.

17.8 Localization Studies

John Doppman's skeptical comment on localization studies "the only localization study in a patient with biochemically proven Primary Hyperparathyroidism is the localization of an experienced parathyroid surgeon" needs correction in view of changes in imaging technology and surgical approach. No doubt we still need a surgeon with appropriate skill to perform successful parathyroidectomy. However, localization studies plus intraoperative adjuncts and technological advances have helped surgeons perform minimally invasive procedures. Newer imaging paved way for remote access procedures as well as open mini-parathyroidectomy limiting the need for extensive bilateral neck exploration for selected patients [2].

17.8.1 Non-Invasive Localization

Ultrasound: A comprehensive imaging of parathyroid using a 5–15 MHz transducer is carried out with the patient in supine position and neck extended. Scanning is done from the Hyoid bone to the sternal notch extending laterally to include possible ectopic gland in the carotid sheath. The thyroid gland is also scanned to look for possible nodules and if present guided FNAC is carried out from suspicious nodules (Fig. 17.1).

This picture shows an enlarged parathyroid gland homogenous, solid hypoechoic lesion which is less echogenic than the thyroid.

Enlarged parathyroids are usually homogenous solid, hypoechoic, and less echogenic than the thyroid. Usually, these adenomas are 8–15 mm in size, oval and as they enlarge become irregular. Cystic parathyroid adenomas present as complex cystic lesions with internal septations, peripheral nodularity and have an echogenic border with the thyroid gland. Parathyroid adenomas are vascular and receive blood supply from polar vessels forming an asymmetric vascular arc around it. This helps to differentiate from lymph nodes which have a central hilar supply. When used alone the sensitivity of ultrasound alone is up to 80% in picking up parathyroid lesions. The diagnostic accuracy is operator dependent and further interfered by locations such as trachea-esophageal groove. Ultrasound imaging has a further advantage of better availability compared to scintigraphy. Ultrasound is useful as an adjuvant anatomical imaging for accurate localization prior to surgery and enables mini- focused parathyroidectomy.

Parathyroid carcinoma is usually more heterogenous, multilobular and shows evidence of invasion to surrounding structures with or without lymphadenopathy. Ultrasound cannot fully assess ectopic locations and multiglandular disease. It also depends on the patient's build and body habitus.

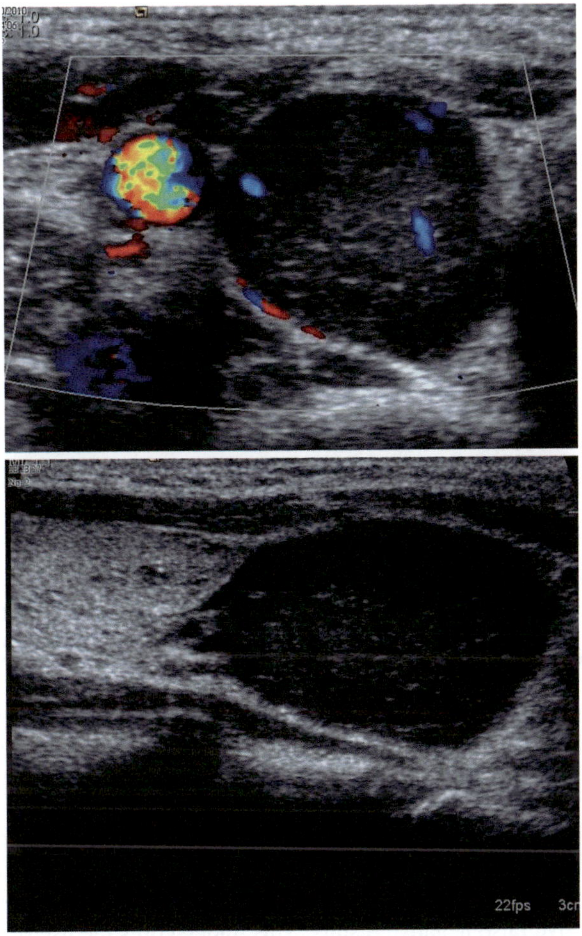

Fig. 17.1 Ultrasound imaging of inferior parathyroid lesion

17.8.2 Scintigraphy

17.8.2.1 Technetium99 Sestamibi Scan

Sestamibi is a lipophilic cation that crosses passively into cells and concentrates in the mitochondria. Adenomas and hyperplastic glands contain mitochondrial rich oxyphil cell that concentrates and retain Sestamibi [4]. This radiotracer has affinity to thyroid as well as to parathyroid but washes off from thyroid earlier. The affinity to hyperactive parathyroid glands is relatively more persistent than to thyroid but normal parathyroid glands are not lighted up with the tracer. The differential affinity is the principle behind dual phase scintigraphy where the delayed scan is done 2–2½ h after injection of tracer (Fig. 17.2).

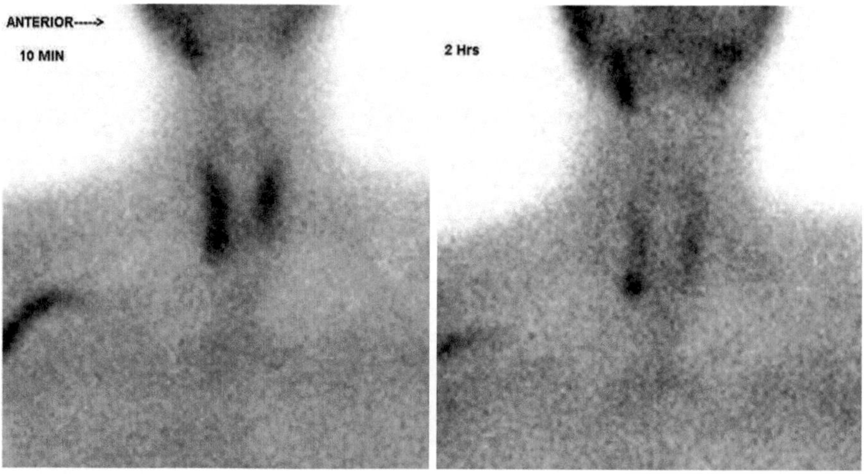

Fig. 17.2 Tc 99 MIBI dual phase scintigraphy

In dual phase single isotope imaging an early image is obtained at 10–15 min after the radiotracer injection in addition to the 90–180-min post injection image. This method utilizes different rates of washout from the thyroid tissue and retention in abnormal parathyroid [6] (Fig. 17.3).

The addition of SPECT (single photon emission computed tomography) allows for three-dimensional imaging thereby improving visualization of abnormal parathyroid glands. This has improved sensitivity of parathyroid detection compared to planar imaging in antero lateral and oblique views.

The sensitivity of Sestamibi dual phase scintigraphy ranges from 61.4% to 100%. Thyroiditis and thyroid neoplastic nodules retain Sestamibi and can result in false positive scans. Sometimes early washout from pathological parathyroid tissue can give false negative results. This underscores the use of SPECT and CT. Size of the adenoma and serum calcium levels also have a bearing on Sestamibi scans. Large adenomas with high PTH values and higher calcium levels are more likely to be lighted up on Sestamibi scans.

Deficiency in vitamin D seems to be a predictive for a positive scan. Calcium channel blockers interfere with the uptake of Sestamibi. Oxyphil cells have higher mitochondrial content than chief cells and hence oxyphil adenomas have a higher positive scan rate.

17.8.2.2 Positron Emission Tomography

It has a sensitivity of 86% and a specificity of 78%. FDG—PET has not been widely used. Recently PET with 11-C methionine that detects amino acid metabolism is being evaluated. The recent introduction Fluro-choline PET has promising results in localization studies od otherwise obscure lesions.

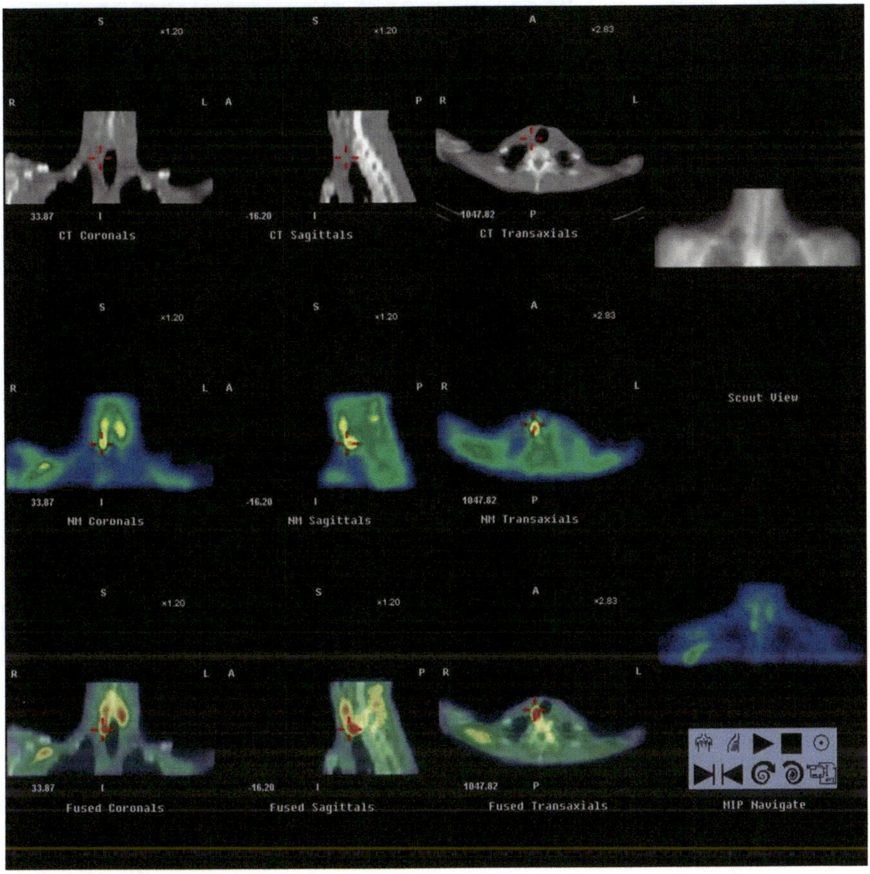

Fig. 17.3 SPECT imaging of right inferior parathyroid adenoma

17.8.3 Computed Tomography

4D CT scan is showing great promise as an imaging modality for locating abnormal parathyroid glands. The advantages of 4 D CT are that precise anatomical information from CT scan is easy to interpret and can direct surgical exploration. Also, ectopic locations are fully assessed (Fig. 17.4).

17.8.4 Magnetic Resonance Imaging

This is less sensitive than other imaging modalities. The primary advantage of MRI is no exposure to radiation and therefore can be used in a pregnant woman with primary hyperparathyroidism (Fig. 17.5).

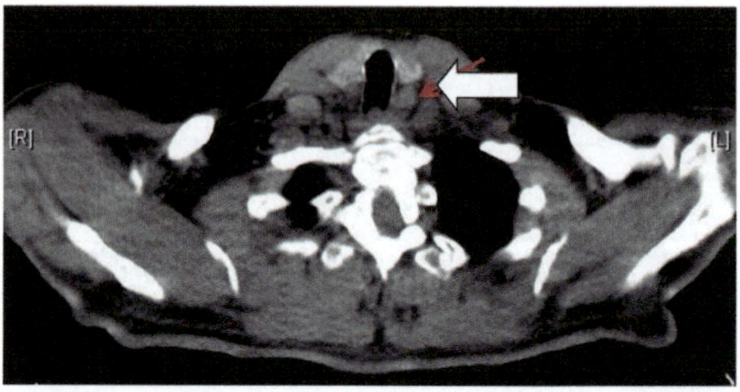

Fig. 17.4 Four D CT showing left inferior parathyroid adenoma

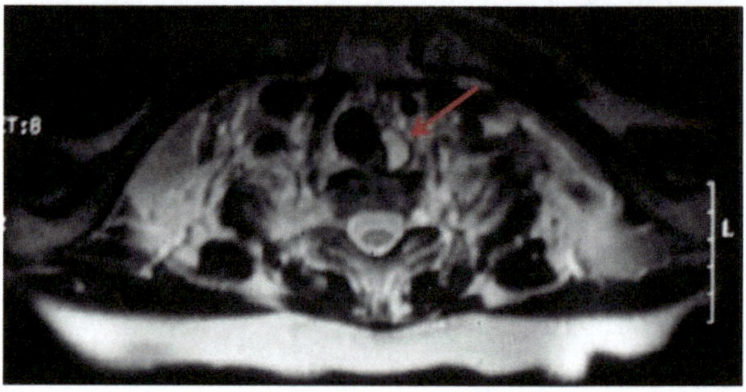

Fig. 17.5 MRI image of left inferior parathyroid adenoma

17.8.5 Invasive Localization

Catheter based localization with arteriography and venous sampling is reserved for remedial patients who fail to localize with noninvasive imaging, discordant imaging, unusual anatomy, or prior neck surgery.

Ultrasound or CT guided FNA combined with rapid PTH assay is a useful adjunct for patients with equivocal imaging in the setting of prior neck surgery or for confirmation of glands in ectopic locations.

17.8.6 Intraoperative Localization

17.8.6.1 Radio Guided Surgery

Patients are injected 10 mci of Sestamibi, 1–2 h prior to surgery. A gamma probe is used to direct surgery. Radio guidance has been reported to provide several advantages during surgery including facilitating speedy identification of glands, especially in ectopic locations.

17.8.6.2 Intraoperative Ultrasound

Intraoperative ultrasound can be a helpful imaging adjunct for identification of abnormal parathyroid glands with equivocal pre-operative imaging and in remedial surgery.

Intra-operative parathyroid aspiration: Aspiration of suspicious lesion and estimation of PTH in the aspirate is suggested as an alternate technique of excision and frozen section biopsy. But this technique needs more validation to estimate the cut off.

17.8.6.3 Methylene Blue

Methylene blue infusion at 7.5 mg/kg is infused over 15 min once the incision is made. Parathyroid gland retains the dye. False positive staining was noted in normal parathyroid glands, thyroid tissue, and lymph nodes. All secretions including saliva, urine will turn blue. Methylene blue is no longer used routinely.

Parathyroid carcinoma (PC): PC is a rare etiology of PHPT and generally considered <1% of all parathyroid lesions. Most of PC are hormonally very active and shows high disease severity indices like serum calcium and serum PTH. Many patients present in metabolic crisis and a palpable neck nodule is not rare. There are no conclusive clinical or biochemical evidences indicating the diagnosis. During surgical procedure a firm or hard adherent lesion is always found but these observations are seldom unique to PC. Similar features are not uncommon in atypical adenoma and even in adenoma. Since only effective treatment with curative intend is complete excision without residual malignancy, wide excision is recommended in patients with suspicious features.

Parathyroidectomy: Parathyroidectomy is the curative surgical procedure for PHPT. Focused parathyroidectomy through a well-planned small incision is the procedure of choice when localization is successful. The operation is planned as a day case procedure in mildly symptomatic or asymptomatic PHPT. The scenario is different in many symptomatic patients with high disease severity indices especially when bone loss is severe. These patients may require early admission and measures to lower serum calcium levels before operation. Vitamin D deficiency is corrected cautiously. Patients with severe bone loss, severe vitamin D deficiency, and advanced age are likely to develop hungry bone syndrome and a few require parenteral supplementation to severe symptoms of hypocalcemia.

References

1. Akerström G, Malmaeus J, Bergström R. Surgical anatomy of human parathyroid glands. Surgery. 1984;95(1):14–21.
2. NIH Conference. Diagnosis and management of asymptomatic primary hyperparathyroidism: consensus development conference statement. Ann Intern Med. 1991;114(7):593–7. https://doi.org/10.7326/0003-4819-114-7-593.
3. Cusano NE, Silverberg SJ, Bilezikian JP. Normocalcemic primary hyperparathyroidism. J Clin Densitom. 2013;16(1):33–9. https://doi.org/10.1016/j.jocd.2012.12.001.
4. Cheung K, Wang TS, Farrokhyar F, Roman SA, Sosa JA. A meta-analysis of preoperative localization techniques for patients with primary hyperparathyroidism. Ann Surg Oncol. 2012;19(2):577–83. https://doi.org/10.1245/s10434-011-1870-5.

5. Chan RK, Ibrahim SI, Pil P, Tanasijevic M, Moore FD. Validation of a method to replace frozen section during parathyroid exploration by using the rapid parathyroid hormone assay on parathyroid aspirates. Arch Surg. 2005;140(4):371–3. https://doi.org/10.1001/archsurg.140.4.371.
6. Coakley AJ, Kettle AG, Wells CP, O'Doherty MJ, Collins RE. 99Tcm sestamibi—a new agent for parathyroid imaging. Nucl Med Commun. 1989;10(11):791–4. https://doi.org/10.1097/00006231-198911000-00003.

Management of PHPT in MEN 1 and MEN 2 Syndrome

18

Sabaretnam Mayilvaganan and P. R. K. Bhargav

18.1 Introduction

Primary hyperparathyroidism (PHPT) is a common endocrine disease, and a few have syndromic associations with Multiple Endocrine Syndrome Neoplasia Type 1 and Multiple Endocrine Neoplasia Type 2. Multiple Endocrine Neoplasia Type 1 (MEN 1) is characterised by neoplasms of parathyroid glands, pituitary, and gastro-pancreatic endocrine cells with phenotype expression of two or more organs. The PHPT is the commonest presentation of MEN 1. MEN 2 is the association of medullary thyroid cancer, pheochromocytoma, and neoplasms parathyroid. Usually, hyperparathyroidism associated with MEN 2 is mild and asymptomatic. PHPT is frequently encountered as expression mutations in exon II, codon 634 of MEN 2. The treatment of choice is excision of the enlarged parathyroid gland [1, 2].

18.2 MEN 1

Multiple endocrine neoplasia Type 1 is a rare hereditary syndrome with a wide range of expressions, including about 20 types of endocrine and non-endocrine tumours. The principal organs are the anterior pituitary gland, parathyroid, and entero-pancreatic endocrine tissue, and phenotype expressions occur in two or more of these locations. The genetic changes in MEN 1 was identified in 1997 and was found positive in about 90% of patients. Genetic testing is now recommended and helps identify asymptomatic carriers who remain on biochemical screening.

S. Mayilvaganan (✉)
Department of Endocrine Surgery, SGPGI, Lucknow, Uttar Pradesh, India

P. R. K. Bhargav
Department of Endocrine & Breast Surgery, Sanjay Gandhi Postgraduate Institute of Medical Sciences (SGPGIMS), Lucknow, Uttar Pradesh, India

C. G. Nair, S. J. Abraham (eds.), *Surgical Management of Thyroid and Parathyroid Diseases*, https://doi.org/10.1007/978-981-97-3774-1_18

18.2.1 Primary Hyperparathyroidism in MEN 1

Primary hyperparathyroidism is the most common form of endocrinopathy in MEN 1 and accounts for 2–4% of all forms of primary hyperparathyroidism. In most occasions, hyperparathyroidism is the earliest occurring endocrine expression of MEN 1. It manifests between 20–25 years of age, about a decade or two earlier than the non-syndromic PHPT. Patients with proven PHPT at a younger age with a positive family history should be screened for MEN 1 syndrome. However, the genotype-phenotype correlation for transmission of PHPT in MEN 1 has yet to be identified.

Nodular hyperplasia involving multiple parathyroid glands is typical of MEN 1; however, adenomas are seen rarely. There is no symmetry in volume or size, volume, and weight. However, each gland expresses a monoclonal lesion. Supernumerary glands can be found in 20% of MEN 1 patients. Rarely parathyroid adenomas located in ectopic locations and parathyroid remnants in the thymus assume great significance.

The treatment is essentially surgery, but the optimal extent of removal is disputed [3, 4]. Subtotal parathyroidectomy is the excision of 3–3½ glands, and total Parathyroidectomy is the removal of all parathyroid glands and autologous parathyroid tissue graft [5, 6]. After total parathyroidectomy, a portion of relatively normal-looking parathyroid is implanted in the brachioradialis muscle of the non-dominant arm. Cervical thymectomy is generally combined with subtotal or total parathyroidectomy to ensure the removal of accessary parathyroid remnants. The so-called total parathyroidectomy may not be a curative procedure, and recurrence is inevitable, so the surgery is aimed at preventing the complications of hyperparathyroidism [7, 8] (Algorithm, Box 18.1 and Fig. 18.1).

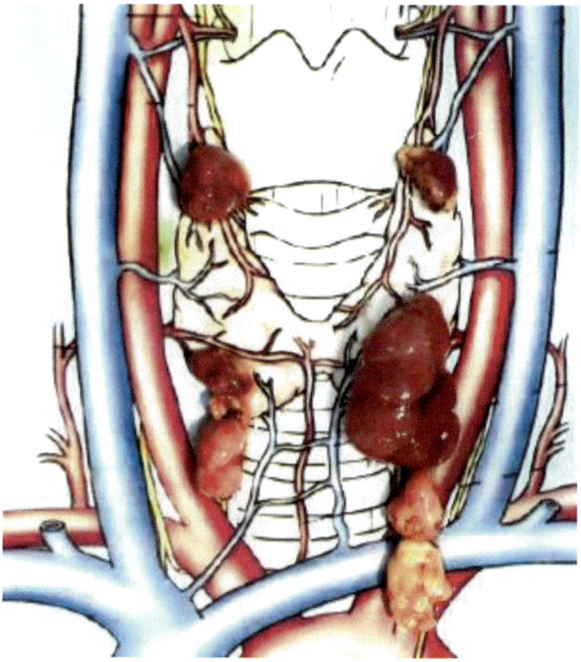

Fig. 18.1 Total parathyroidectomy with auto transplantation for MEN 1

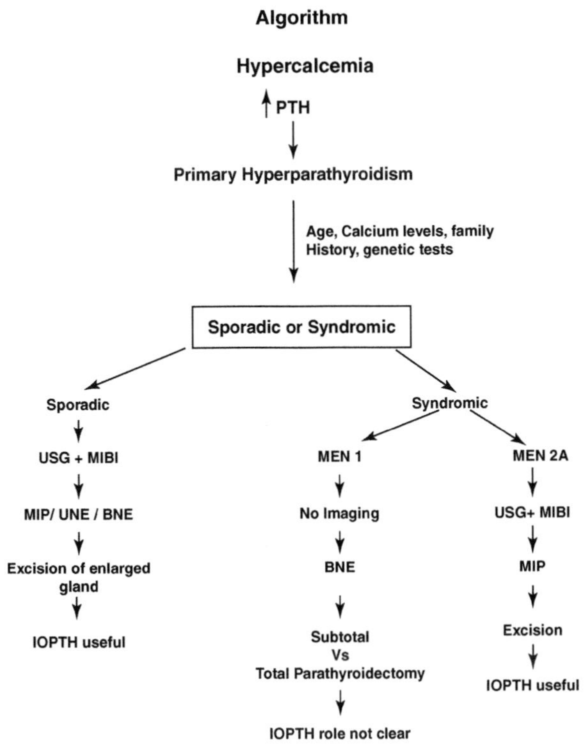

18.3 MEN 2

Multiple endocrine neoplasia Type 2A is an autosomal dominant syndrome [9] the phenotype includes Pheochromocytoma (usually bilateral synchronous), medullary thyroid carcinoma, and PHPT. The majority of patients have medullary thyroid carcinoma as the initial manifestation, and the pheochromocytoma may be diagnosed at a later stage. PHPT is seen in 15% of MEN 2A and is asymptomatic. Mutations of the Rearranged During Transfection (RET) proto-oncogene have been documented in MEN 2, and the mutation analysis should be considered in suspected MEN 2 patients. The clinical expression developing each of the three pathognomonic tumours in MEN 2A is based on the codon-specific variant, and it is most penetrant in the highest and high-risk groups.

MEN 2B phenotype includes pheochromocytoma, MTC, mucosal neuromas, thickened corneal nerves, intestinal ganglioneuromas, and marfanoid body habitus. PHPT is usually not present in these individuals. MEN 2B is primarily associated with mutations in codon 918 [10].

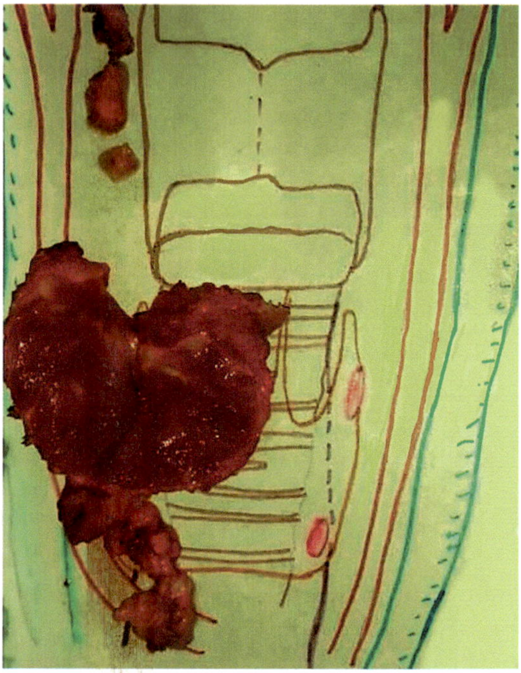

Fig. 18.2 Total thyroidectomy for medullary thyroid carcinoma and parathyroid adenoma

18.3.1 PHPT in MEN 2A

PHPT is associated with 15–20% of MEN 2A and is diagnosed more frequently with mutations in exon II, codon 634, and the RET gene [11]. The genotype-phenotype correlation is more characteristic of these mutations. The peak age for clinical manifestations is the 3rd decade, with symptoms usually mild. The PHPT is mostly detected on laboratory evaluations, and the treatment is surgical. The surgical treatment again is an occasion for the enlarged parathyroid gland (adenoma) only. The Intensity of the screening should be higher in those carrying the ret gene mutation, especially with exon 11 (Algorithm, Box 18.2 and Fig. 18.2) [12].

18.4 Conclusion

Subtotal parathyroidectomy, either three glands or 31/2 gland excision, has the edge over total parathyroidectomy since a similar recurrence rate is found. Persistent or prolonged hypoparathyroidism is more frequent with total parathyroidectomy. Clinically evident PHPT is rare associated with MEN 2A. Most index cases presenting with PHPT as the first manifestation have synchronous MTC and are often node-positive, and excision of the enlarged gland is the treatment of choice.

Box 18.1 Salient Features of PHPT in MEN 1

PHPT in MEN I

- Most common endocrinopathy
- 100% penetrance by 50 years of age
- Hypercalcemia is mild
- Hyperparathyroidism induced hypercalcemic crisis rare
- Earlier onset
- Equal male: female ratio
- Total parathyroidectomy + AT Vs subtotal Parathyroidectomy

Box 18.2 Salient Features of PHPT in MEN 2A

PHPT in MEN 2A

- PHPT is generally diagnosed at or after thyroidectomy
- Milder, asymptomatic and caused by single enlarged gland
- Germline mutation at codon 634—highest association

Box 18.3 Take Home Message

MEN Patients

- Risk for multiple conditions
- Complex management
- Lifelong follow up
- Genotype-Phenotype correlation
- New Paradigm in Prophylactic Surgery
- Risk of disease in relatives and counseling

References

1. Giusti F, Tonelli F, Brandi ML. Primary hyperparathyroidism in multiple endocrine neoplasia type 1: when to perform surgery? Clinics. 2012;67:141–4.
2. Lourenço DM Jr, Coutinho FL, Toledo RA, Gonçalves TD, Montenegro FL, Toledo S. Biochemical, bone and renal patterns in hyperparathyroidism associated with multiple endocrine neoplasia type 1. Clinics. 2012;67:99–108.
3. Tonelli F, Giudici F, Cavalli T, Brandi ML. Surgical approach in patients with hyperparathyroidism in multiple endocrine neoplasia type 1: total versus partial parathyroidectomy. Clinics. 2012;67:155–60.
4. Nilubol N, Weinstein LS, Simonds WF, Jensen RT, Marx SJ, Kebebew E. Limited parathyroidectomy in multiple endocrine neoplasia type 1-associated primary hyperparathyroidism: a setup for failure. Ann Surg Oncol. 2016;23(2):416–23.
5. Coutinho FL, Lourenco DM Jr, Toledo RA, Montenegro FL, Toledo S. Post-surgical follow-up of primary hyperparathyroidism associated with multiple endocrine neoplasia type 1. Clinics. 2012;67:169–72.

6. Fyrsten E, Norlén O, Hessman O, Stålberg P, Hellman P. Long-term surveillance of treated hyperparathyroidism for multiple endocrine neoplasia type 1: recurrence or hypoparathyroidism? World J Surg. 2016;40(3):615–21.
7. Montenegro FL, Lourenço Junior DM, Tavares MR, Arap SS, Nascimento Junior CP, MassoniNeto LM, D'Alessandro A, Toledo RA, Coutinho FL, Brandao LG, Cordeiro AC. Total parathyroidectomy in a large cohort of cases with hyperparathyroidism associated with multiple endocrine neoplasia type 1: experience from a single academic center. Clinics. 2012;67:131–9.
8. Versnick M, Popadich A, Sidhu S, Sywak M, Robinson B, Delbridge L. Minimally invasive parathyroidectomy provides a conservative surgical option for multiple endocrine neoplasia type 1–primary hyperparathyroidism. Surgery. 2013;154(1):101–5.
9. Alevizaki M, Saltiki K. Primary hyperparathyroidism in MEN2 syndromes. In: Medullary thyroid carcinoma. Cham: Springer; 2015. p. 179–86.
10. Magalhães PK, Antonini SR, de Paula FJ, de Freitas LC, Maciel LM. Primary hyperparathyroidism as the first clinical manifestation of multiple endocrine neoplasia type 2A in a 5-year-old child. Thyroid. 2011;21(5):547–50.
11. Alevizaki M. Management of hyperparathyroidism (PHP) in MEN2 syndromes in Europe. Thyroid Res. 2013;6(S1):S10.
12. Machens A, Lorenz K, Dralle H. Peak incidence of pheochromocytoma and primary hyperparathyroidism in multiple endocrine neoplasia 2: need for age-adjusted biochemical screening. J Clin Endocrinol Metab. 2013;98(2):E336–45.

Hyperparathyroidism of Renal Origin 19

A. Vimala, Ranjani Ravi, and Jinsi N Babu

Hyperparathyroidism is defined as increased production of parathormone by parathyroid glands and based on pathophysiology is classified as: primary hyperparathyroidism, secondary hyperparathyroidism, tertiary hyperparathyroidism, and refractory hyperparathyroidism. This chapter essentially deals with secondary, tertiary, and refractory hyperparathyroidism related to renal dysfunction.

19.1 Secondary Hyperparathyroidism (SHPT)

Deteriorating renal function causes phosphate retention and hypocalcemia. Parathyroid hyperactivity is a secondary phenomenon to maintain calcium/phosphate metabolism. This is achieved by several feedback loops between the kidney, the parathyroid glands, bone, and intestine as shown in Fig. 19.1.

The normal homeostasis of calcium is based on an interplay among kidneys, bones, and intestines and is closely linked to phosphate metabolism and vitamin D. Serum calcium levels may be disturbed by changes in any of the homeostatic mechanisms and may result in changes in any of the following abnormalities singly or in combination.

Hyperphosphatemia, hypocalcemia, high serum fibroblast growth factor 23 (FGF 23), and decreased 1,25OH vitamin D stimulate parathyroid cells to synthesis parathyroid hormone. Common causes of secondary hyperparathyroidism are noted in Table 19.1 [1].

A. Vimala (✉) · R. Ravi · J. N. Babu
Cosmopolitan Hospitals, Thiruvananthapuram, Kerala, India

C. G. Nair, S. J. Abraham (eds.), *Surgical Management of Thyroid and Parathyroid Diseases*, https://doi.org/10.1007/978-981-97-3774-1_19

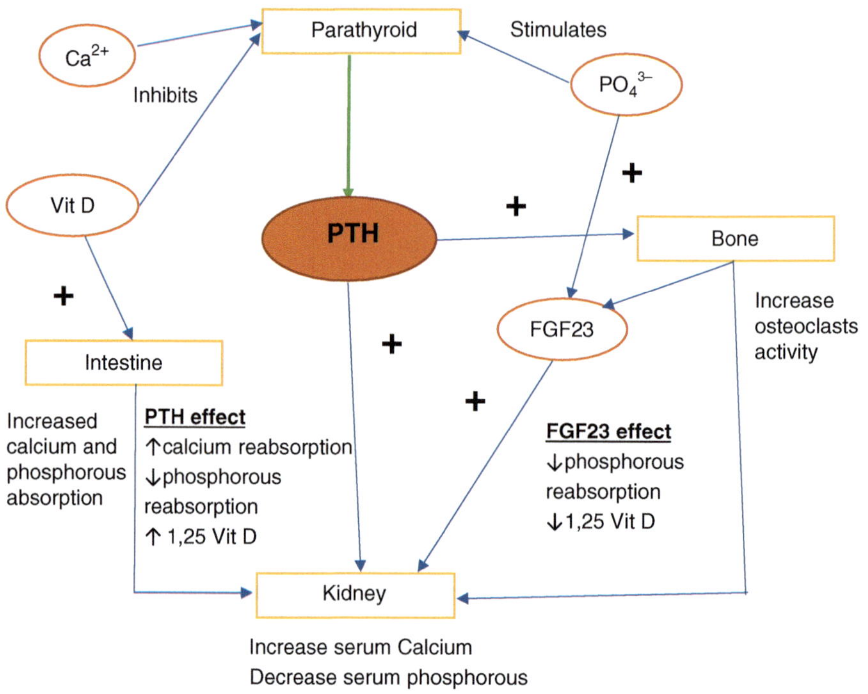

Fig. 19.1 Feedback loops involving *PTH* parathyroid hormone; *FGF-23* fibroblasts growth factor 23 [1]

Table 19.1 Causes of secondary hyperparathyroidism

Causes	Remarks
Chronic kidney disease	Abnormalities are seen earlier in tubulointerstitial disease
Vit D insufficiency	25(OH) vitamin D < 20 mcg/L
Drugs	Bisphosphonates, anticonvulsants, frusemide
Gastrointestinal cause	Celiac disease, cystic fibrosis
Hypercalciuria due to renal leak	Patients may be normocalcemic

19.2 Secondary Hyperparathyroidism of Renal Origin (SHPT)

Chronic kidney disease (CKD) is commonly associated with disorders of mineral and bone metabolism and is grouped as Chronic Kidney Disease-Mineral Bone Disorder (CKD-MBD). Secondary hyperparathyroidism is one of the manifestations of CKD- MBD and is the commonest cause of secondary hyperparathyroidism.

Renal dysfunction is classified based on estimated glomerular filtration rate (eGFR) and the KDIGO grading system of the renal functional stage is noted in Table 19.2.

Table 19.2 KDIGO CKD staging according to eGFR (mL/min) [2]

CKD stage		GFR categories mL/min/1.73 m^2
G1	Normal or high	≥90
G2	Mildly decreased	60–89
G3a	Mild to moderately decreased	45–59
G3b	Moderately to severely decreased	30–44
G4	Severely decreased	15–29
G5	Kidney failure	<15

These stages can be further classified into A1, A2, A3 based on albumin excretion

eGFR (EPI) = 141 * min (Scr/κ, 1)$^{\alpha}$ * max(Scr/κ, 1)$^{-1.209}$ * 0.993^{Age} * 1.018 [if female] * 1.159 [if black] (Scr is serum creatinine (mg/dL), κ is 0.7 for females and 0.9 for males, α is −0.329 for females and −0.411 for males, min indicates the minimum of Scr/κ or 1, and max indicates the maximum of Scr/κ or 1).

Subclinical abnormalities such as loss of klotho, increased FGF23 secretion, decreased bone formation rates, and vascular calcification occur from stage G2 onwards. But clinical manifestations and biochemically detectable abnormalities in serum phosphorus, calcium, PTH, and vitamin D become evident only when the glomerular filtration rate (GFR) falls below 40 mL/min, In CKD stage 5 (eGFR < 15 mL/min/1.73 m^2), almost all patients have evidence of bone disease. Approximately 80% of incident dialysis patients have evidence of coronary artery calcification. Hence, guidelines for CKD-MBD advise monitoring for the same for patients from CKD stage III onwards. According to KDIGO (Kidney Disease Improving Global Outcomes) 2012 guidelines, CKD is staged based on eGFR (eGFR is estimated Glomerular Filtration Rate) calculated by the CKD-EPI equation.

19.3 Prevalence of Secondary Hyperparathyroidism

In 1982, Memmos D et al. reported hyperparathyroidism in 90% of patients on dialysis [3]. Chiranjee Lal Dayma et al. from Rajasthan reported a 72% prevalence in CKD patients [4]. In our center, 90% prevalence of secondary hyperparathyroidism is seen in dialysis-dependent patients (unpublished).

Pathophysiology. The balance of serum calcium and serum phosphorous is grossly altered in CKD. Secondary hyperparathyroidism is adaptive parathyroid gland hyperplasia which maintains calcium phosphate balance. Increased production of PTH is secondary to stimuli such as hyperphosphatemia, hypocalcemia, and increased FGF23.

The serum PTH production in these patients is increased by three mechanisms and is shown in Fig. 19.2.

1. **Hyperphosphatemia:** Phosphate homeostasis is maintained by glomerular filtration and tubular secretion. With the decline in GFR in CKD, filtration and hence excretion of phosphate is decreased resulting in phosphate retention. During early

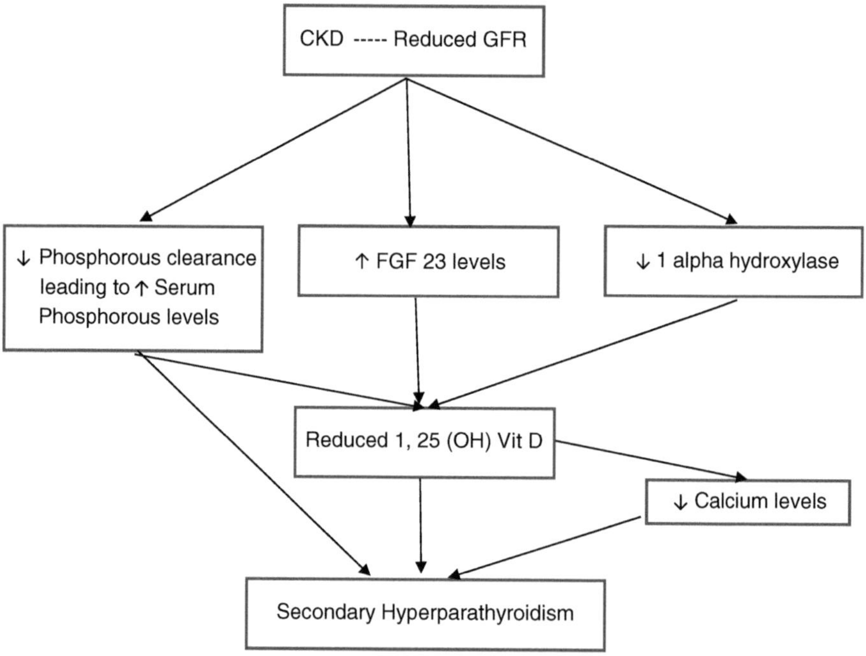

Fig. 19.2 Showing mechanisms of hyperparathyroidism in chronic kidney disease

phases of renal dysfunction, phosphate reabsorption is decreased in remaining normal nephrons as an attempt to lower mounting serum phosphate levels. This effect is mediated by both FGF23 and PTH. In advanced stages 4 and 5 of CKD, phosphate homeostasis cannot be maintained by the above mechanism and patients develop hyperphosphatemia. Hyperphosphatemia suppresses the renal hydroxylation of inactive 25-hydroxy vitamin D to 1, 25-hydroxy vitamin D (calcitriol).

2. **Hypocalcemia:** Hypocalcemia results secondary to phosphate accumulation and decreased production of calcitriol by kidneys and is evident in stage 5 of CKD. Hypocalcemia has a negative feedback effect on calcium-sensing receptors (CaSR) on parathyroid cells leading to cellular hyperplasia excessive synthesis and secretion of PTH.
3. **FGF 23:** Reduction in GFR results in an increase in FGF 23 from the bones which directly acts on parathyroid cells as well as indirectly by suppressing calcitriol synthesis.

19.3.1 Bricker's Trade-off Hypothesis

Bricker proposed the following sequence. There is increased reabsorption of phosphate from the GI tract. An increase in phosphate leads to a reduction in calcium as it forms complexes with calcium. This is sensed by parathyroid which increases

PTH response. Increased PTH restores calcium balance and decreases phosphate reabsorption. A trade-off thus occurs in which SHPT is the price paid for normal calcium and phosphorous.

Low calcitriol stimulates parathormone production. So, in end-stage renal diseases, parathyroid glands are stimulated by various factors which include hyperphosphatemia, hypocalcemia, low serum calcitriol, and elevated FGF 23. SHPT is a reversible condition and overactive glands revert to the normal state when the stimuli are removed.

Monitoring in CKD for secondary hyperparathyroidism (Table 19.3):

- Serum calcium and phosphorous should be monitored every 6–12 months in CKD stage III, every 3–6 months in stage CKD stage 4, and every 1–3 months on CKD stage V including G 5 D (dialysis).
- Serum PTH should be monitored every 6–12 months in CKD stage IV and every 3–6 months in CKD stage V. The value of PTH in CKD VD should be maintained between two to nine times the upper limit of PTH assay.
- Serum alkaline phosphatase (ALP) is an indirect indicator of CKD MBD; however, its use as an indicator and to monitor the same has not been validated or recommended. Serum ALP is an enzyme that removes phosphate from protein and nucleotides. Other than bone it is present in the liver, intestine, and placenta. Although ALP levels reflect bone turnover and are usually raised in secondary hyperparathyroidism of renal origin, ALP is not a specific indicator for the same. Bone-specific ALP may be a better marker however further studies are required to validate the same.

The swing of PTH, serum calcium, and phosphorous with a decrease in eGFR is shown in Fig. 19.3.

Table 19.3 Showing appropriate monitoring of parameters in various CKD stages

Monitoring of	CKD stage III (eGFR- 30–59 mL/min)	CKD stage IV (eGFR- 15–29 mL/min)	CKD stage V/V D (eGFR <15 mL/min)	Remarks
Calcium and phosphorous	Every 6–12 months	3–6 months	1–3 months	Phosphorous should be maintained in the normal range and calcium in the low normal range
Parathyroid hormone levels		6–12 months	3–6 months	In CKD V, iPTH should be maintained 2–9 times the upper limit of normal
Alkaline phosphatase	Serum alkaline phosphatase is an indirect indicator of CKD MBD; however, its use as an indicator and to monitor the same has not been validated or recommended			

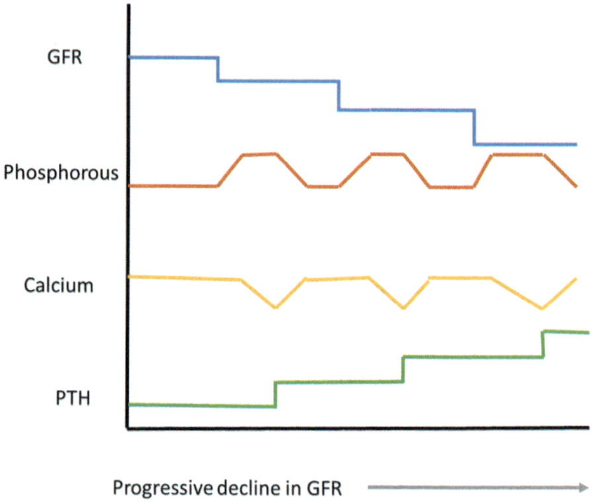

Fig. 19.3 Showing change in various laboratory parameters with the decline in GFR

19.3.2 Consequences of Secondary Hyperparathyroidism

Apart from bone loss secondary hyperparathyroidism can lead to soft tissue and vascular calcifications in patients with chronic kidney disease. The presence of CKD MBD is a strong predictor of increased cardiovascular morbidity and mortality. Hence, the treatment of secondary hyperparathyroidism should be initiated in the early stages itself.

19.3.3 Medical Treatment

The treatment of secondary hyperparathyroidism is aimed at treating the trigger for PTH secretion.

Treat vitamin D deficiency: Correction of vitamin D deficiency is recommended except when associated with severe hyperphosphatemia or hypercalcemia [5].

Control hyperphosphatemia: Serum levels of phosphate should be maintained between 3.5 and 5.5 mg/dL.

Measures to Lower Serum Phosphate:

1. Dietary restriction of phosphate.
2. Dietary regulations alone are not effective in the control of hyperphosphatemia. Phosphate binders are recommended in patients with persistently elevated serum phosphate >5.5 mg/dL in addition to dietary restriction. Phosphate binders are calcium-based—calcium acetate, calcium carbonate, non-calcium binders such as sevelamer carbonate, sevelamer hydrochloride, lanthanum carbonate, nicotinic acid, ferric citrate, and tenapanor.

Correction of hypocalcemia: Maintaining serum calcium levels to a low normal is done by supplementation when symptomatic hypocalcemia is encountered. The serum corrected calcium is maintained above 7.5 mg/dL. Calcium supplements with dihydroxy vitamin D are done to treat hypocalcemia. Adverse effects of calcium supplementation are hypercalcemia or positive calcium balance without hypercalcemia both of which can contribute to cardiovascular mortality.

Treatment of persistent progressive parathormone levels (PTH > 150–200 pg/mL) if upper limit of normal is 65 pg/mL.

Patients with renal dysfunction and elevated PTH are given calcitriol (1-25OHD-0.25 μg) thrice weekly especially in patients with hypovitaminosis but better avoided when associated with hypercalcemia.

19.4 Tertiary Hyperparathyroidism

Secondary hyperparathyroidism is an adaptive phenomenon and resolves to the normal stage when the precipitating factors are removed. Prolonged stimulation of parathyroid cells due to increased phosphate, low calcitriol, and hypocalcemia results in nodular or diffuse hyperplasia. These hyperplastic nodules do not undergo regular involution. Rarely these nodules can undergo a malignant transformation as well. In patients with tertiary hyperparathyroidism decreased expression of CaSR and VDR result in a lack of suppression of PTH by increased calcium levels or vitamin D analogs. In contrast to SHPT which is a reversible condition, tertiary hyperparathyroidism is irreversible. Tertiary hyperparathyroidism is a progression from SHPT. The parathyroid glands are autonomous and are no longer regulated by calcium levels. This results in hypercalcemia and variable phosphate levels. At our center, 12.5% of dialysis patients have tertiary hyperparathyroidism. We have taken hypercalcemia as corrected calcium > 9.5 mg/dL. This is after excluding calcium supplements and vit D as a cause for hypercalcemia. [Corrected calcium = serum calcium + 0.8 * (4 − serum albumin)] Example: if measured Ca is 7.5 mg/dL and measured Albumin is 3 g/L, corrected Ca = 7.5 + 0.8 × (4 − 3) =7.5 +0.8 = 8.3 mg/dL.

Treatment of Tertiary Hyperparathyroidism:

1. Calcitriol or other vitamin D analog supplementation after correcting hyperphosphatemia – Vitamin D supplementation suppresses serum parathyroid levels by reducing production as well as by increasing serum calcium levels.
2. Calcimimetics: The activation of calcium-sensing receptor (CaSR) decreases PTH secretion. Calcimimetics like cinacalcet increase the sensitivity of the CASR to calcium thereby reducing plasma PTH concentration and in turn reducing calcium and phosphorous levels. However, although they do cause a decrease in PTH levels, they have not been shown to have any cardiovascular mortality benefit [6].

19.5 Refractory Hyperparathyroidism

Definition: Severe persistent progressive elevation of PTH that cannot be treated adequately by medical therapy including VIT D analogs and calcimimetics without causing significant hyperphosphatemia or hypocalcemia. PTH levels should be more than 800 pg/mL.

Signs and Symptoms:

1. Bone and joint pain
2. Fractures
3. Proximal muscle weakness
4. Extra-skeletal vascular calcifications
5. Calciphylaxis
6. Pruritis
7. Hypercalcemia and hyperphosphatemia

19.6 Indications for Surgery in Refractory Hyperparathyroidism

The failure to respond to medical management is the main indication for surgery. This poor response to drugs and persistent rise in PTH levels is usually associated with monoclonal proliferation and nodular hyperplasia of the parathyroid gland. There may be a decrease in CASR as well.

- Persistently raised serum parathyroid hormone for 6 months despite maximum medical therapy.
- If the patient has symptomatic hyperparathyroidism with a PTH level of more than 800 pg/mL.
- Absolute value of serum PTH value > 1000 pg/mL.
- Persistent hypercalcemia or hyperphosphatemia, tissue or vascular calcification, worsening osteodystrophy.

19.6.1 Reasons Why Parathyroidectomy Is Not Done for All Patients with Hyperparathyroidism

1. Although parathyroidectomy seems a lucrative treatment for persistent hyperparathyroidism, it is not always a definitive treatment. One-third of these patients may still have persistent parathyroid levels 1-year post surgery [7].
2. Around 10% may experience extremely low levels of PTH [7].
3. Also 30-day mortality of 2% has been reported in this elective surgery [8].

19.6.2 Preoperative Management

Before surgery, patients with severe renal failure who are prone to fluid overload and patients already on dialysis should be could be considered for dialysis and fluid removal to create adequate space and optimize the patients for surgery.

19.6.3 Post-operative Management

1. Intensive monitoring of serum calcium levels as hypocalcemia is an immediate sequela.
2. Supplementation of IV calcium gluconate and oral calcium carbonate, magnesium, and calcitriol as per requirement.
3. Monitor for fluid overload since most of these patients are in severe renal failure or on hemodialysis.
4. Requirement of IV calcium is 1–1.5 mg/kg body weight/h. This requires a large volume of IV fluids which can result in fluid overload. Dialysis support with high calcium dialysate may be required.

19.6.4 Hungry Bone Syndrome

Hypocalcemia post parathyroidectomy or thyroidectomy despite normal or even elevated parathyroid hormone levels. The same can be seen in end-stage kidney disease on calcimimetics.

Clinical features are those of hypocalcemia (tetany, seizures, laryngeal spasm, and cardiac arrhythmias). Associated hypophosphatemia, hypomagnesemia, hyperkalemia may be present.

Treatment—supplementation of IV or oral calcium to maintain ionized calcium in the normal range. In dialysis patients, high calcium dialysate may be used. In CAPD patients, calcium gluconate may be added to each bag.

Prevention—Preoperative administration of calcitriol if they are not on calcitriol.

Although the risks of surgery are greater in these patients with chronic kidney disease, long-term outcomes from observational studies in parathyroidectomy cohorts show favorable surgical outcomes.

19.7 Post-transplant Hyperparathyroidism

Persistent hyperparathyroidism may occur in up to 50% of a kidney transplant. It has been associated with fractures, increased mortality, and decreased allograft survival. The primary manifestations include hypercalcemia and hyperphosphatemia. Symptoms observed in non-transplant chronic kidney diseases such as bone pain, pruritis, and myopathy are rare. Patients with a PTH level that is 2–3 times the upper limit of normal are considered to have persistent hyperparathyroidism.

In patients with mild to moderate hypercalcemia (1 mg more than the upper limit of normal), cinacalcet may be given. In patients with severe hypercalcemia > 12 mg/dL or symptomatic hypercalcemia, subtotal parathyroidectomy is advised.

In patients with mild to moderate hypophosphatemia, if the PTH level is high, first PTH is treated. If phosphate level is still low, despite correction of PTH, phosphorus-rich feeds are given.

In patients with severe hypophosphatemia, irrespective of PTH levels, phosphorous should be supplemented.

Take-Home Points

- Since abnormalities of calcium, phosphorous, and PTH can be detected in stage III CKD itself, appropriate monitoring should be instituted and measures are taken to prevent the development of hyperparathyroidism.
- Hyperparathyroidism is a strong predictor of cardiovascular mortality and morbidity.
- The main indication of surgery in refractory hyperparathyroidism is a failure of medical management.
- Although the surgical risk is greater in CKD, parathyroidectomy in patients resistant to medical management may have acceptable outcomes.

References

1. Saliba W, El-Haddad B. Secondary hyperparathyroidism: pathophysiology and treatment. J Am Board Family Med. September. 2009;22(5):574–81. https://doi.org/10.3122/jabfm.2009.05.090026.
2. Kidney Disease: Improving Global Outcomes (KDIGO) Acute Kidney Injury Work Group KDIGO Clinical Practice Guideline for Acute Kidney Injury. Kidney Int Suppl. 2012;2:1–138.
3. Memmos DE, Williams GB, Eastwood JB, Gordon EM, et al. The role of parathyroidectomy in the management of hyperparathyroidism in patients on maintenance haemodialysis and after renal transplantation. Nephron. 1982;30:143–8.
4. Dayma CL, et al. Study of prevalence of secondary hyperparathyroidism in chronic renl failure in hadoti region. IJRMS. 2019;7(8).
5. KDIGO 2012 Clinical Practice Guideline for the Evaluation and Management of Chronic Kidney Disease Kidney Int Suppl. 2013;3:5.
6. The EVOLVE Trial Investigators. Effect of Cinacalcet on cardiovascular disease in patients undergoing dialysis. N Engl J Med. 2012;367:2482–94. 1056/NEJMoa1205624.
7. Wetmore JB, Liu J, Do TP, et al. Changes in secondary hyperparathyroidism-related biochemical parameters and medication use following parathyroidectomy. Nephrol Dial Transplant. 2016;31:103–11.
8. Ishani A, Liu J, Wetmore JB, et al. Clinical outcomes after parathyroidectomy in a nationwide cohort of patients on hemodialysis. Clin J Am Soc Nephrol. 2015;10:90–7.

Surgical Treatment of Hyperparathyroidism: An Overview

20

C. Gopalakrishnan Nair

20.1 Introduction

Parathyroid glands are tiny structures, usually four in number, and located in the perithyroidal region and are pivotal in regulating calcium homeostasis. The very candid initial description of parathyroid glands was given by Ivar Sandström in 1879 [1]. But the English Anatomist Sir Richard Owen (1850) observed a few unusual structures on animal dissection on an Indian rhinoceros which he described as a small compact yellow glandular body attached to the thyroid gland at the point where the veins emerge [2]. Halsted and Evans, in 1907, gave a detailed description of the human parathyroid blood supply [3].

But the function of these tiny structures was not clearly known until Harald Salvesen firmly established the relationship of the parathyroid gland to calcium metabolism. "The parathyroids control the calcium level of the blood and by doing so they influence the function not only of the muscle and nerve tissues but probably of all the organs" [4]. William J MacCallum, a pathologist in John Hopkins, convincingly proved that tetany was a direct result of calcium deficiency [5, 6].

The characteristics bone changes of known as osteitis fibrosa cystica were described by von Recklinghausen in 1891 but was not attributed to parathyroid dysfunction [7]. In 1907, Jakob Erdheim reported that patients dying of advanced skeletal disease frequently exhibited enlarged parathyroid glands [8]. Subsequently, similar anecdotal reports of such association were seen in the literature. But general contention was that the parathyroid enlargement was compensatory to bone changes and the causal association was not thought of. In 1915, Friedrich Schlagenhaufer had observed that only one of the four glands was enlarged [9], and proposed that

C. G. Nair (✉)
Professor of Surgery, Endocrine Surgery Division, Amrita Institute of Medical Sciences and Research Centre, Kochi, Kerala, India

C. G. Nair, S. J. Abraham (eds.), *Surgical Management of Thyroid and Parathyroid Diseases*, https://doi.org/10.1007/978-981-97-3774-1_20

the change in the parathyroid glands was leading to the bone disease and suggested removal of the diseased gland. This paved way to therapeutic parathyroidectomy.

J. B. Collip extracted the essence from ox parathyroid and he essentially believed it to contain an active hormone which he named parathyrin [10]. But he is more remembered for his role in the discovery of insulin than for his role in the discovery of parathyroid hormone. The parathyroid hormone was isolated and purified from bovine parathyroid tissue by Rasmussen and Craig in 1959. They also defined the polypeptide structure of the hormone [11].

Anecdotal stories of therapeutic parathyroidectomy were recorded in the history of surgery during the early phase of the 20th century. The story of the first therapeutic parathyroidectomy by Felix Mandl in 1925 typifies the complexities of management of hyperparathyroidism and many still prevail even after a century. The 34-year-old male had bone pain features suggestive of osteopenia was diagnosed with osseous tuberculosis. After detection of hypercalcemia, he was treated with parathyroid extract by Felix Mandl himself. But later he ventured to explore his neck and removed a parathyroid lesion. This dramatically relieved his problem and remained healthy for 5 years and developed recurrence of symptoms. Felix Mandl re-explored but failed to identify a second lesion and the patient died of renal failure. Even autopsy did not locate a second lesion [1].

There is evidence that the first intentional removal of a parathyroid tumor was probably carried out at the Middlesex Hospital, London, UK, by Sir John Bland-Sutton at least a decade earlier [12]. Sir John Bland-Sutton described post mortem evidence of parathyroid tumor in 1886 in his book "Tumours Innocent and Malignant." He surgically removed a parathyroid cyst in 1909, and then carried out an intentional parathyroidectomy for a parathyroid tumor sometime before 1917 [13].

The multiplicity of parathyroid glands and lack of stable positioning of the glands compelled the early surgeons to perform wide exposure and tedious dissections. Technological advances gave a better idea of the biological nature of the disease, diagnosis, and treatment of primary hyperparathyroidism.

20.2 Major Events that Influenced the Management of PHPT

20.2.1 Parathyroid Hormone Assay

A year after the structure of parathyroid hormone was described, Berson et al. developed the first immunoassay technique to estimate parathyroid hormone [14]. But the assay technique was hugely influenced by the pattern of disintegration of the hormone. The active form of parathyroid hormone is comprised of 84 amino acids chain as that is stored, secreted, and functions in the body. The PTH is cleared from the blood in two phases with an early rapid phase (usually <5 min) and a late slow phase (several days). The half-life of human PTH during the early rapid phase varies depending on the study and ranges from a low of 1.68 min [12] to a high of 21.5 ± 10 [15]. Estimation of PTH was a challenge since it circulates at extremely low levels in the presence of a much greater number of PTH fragments [16]. The first-generation radioimmunoassay had poor sensitivity and specificity as they frequently detected inactive fragments also. The second-generation immunoradiometric assays were developed in 1987 and were considered to be specific for intact PTH. But later this technique also was found to detect a few fragments also [17]. The third-generation assays are more specific to capture intact PTH alone but have not gained wide popularity probably due to lack of universal availability [18]. The routine second-generation assays require approximately 1 h to complete and are widely used for diagnostic purposes.

20.2.2 Intra-Operative PTH Assay

The parathyroid hormone decays rapidly in the initial phase and logically the serum level should decrease when the source of synthesis is removed. Nussbaum SR et al. (1988) showed the advantage of a sensitive immunoradiometric assay (IRMA) of PTH during parathyroidectomy and concluded that a 40% decline of serum level from the baseline indicated successful parathyroidectomy [19]. But the second-generation assays required 1 h to give the result and so were not suited for intraoperative use.

But the introduction of the rapid chemiluminescent technique of PTH measurement by Nussbaum SR et al. and subsequent refining and establishing the proper methodology by Irvin et al. from Miami are major landmarks in the history of parathyroidectomy [20, 21]. Later Irvin and Deriso improved the technique by immunochemiluminometric assay reducing estimation time to about 10 min [20]. The rapid assay technique was used as a tool to determine the efficacy of parathyroidectomy in the complete removal of hyperfunctioning parathyroid lesions. The basic principle behind the newly postulated technique is that PTH decays in 3–5 min time and when the primary source of uncontrolled production is removed iPTH level declines significantly.

Essentially few criteria need to be well defined. What should be the baseline value to correlate? Second, what should be a time of interval for the perceptible decay of the hormone? When to draw the blood for baseline assay; before induction of anesthesia, after induction, or before excision of the lesion? So, post-excision blood draws may be after 10 min or after 15 min. The following criteria differ marginally in timing the blood draw and in defining the extent of fall.

Irvin III et al. from the University of Miami School of Medicine were the first to propose the use of iPTH assay as an intraoperative adjunct to predict the successful removal of parathyroid lesions. Six ml of peripheral venous blood samples were taken after induction of anesthesia, just before excision of tumor 5, 10, and 20 (optional) minutes after excision of the tumor. The decline of serum iPTH level more than 50% of the highest pre-excision at 10 min post excision blood sample signals successful removal of hyperfunctioning parathyroid lesion [21]. This popular criterion is known as the Miami criterion. Miami criterion defined by Irvin et al. "the most accurate criterion for predicting the outcome of parathyroidectomy is a drop of 50% or more from the highest parathormone level from either the pre-incision or pre-excision level, measured 10 min after the suspected gland is excised (Miami criterion)" [22].

But in a large series of 1158 patients, IOPTH assay with a 50% decline at 10 min failed to identify 16% of multiglandular diseases [23]. Hence, much stricter protocols and criteria were postulated. There are unsolved controversies related to the timing of pre-operative as well as a postoperative blood draw. The preoperative blood draw may be before induction of anesthesia or before incision. There are varying protocols in post-excision blood draw also.

Halle criterion: The Halle criterion dictates a post-excisional blood draw after 15 min and insists on the decline of iPTH to a lower level of normal ranges (<35 ng/L).

Rome criterion: The criterion is essentially a predictor of operative failure rather than successful removal of the culprit lesion. PTH drop of <50% from the highest basal value within 20 min after excision and/or a residual PTH level at 20 min higher than the reference range (65 ng/L) and/or a significant increase (>7.5 ng/L) from PTH at 10 min to PTH at 20 min was or were considered a predictor of persistent hyperfunctioning parathyroid tissue requiring further surgical exploration [24].

Vienna Criterion: This criterion emphasizes the baseline value defined as right after the induction of anesthesia but before the incision is made. It is defined as a 50% or more defined baseline value of iPTH 10 min after surgery predictive of successful removal of the hyperfunctioning parathyroid lesion [25].

However, the Miami criterion is widely practiced due to its simplicity.

20.2.3 Localization Studies

The surgical procedures for hyperparathyroidism are greatly influenced by the complex anatomy of parathyroid glands. The number, position, and size of parathyroid glands are variable and approximately 80% of them show bilateral symmetry in position. Of the two pairs, the 4th parathyroid or the superior glands are more constant in position. These glands are closely related to the posterior aspect superior pole of the thyroid and cover the terminal portion of the recurrent laryngeal nerve. This relationship gains more significance when pre-laryngeal branching of RLN is present. The lesion has a higher tendency to dissect the trachea-esophageal grove. The primary arterial supply of this lesion is by direct end artery from the inferior thyroid artery or indirectly from the arterial anastomotic on the posterior aspect of the lobe. Careful ligation is absolutely essential since controlling a bleeding vessel close to RLN may be troublesome.

The inferior glands are more wandering and unstable in position probably because they share their origin from the 3rd pharyngeal pouch with that of the thymus. The longer the course of migration higher is the chance of waywardness. Traditionally the inferior glands are situated in 1 cm vicinity of the lower pole anterior to the course of recurrent laryngeal nerve. The wandering gland may migrate to the superior mediastinum along with the thymus or less commonly follow the trachea to the posterior mediastinum. Parathyroid lesions are rarely detected in the carotid sheath probably since the artery of the third pharyngeal pouch becomes the common carotid artery and part of the internal carotid artery.

The earliest case series of parathyroid tumors Castleman showed hyperparathyroidism was not only associated with the solitary parathyroid adenoma, but also with multiple adenomas and with glandular hyperplasia [26]. So total neck exploration became a routine procedure and a single-center experience of 379 surgically treated PHPT patients had a 99.5% success rate with insignificant rates of complications [27].

But total neck exploration was laborious and the success rate was dependent on the experience of the surgeon. The incidences of permanent hypocalcemia probably due to vascular damage to normal parathyroid glands and recurrent laryngeal injuries were high. A change in approach emerged following the introduction of localization procedures. Scan-directed minimally invasive parathyroidectomy was found equally effective with fewer incidences of postoperative hypocalcemia and recurrent laryngeal nerve palsy [28]. But this journey was long till successful localization techniques were available.

The remark by interventional radiologist John L. Doppman in 1986 that "In my opinion, the only localizing study indicated in a patient with untreated 1°HPTH is to localize an experienced parathyroid surgeon," reflects the initial attitude toward routine preoperative parathyroid imaging [29]. In the last three decades, advances in imaging technology have influenced surgeons dramatically to modify the procedures.

Selective angiography and venous catheterization and sampling for PTH were laborious techniques and were used initially with 60–85% sensitivities. Newer less aggressive but accurate imaging techniques have replaced such laborious procedures.

The use of radiopharmaceuticals in parathyroid localization was initially described in the late 1970s. Thallium-201/^{99m}Tc-pertechnetate subtraction scan was found useful in parathyroid localization in 1983 [30]. Thallium concentrates in thyroid and parathyroid glands whereas Tc pertechnetate concentrates in thyroid alone. The principle is to remove the images acquired by Tc pertechnetate from the Thallium-201 images when both are acquired serially from still patients. All subtraction essentially requires extreme patient compliance in remaining still to avoid motion artifacts [31].

99mTc-sesta-(2-methoxy-isobutyl-isonitrile) popularly known as Tc-MIBI was introduced as a tracer for parathyroid lesions in 1989. MIBI is administered intravenously (185–900 Mbq) and the tracer accumulates in functioning thyroid follicular cells and hyperactive parathyroid cells. The radiopharmaceutical is concentrated in mitochondria of many normal tissues like cardiac and thyroid cells. The material washes out differentially from tissues as early as thyroid follicular cells but is delayed from abnormal parathyroid cells. Sestamibi concentration is related to baseline PTH level, the weight of lesion, cell cycle phase in the lesion. When serum PTH level falls below 200 pg/ml the accuracy falls to 81% while it climbs up to 100% when PTH level is >400 pg/ml. Increase perfusion of the gland and functional activity as well as abundant mitochondria-rich oxyphil cells are the key component of successful MIBI uptake. The sensitivity of the MIBI scan is influenced by the size and cellularity of the lesions. Some observations suggest a positive correlation with the oxyphil cell population of the tumor [32, 33]. 99mTc-tetrofosmin behaves in a similar fashion and can be used instead of Tc-MIBI. But 99mTc-tetrofosmin may clear more slowly from the thyroid gland.

The technique of conventional MIBI dual-phase parathyroid scintigraphy involves injecting the tracer intravenously and taking neck planar images at 10–20 min and late images at 2 h and 3 h [30]. Incorporating single-photon emission computed tomography (SPECT) has greatly enhanced the accuracy of dual-phase scans. The reported sensitivity of the MIBI scan ranges from 80% to 100% [34]. Two rare issues that may interfere with the accuracy of dual-phase scans are concurrent thyroid neoplasm and parathyroid lesions with decreased affinity to the tracer. Parathyroid scintigraphy was found to have high sensitivity in symptomatic diseases probably due to the large tumor volume and high metabolic activity.

Ultrasound imaging has improved with the addition of newer highly sensitive probes. The parathyroid lesions classically appear as hypoechoic vascular lesions separate from the thyroid in the characteristic locations. The smaller lesions could be confused with lymph nodes. The lesions in ectopic locations such as mediastinum, carotid sheath, and tracheo-esophageal grove are not imaged with ultrasound. But the US is widely used as second imaging for anatomical localization complementing parathyroid scintigrams. The US done by the surgeon helps to locate the lesion precisely to plan a suitable incision for mini-exploration.

When scintigraphy fails to localize the lesion higher imaging options may be selected. Four-dimensional computed tomography (4DCT) is a relatively new addition to conventional CT. The respiratory movements distort the images with changes in the center of the target. To overcome this issue 3D scan images are captured at different phases and tagged with breathing signals. 4D CT is now employed in scintigraphy negative parathyroid lesions.

Still, a newer entry is fluro-choline PET imaging. Upregulation of choline kinase activity is noted in adenoma or hyperplasia and this leads to increased choline uptake. 18 F-Fluro-choline (FCH) is used as a tracer in PET imaging. A large series of PHPT Fluro-choline PET showed 94.1% sensitivity and could localize accurately 86% MIBI negative patients.

MRI is used generally as a supplementary anatomical localization modality in mediastinal lesions.

Based on improvements in diagnosis and localization studies the surgical procedure was modified. Instead of bilateral neck exploration, exploration targeting the scan-directed abnormality became popular. The following terminologies are widely used to categorize the type of surgical procedures:

(1) Unilateral exploration; (2) Focused parathyroidectomy; (3) Mini-parathyroidectomy; (4) Radio guided parathyroidectomy; (5) Remote access parathyroidectomy.

Total neck exploration is the procedure to expose and explore the normal anatomical locations of all four parathyroid glands, and also aberrant locations in case of failure to identify the glands. This procedure yields excellent results in trained hands and does not necessitate localization. Unilateral neck exploration or limited neck exploration denote exposure of one side to ensure normality in size and appearance of the remaining gland.

During the 1990s, minimally invasive, targeted parathyroidectomy gained attention among surgeons [35, 36].

Focused parathyroidectomy is a widely practiced procedure whereby the scan-directed lesion alone is removed. Since 80–85% PHPT is due to a single hyperfunctioning gland the focused removal is curative in those patients. This procedure could be done through a small incision (2–3 cm) if the lesion is precisely localized. Mini-parathyroidectomy is attempted when concordance in parathyroid scintigraphy and anatomical imaging like ultrasound is reached.

20.2.3.1 Radio-Guided Parathyroidectomy

Further advances in technology helped the surgeons to use a gamma probe in the operation theatre. The first use of radiation emitter during the operation was in 1949 by Selverstone et al. in intraoperative localization of brain tumor [37].

But 7 years later in 1956 gamma detection probe system was introduced to localize thyroid cancer recurrence. The lesion was successfully removed using a gamma probe [38]. The basic principle in this adjunct is radiolabeled antibodies concentrate in tissues which showed an affinity and portable gamma emission detector

Radiolabeled antibodies concentrate in tissues that exhibit affinity to them and a gamma camera is used to detect them. Sophisticated handheld probes are available now and can be used as an intraoperative adjunct for precise localization.

Tc99 MBI is widely chosen for localization of hyperfunctioning parathyroid gland. The infused radiopharmaceutical is widely distributed in circulation and induces widespread background radiation which is perceptible by the probe. According to Murphy and Norman, lymph nodes, normal parathyroid tissue, and fat have about 2.2% higher radioactivity compared to the background, whereas thyroid tissue expresses about 5.5% higher radioactivity. On the basis of the data from 345 patients and a total of 1290 specimen, they concluded thyroid and hyperplastic parathyroid glands showed 5.5% and 7.5% higher radioactivity but never more than 16%. But adenomas used express still higher radioactivity with a range of 18–136%. They concluded that 20% higher radioactivity is confirmatory and eliminates the need for frozen sections [39]. Now well electronically collimated Bluetooth-enabled probes are available and are very handy for intraoperative use.

The protocols for the administration of tracer differs. Norman et al. described a same-day protocol where a diagnostic dual-phase scan is performed with IV administration of 20–25 mCi (6740-925 MBq) of MIBI. The patient is taken immediately for exploration aided by a gamma probe [40]. But this protocol does not offer an efficient patient selection and so Flynn et al. described an approach where diagnostic scan to be done earlier and second similar dose 60–90 min before the surgical procedure [41, 42].

The probe is used to identify the hot spot and the incision is planned directly over it or at a cosmetically acceptable nearby site and the line is chosen. The count detected over the suspected lesion is 1.5 times higher than the background count and the lesion is completely removed. To ensure accuracy the background radioactivity count to tumor bed count is measured and the ratio becomes 1. The excised tumor retains radioactivity and the ex vivo count is higher than the background radioactivity count. Ratio > 1.2 indicates complete removal. Hyperfunctioning adenomas retain higher radioactivity, almost 50% more than that of background.

References

1. Sandström IV. On a new gland in man and several mammals—glandulae parathyroideae. Upsala Läk Förenings Förh. 1879–1880;15:441–71.
2. Owen R. On the anatomy of the Indian rhinoceros (Rh. unicornis, L). Trans Zool Soc Lond. 1862;4:31–58.
3. Halsted WS, Evans HM. The parathyroid glandules. Their blood supply and their preservation in operation upon the thyroid gland. Ann Surg. 1907;46(4):489–506.
4. Salvesen HA. Studies of the physiology of the parathyroids. Acta Med Scand. 1923;(Suppl. 6):1–159.
5. MacCallum WJ, Voegtlin C. On the relation of the parathyroid to calcium metabolism and the nature of tetany. Bull Johns Hopkins Hosp. 1908;19:91–2.
6. MacCallum WJ, Davidson CF. Further notes on the function of the parathyroid glands. Med News. 1905;86:625–33.
7. von Recklinghausen FD (1891) Festschrift fur Rudolf Virchow Reimer. Berlin. Die fibröse oder deformierende Ostitis, die Osteomalacie und die osteoplastische Karzinose in ihren gegenseitigen Beziehungen; pp. 1–89.
8. Erdheim J. Uber die Dentinverkalkung in Nagezahn bei der Epithelkörerchen transplantation. Frankfurt Z Path. 1911;7:295–342.
9. Schlagenhaufer F. Zwei Fälle von Parathyroideatumoren. Wien Klin Wschr. 1915;28:1362.
10. Collip JB, Leitch DB. A case of tetany treated with parathyrin. Can Med Assoc J. 1925;15(1):59–60.
11. Rasmussen H, Craig LC. The parathyroid polypeptides. Recent Prog Horm Res. 1962;18:269–95.
12. Dorairajan N. Pradeep PV Vignette hyperparathyroidism: glimpse into its history. Int Surg. 2014;99(5):528–33.
13. Delbridge LW, Palazzo FF. First parathyroid surgeon: Sir John Bland-Sutton and the parathyroids. ANZ J Surg. 2007;77:1058–61.
14. Berson SA, Yalow RS, Aurbach GD, Potts JT Jr. Immunoassay of bovine and human parathyroid hormone. Proc Natl Acad Sci U S A. 1963;49:613–7.
15. Carter AB, Howanitz PJ. Intraoperative testing for parathyroid hormone: a comprehensive review of the use of the assay and the relevant literature. Arch Pathol Lab Med. 2003;127(11):1424–42.
16. Brasier AR, Wang C, Nussbaum SR. Recovery of parathyroid hormone secretion after parathyroid adenomectomy. J Clin Endocrinol Metab. 1988;66:495–500.
17. Nussbaum SR, Zahradnik RJ, Lavigne JR, Brennan GL, Nozawa-Ung K, Kim LY, Keutmann HT, Wang CA, Potts JT Jr, Segre GV, Au Cole DE, Webb S, Chan PC. Clin Chem. 1987;33(8):1364–7.
18. Cole DE, Webb S, Chan PC. Update on parathyroid hormone: new tests and new challenges for external quality assessment. Clin Biochem. 2007;40(9–10):585–90.
19. Nussbaum SR, Thompson AR, Hutcheson KA, Gaz RD, Wang CA. Intraoperative measurement of parathyroid hormone in the surgical management of hyperparathyroidism. Surgery. 1988;104(6):1121–7.
20. Irvin GL 3rd, Deriso GT 3rd. A new, practical intraoperative parathyroid hormone assay. Am J Surg. 1994;168(5):466–8.
21. Irvin GL III, Carneiro DM, Solorzano CC. Progress in the operative management of sporadic primary hyperparathyroidism over 34 years. Ann Surg. 2004;239(5):704–11.
22. Irvin GL III, Solorzano CC, Carneiro DM. Quick intraoperative parathyroid hormone assay: surgical adjunct to allow limited parathyroidectomy, improve success rate, and predict outcome. World J Surg. 2004;28:1287–92.
23. Siperstein A, Berber E, Barbosa GF, Tsinberg M, Greene AB, Mitchell J, Milas M. Predicting the success of limited exploration for primary hyperparathyroidism using ultrasound, sestamibi, and intraoperative parathyroid hormone: analysis of 1158 cases. Ann Surg. 2008;248(3):420–8.

24. Riss P, Kaczirek K, Heinz G, Biglmayer C, Niederle B. A "defined baseline" in PTH monitoring increases surgical success in patients with multiple gland disease. Surgery. 2007;142:398–404.
25. Lombardi CP, Rafaelli M, Traini E, Di Stasio E, Carrozza C, De Crea C, Zuppi C, Bellantone R. Intraoperative PTH monitoring during parathyroidectomy: the need for stricter criteria to detect multiglandular disease. Langenbecks Arch Surg. 2008;393:639–45.
26. Castleman B, Mallory TB. The pathology of the parathyroid glands in hyperparathyroidism: a study of 25 cases. Am J Pathol. 1935;11(1):1–72.17.
27. van Heerden JA, Grant CS. Surgical treatment of primary hyperparathyroidism: an institutional perspective. World J Surg. 1991;15(6):688–92.
28. Singh Ospina NM, Rodriguez-Gutierrez R, Maraka S, et al. Outcomes of parathyroidectomy in patients with primary hyperparathyroidism: a systematic review and meta-analysis. World J Surg. 2016;40(10):2359–77.
29. Doppman JL. Reoperative parathyroid surgery; localization procedures. Prog Surg. 1986;18:117–32.
30. Coakley AJ, Kettle AG, Wells CP, O'Doherty MJ, Collins RE. 99Tcm Sestamibi—a new agent for parathyroid imaging. Nucl Med Commun. 1989;10:791–4.
31. Young AE, Gaunt JI, Croft DN, Collins RE, Wells CP, Coakley AJ. Location of parathyroid adenomas by thallium-201 and technetium-99m subtraction scanning. Br Med J. 1983;286:1384–6.
32. Carpentier A, Jeannotte S, Verrault J, et al. Preoperative localization of parathyroid lesions in hyperparathyroidism: relationship between technetium-99m-MIBI and oxyphil cell count. J Nucl Med. 1998;39:1441–4. 2007; 48:1084
33. Torregrosa F, Fuster JV, D. Biological factors influencing parathyroid localization. Nucl Med Commun. 2003;24(2):121–1241089.
34. Mitchell BK, Merrell RC, Kinder BK. Localization studies in patients with hyperparathyroidism. Endocr Surg. 1995;75:483–98.
35. Udelsman R, Donovan PI, Sokoll LJ. One hundred consecutive minimally invasive parathyroid explorations. Ann Surg. 2000;232(3):331–9.
36. Scott NA, Reeve TS, Parkyn RF, et al. Systematic review of minimally invasive parathyroidectomy: update and re-appraisal. 2001. In: Database of abstracts of reviews of effects (DARE): quality-assessed reviews. York (UK): Centre for Reviews and Dissemination (UK); 1995. Available from: https://www.ncbi.nlm.nih.gov/books/NBK68828.
37. Selverstone B, Sweet WH, Robinson CV. The clinical use of radioactive phosphorus in the surgery of brain tumors. Ann Surg. 1949;130(4):643–51.
38. Harris CC, Bigelow RR, Francis JE, Kelly GG, Bell PR. A Csi(Ti)-crystal surgical scintillation probe. Nucleonics. 1956;14:102–8.
39. Murphy C, Norman J. The 20% rule: a simple, instantaneous radioactivity measurement defines cure and allows elimination of frozen sections and hormone assays during parathyroidectomy. Surgery. 1999;126(6):1023–9.
40. Norman J, Chheda H. Minimally invasive parathyroidectomy facilitated by intraoperative nuclear mapping. Surgery. 1997;122:998–1003.
41. Flynn MB, Bumpous JM, Schill K, K.M. McMasters minimally invasive radioguided parathyroidectomy. J Am Coll Surg. 2000;191:24–31.
42. Rubello D, Mariani G, Pelizzo MR, Italian Study Group of Radioguided Surgery and ImmunoScintigraphy. Minimally invasive radioguided parathyroidectomy on a group of 452 primary hyperparathyroid patients: refinement of preoperative imaging and intraoperative procedure. Nuklearmedizin. 2007;46:85–92.
43. Taillefer R, Boucher Y, Potvin C, Lambert R. Detection and localization of parathyroid adenomas in patients with hyperparathyroidism using a single radionuclide imaging procedure with Tc-99m sestamibi (double phase study). J Nucl Med. 1992;33:1801–7.

21 Parathyroidectomy: Surgical Techniques

C. Gopalakrishnan Nair and Riju Ramachandran

21.1 Preoperative Workup and Preparation

The majority of Indian patients are symptomatic unlike those frequently encountered in North American and European countries. Hypercalcemia used to be severe among symptomatic patients. Mild hypercalcemia (<12 mg/dl) does not require active correction before surgery. Hypercalcemia above 12 mg/dl requires correction before surgery. Volume expansion by administration of normal saline 200–300 ml/h (4–6 L in 24 h) brings down the calcium and is the first line of treatment. The volume expansion should be carefully monitored in the elderly and in those with cardiac and renal dysfunction. Intravenous bisphosphonates (Zoledronic acid 4 mg slowly over 20 min, or Pamidronate 30–90 mg at 20 mg per h is effective in bringing down the serum calcium level but usually by 3–4 days. The moderate (12–14 mg/dl) and severe hypercalcemia (>14 mg/dl) require close observation in high dependency wards. Subcutaneous calcitonin (4–8 units/kg body weight) administered twice daily or even 6th hourly is effective in controlling hypercalcemia. Calcitonin reduces the serum calcium concentration by increasing renal calcium excretion and, more importantly, by decreasing bone resorption via interference with osteoclast function. Calcitonin brings down the calcium level by 1–2 mg in 4 h and is safe along with intravenous saline. But the effect is short-lived and limited to 48 h indicating the development of tachyphylaxis, perhaps due to receptor downregulation. More sustained but slow control of hypercalcemia is obtained by intravenous administration of Bisphosphonates. Zoledronic acid (4 mg) administered as a slow infusion effectively brings down hypercalcemia. Bisphosphonates are preferentially

C. G. Nair (✉)
Professor of Surgery, Endocrine Surgery Division, Amrita Institute of Medical Sciences and Research Centre, Kochi, Kerala, India

R. Ramachandran
Department of Surgery, AIMS, Kochi, Kerala, India

C. G. Nair, S. J. Abraham (eds.), *Surgical Management of Thyroid and Parathyroid Diseases*, https://doi.org/10.1007/978-981-97-3774-1_21

incorporated into sites of active bone remodeling in conditions characterized by accelerated skeletal turnover. Bisphosphonates inhibit bone resorption by internalization of osteoclasts. But the action is not limited to osteoclasts alone but also bone remodeling also. The effects last for 3–6 months. About one-third of patients may have acute phase reactions of fever, myalgia, and joint pain lasting for a few days.

Calcimimetic drugs like Cinacalcet (30 mg twice daily) are used in secondary HPT of renal origin. Cinacalcet was introduced for clinical use in 2004 and is occasionally used in PHPT but is not an effective substitute for parathyroidectomy. The drug is widely used in patients with parathyroid dysfunction of renal origin.

The cardiovascular system is not a direct target organ in PHPT. But associated cardiac abnormalities such as diastolic myocardial dysfunction, left ventricular hypertrophy, hypertension, autonomic imbalance, metabolic disturbances, and endothelial vasodilatory dysfunction are not uncommon [1, 2]. Mild forms of PHPT as seen in asymptomatic disorder without cardiac risk factors may have normal cardiac morphology and function. But patients with more severe disease, as manifested by higher serum calcium levels, may have an increased risk of morbidity and mortality from cardiac diseases. The incidence of hypertension is also high among PHPT patients. So, it is always advisable to do a complete cardiac assessment prior to parathyroidectomy.

Renal dysfunction is not rare as part of metabolic dysfunction in severe diseases and usually gets corrected by measures taken to correct hypercalcemia. Obstructive uropathy with or without renal dysfunction merits early urological intervention before parathyroidectomy.

Vocal cord assessment is mandatory prior to exploration.

Patients who have severe bone demineralization are likely to develop hungry bone syndrome during the postoperative period. Many such patients have vitamin D deficiency of insufficiency but active correction may exacerbate the hypercalcemia transiently. Patients who have shreds of evidence of bone loss are given 1–25 OH vitamin D preparation on the day before operation and continued after the operation before evident hypocalcemia manifests.

21.1.1 Preoperative Counselling

Just like any other interventional procedure patients should be informed about the details of the disease, need for operation, and likely consequences of the procedure. Awareness of parathyroid glands and diseases related to them is sparse among the general public. The vivid functions of the parathyroid hormone may have to be explained since many fail to understand the logic of operating on the neck for renal stones or fragility fractures.

For focused procedure is done through a small strategically placed incision and the resultant scar could be inconspicuous. But unlike this popular concept occasionally the scar hypertrophy is troublesome. Concern about an ugly scar on the exposed area of the neck is high and hence surgeon should take an effort to reduce this anxiety.

The concern of hungry bone syndrome in patients with considerable bone mineral loss is high. The prolonged need for calcium and dihydroxy vitamin D supplementation is to be explained. The surgeon should ensure adequate compliance regarding the timely intake of tablets. The probable chances of distressing symptoms of hypocalcemia are explained adequately. Those rare patients with a lack of compliance may require unexpected hospital visits for intravenous calcium administration. These patients require a monthly visit to laboratory estimation serum calcium, serum PO_4 and alkaline phosphatase for dose adjustment of supplementation.

Vitamin D deficiency is usual in PHPT patients and adequate supplementation is necessary for the postoperative phase.

Recurrent laryngeal nerve injury is a rare complication of parathyroidectomy. Preoperative assessment of the larynx to assess the movement and function of the vocal cord is always indicated. In case of such catastrophic event of palsy or paresis of video-laryngoscopy is done to assess the extent of the damage. The details of management protocol are given elsewhere and have to be briefed to the patient before surgery.

21.1.2 Desirable Additional Supports

A portable ultrasound machine in the operation theatre is of great help during exploration when preoperative imaging was inconclusive.

An intraoperative gamma probe is available only in selected institutions but supports surgeons efficiently in patients who are undergoing re-explorations. Gamma probe also helps to identify rare parathyroid adenomas which retain tracer for short time only ("rapidly washing out adenomas").

Laboratory service available close to operation theatre with facilities to do rapid PTH assays is desirable. Intraoperative PTH assay is essential when focused parathyroidectomy is planned.

The pathology laboratory equipped to do frozen section studies helps to differentiate structures such as thyroid remnants, cysts of the thymus, and lymph nodes.

1. Anesthesia—The procedure may be done under local anesthesia given as cervical plexus block and infiltration at the line of incision. But many of the parathyroid lesions of patients in symptomatic diseases of Asiatic countries are bigger in size compared to those of asymptomatic PHPT patients. Hence, these lesions used to have a complex relation to vital structures and are highly vascular. So general anesthesia with adequate muscle relaxation is preferred if the general condition is satisfactory (Fig. 21.1).
2. When facility for intraoperative PTH assay is available 4 ml blood is drawn before induction and administration of anesthetic agents. The transporting team is instructed to transfer quickly to the laboratory protected in icepack to minimize disintegration of the hormone (Fig. 21.2).

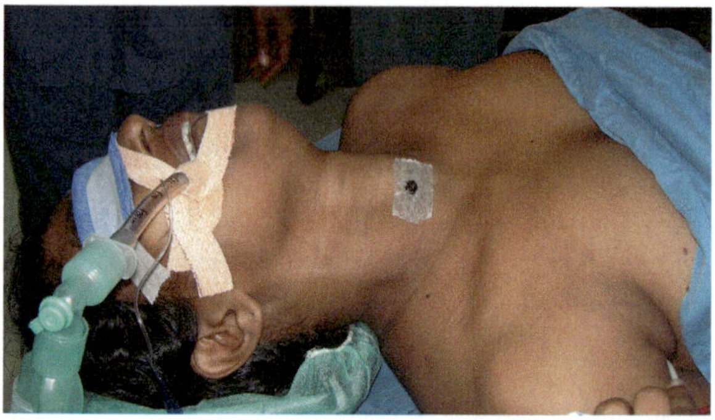

Fig. 21.1 The patient positioned after completing administration of general anesthesia

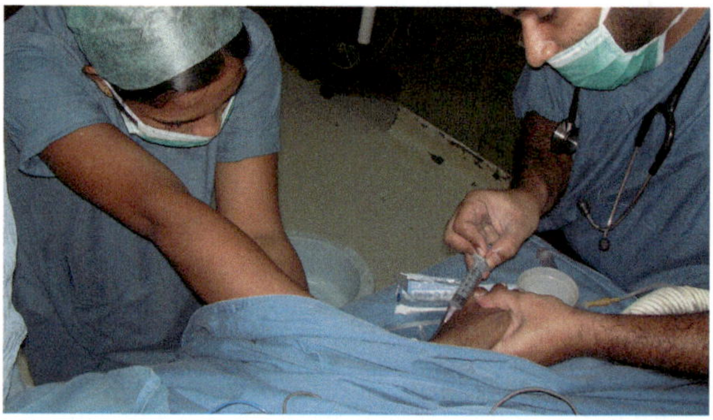

Fig. 21.2 Blood draw for IOPTH assay

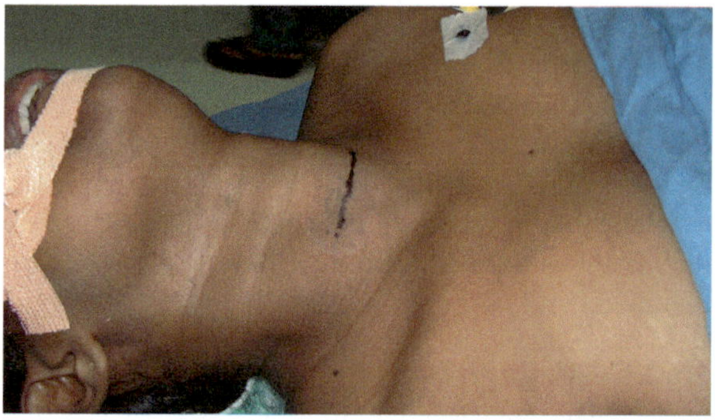

Fig. 21.3 The incision marked

3. Position: The patient is positioned supine with a support behind the shoulder blades. The head is steadied on a ring. It is very important to avoid excessive extension strain on the cervical spine (Figs. 21.3, 21.4 and 21.5).
4. If an ultrasound machine is available reassessment of anatomical localization is done and the location is marked with a permanent marker pen. This procedure may be done in the ultrasound room if the facility is not available in the operation theatre.
5. Neck is painted and the patient is draped (Fig. 21.6).

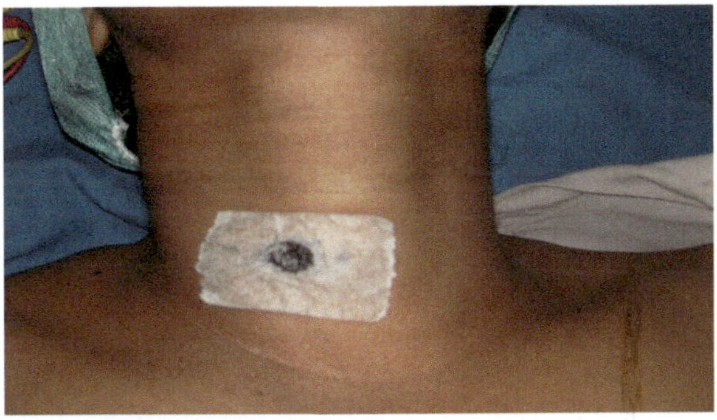

Fig. 21.4 Patient positioned

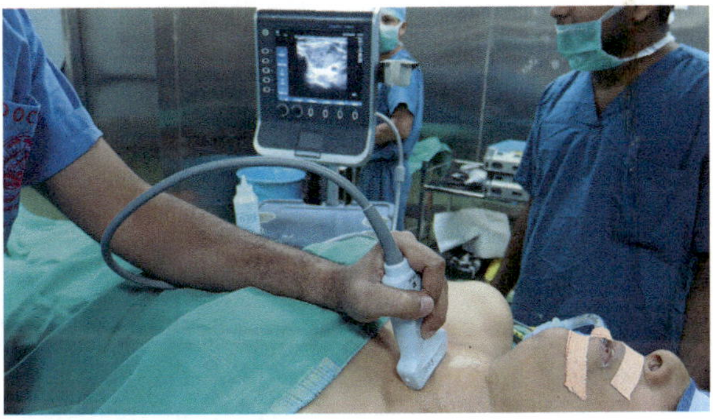

Fig. 21.5 Surgeon performed localization

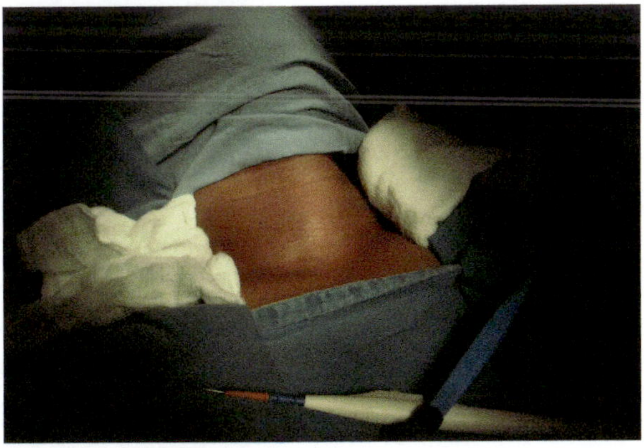

Fig. 21.6 Patient painted and draped

6. Operating surgeon stands on the side of the lesion and the first assistant on the other side. The scrub sister is on the left side of the table with the instrument spread over Mayo's trolley.
7. Energy sources for hemostasis are connected to the patient. Suction equipment is kept ready and connected with the fine-tip with control. The intensity of energy sources is set to the minimum after the initial steps.

8. Incision is planned based on the location of the lesion and the chosen type of procedure. For lesions from inferior parathyroid 2–3 cm transverse incision at the marked area is sufficient. If the ultrasound imaging indicates a large lesion close to the trachea-esophageal groove larger incision is planned. If total neck exploration is planned the incision shall be as for thyroidectomy. A five- to six-centimeters-long transverse incision given on the lower part of the front of the neck reaching equally on both sides from the midline, gives good exposure (Figs. 21.7 and 21.8).

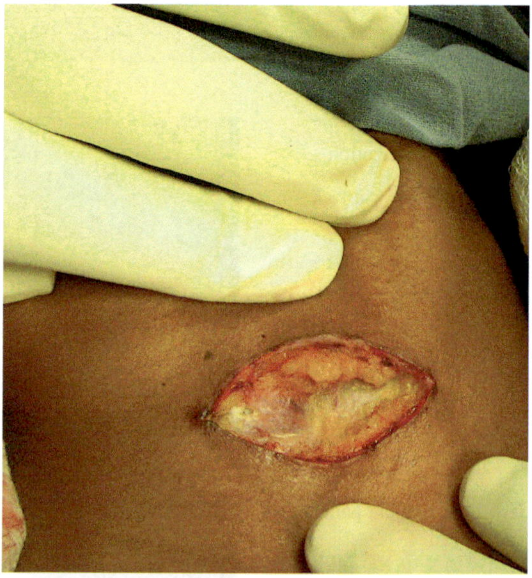

Fig. 21.7 The incision

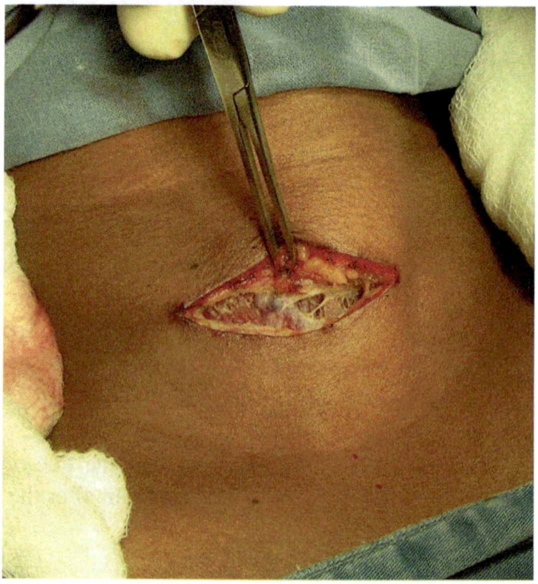

Fig. 21.8 The subcutaneous tissue and investing layer divided

9. Incision is deepened to expose the investing layer of deep fascia. For focused parathyroidectomy, further exposure is done by the division of strap muscles. The investing layer of the deep fascia is divided transversely and strap muscles are exposed. The median raphe is divided and a blunt hemostat is introduced lifting the strap muscles from the anterior surface of the thyroid. The anterior jugular vein is a potential source of bleeding and should be controlled effectively. The strap muscles are divided with bipolar diathermy for controlling bleeding from cut edges (Figs. 21.9, 21.10, and 21.11).
   When total neck exploration is planned upper and lower skin flaps are made to expose the investing layer of fascia. The thyroid gland is exposed by the traditional method of dividing investing layer of fascia vertically in the midline.
10. But inferior parathyroid glands are generally located in the vicinity of the lower pole of thyroid lobes. Adequate mobilization of the thyroid lobe with minimal damage to its blood supply is necessary to locate the lesion. The parathyroid gland is mostly seen anterior to RLN on the side of the trachea or rarely embedded in vestigial remnants of the cervical thymus.
    Division of strap muscles exposes the surface of the thyroid gland and dissection is extended laterally between the Sternocleidomastoid muscle and lateral aspect of the thyroid. Middle thyroid vein and rarely Kocher's vein are encoun-

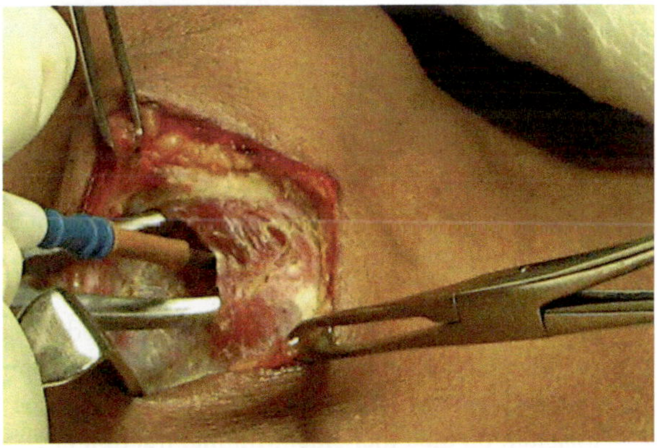

Fig. 21.9 Division of the strap muscles

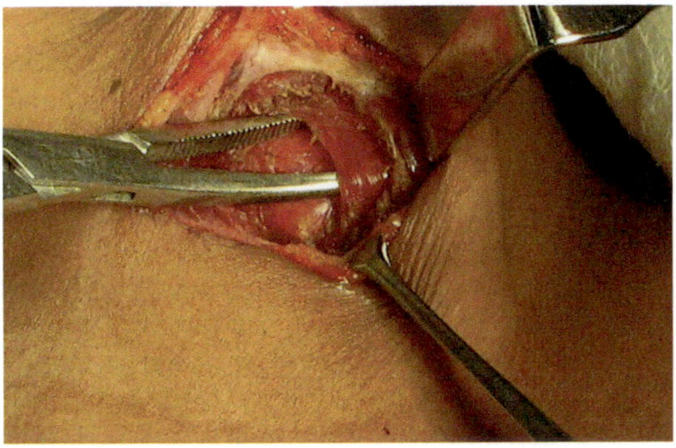

Fig. 21.10 The division of strap muscle continues exposing the thyroid

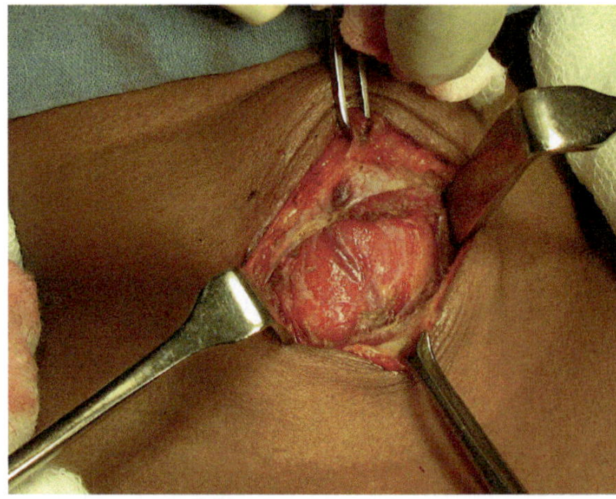

Fig. 21.11 The thyroid lobe exposed

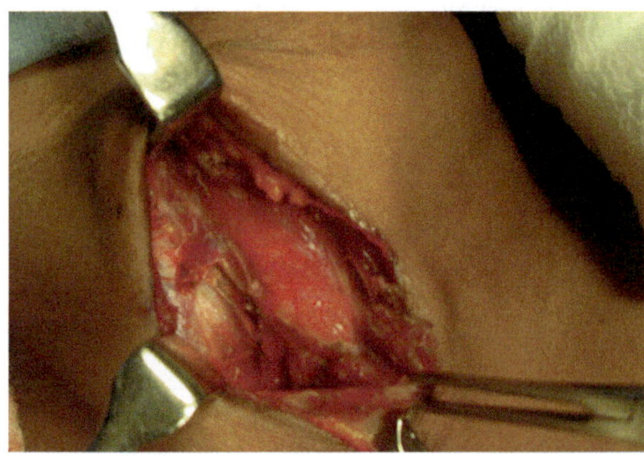

Fig. 21.12 Exposure of lobe of thyroid

tered during the dissection and are secured effectively. The thyroid lobe is mobilized medially exposing the trachea-oesophageal grove. A lightweight Babcock tissue holding forceps is applied and the lobe is held medially by the assistant on the other side. Minimal traction is given to avoid injury of the thyroid gland since subsequent bleeding adds to the difficulty in dissection and identification of parathyroid lesions and vital structures. A Langenbeck retractor is introduced to expose the lateral aspect of the lobe. This retractor is held by the second assistant on the side of the operating surgeon (Fig. 21.12).

11. The inferior or 3rd parathyroid lesions may be easily visible now close to the lower pole of the thyroid gland. The inferior parathyroid lesion is occasionally placed close to the lower pole and discretely visible in the early phase of dissection. These lesions are usually anterior to recurrent laryngeal nerve and mostly derive their blood supply from the inferior thyroid artery. Usually, adenomas are chocolate brown in color and are very soft inconsistency. The thyroid lobe is held medially and the Langenbeck retractor holds the SCM muscle laterally exposing the trachea-esophageal grove. Loose areolar tissue is gently separated exposing the recurrent laryngeal nerve (RLN) (Figs. 21.13, 21.14, and 21.15).

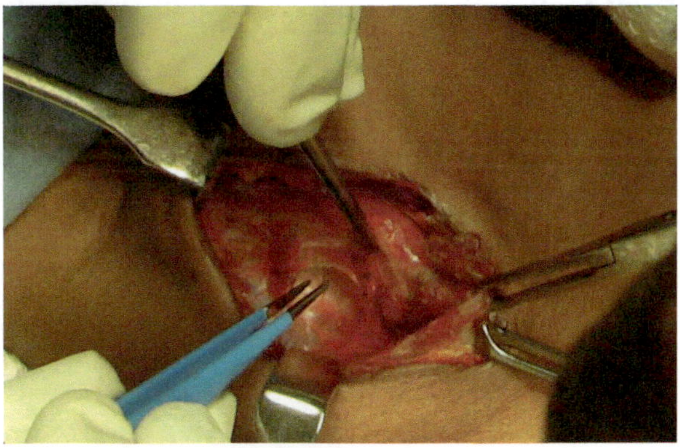

Fig. 21.13 Dissection in tracheoesophageal grove

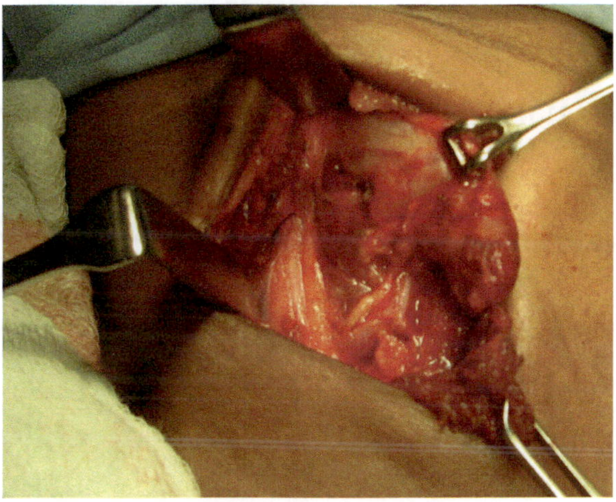

Fig. 21.14 The RLN exposed and the large parathyroid lesion exposed

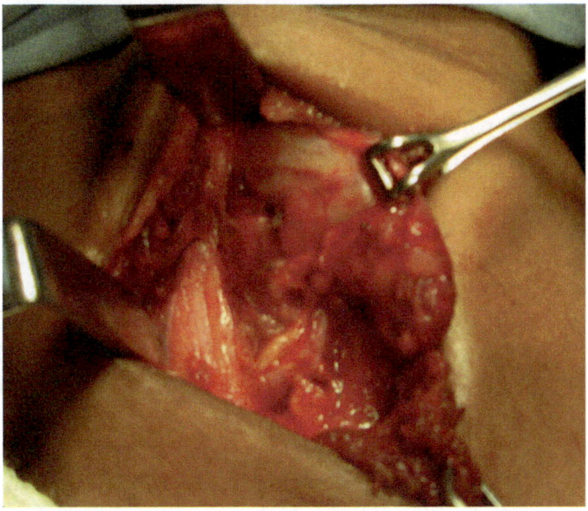

Fig. 21.15 The soft tissue is gently separated

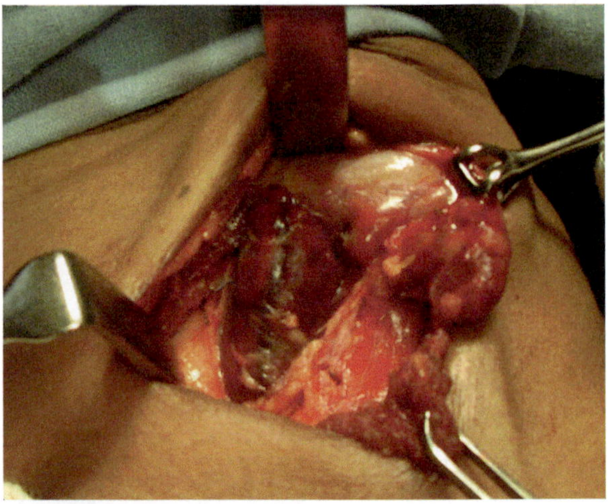

Fig. 21.16 Dissection continues

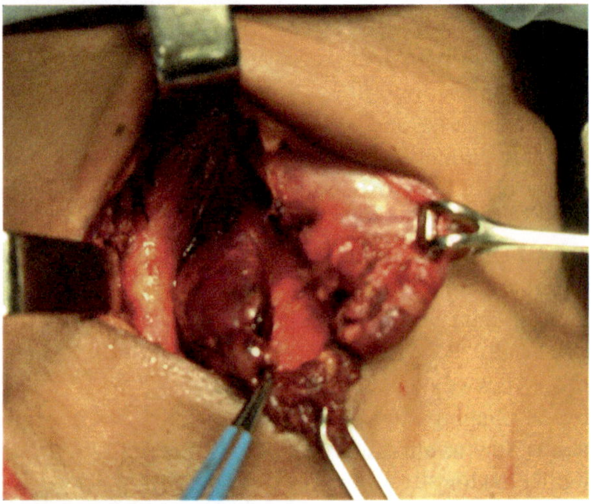

Fig. 21.17 A large parathyroid lesion being removed

The parathyroid lesion is identified in the 1–2 cm vicinity of the lower pole of the thyroid lobe usually anterior to RLN. The lesion traditionally appears as a brown colored soft vascular nodule. Instruments applied to the lesion may rupture it. Tumor tissues thrown off to soft tissue may lead to parathyromatosis which is a rare intractable cause of recurrence. The neighboring soft tissue is separated from the lesion rather than the lesion is separated from the soft tissue. DeBakey forceps applied to a tuft of soft close to adenoma helps to steady the lesion while dissection is continued (Figs. 21.16, 21.17, 21.18, and 21.19).

The inferior gland is highly variable in position and may extend to the superior mediastinum. Lesions placed below the level of innominate vessels may require median sternotomy.

Occasionally the lesion is felt firm and adherent to neighboring soft tissue and or to the thyroid gland. Preoperative clinical features such as palpable neck

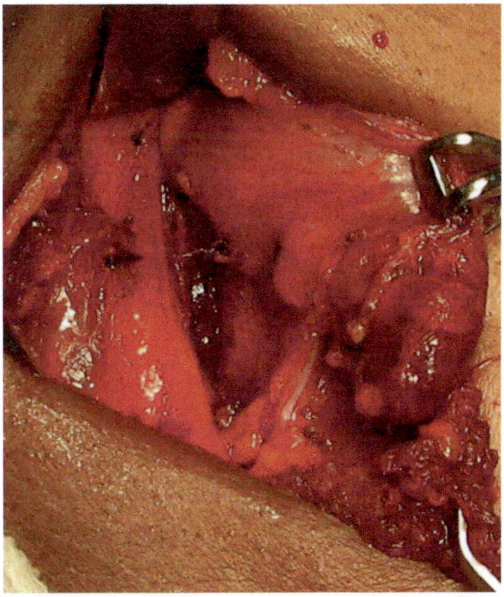

Fig. 21.18 Inspection of the field for bleeding

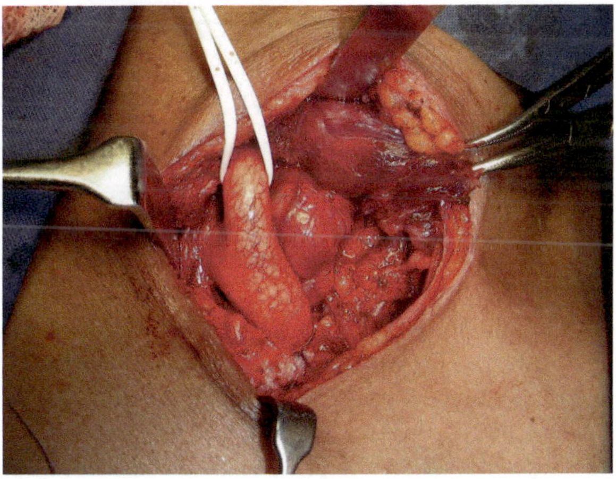

Fig. 21.19 Exposure of a lesion inside the carotid sheath.

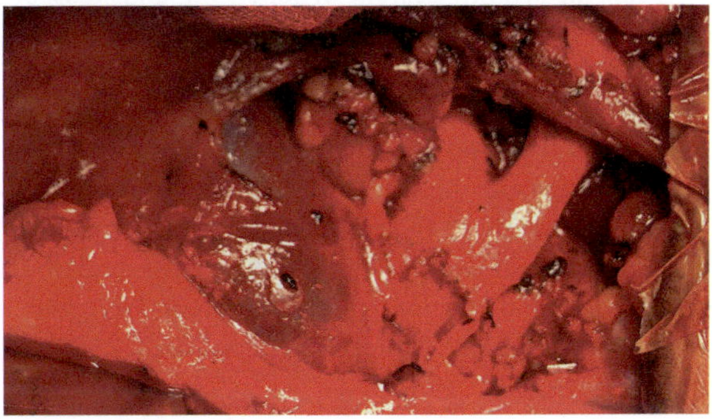

Fig. 21.20 Parathyroid lesion at the origin of carotid artery right side

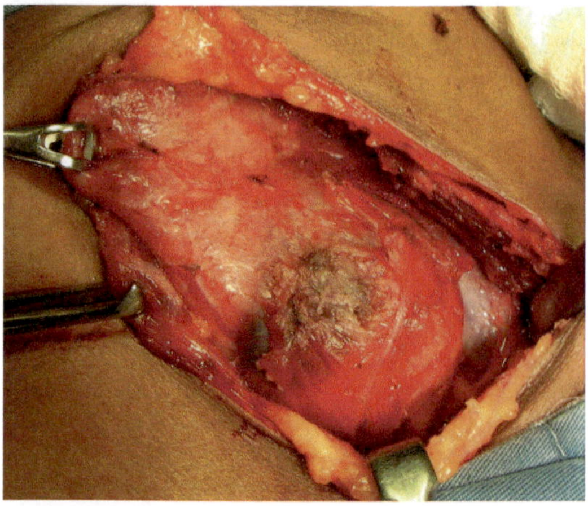

Fig. 21.21 Right inferior parathyroid adherent to thyroid gland

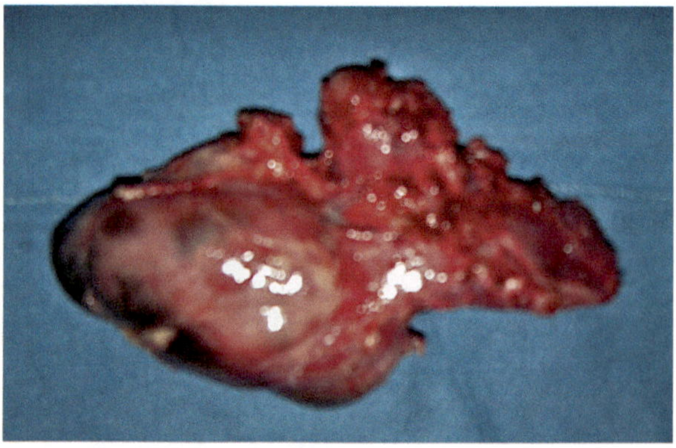

Fig. 21.22 Right hemithyroidectomy with adherent parathyroid lesion

swelling, high disease severity indices, and operative features such as a firm or hard adherent lesion favor diagnosis of carcinoma. The only effective treatment of parathyroid carcinoma is wide excision. So wide excision with ipsilateral thyroid lobe is recommended when PC is suspected (Figs. 21.21 and 21.22).

12. Superior parathyroid gland is more constant in a position close to the posterior border of the thyroid gland. But the recurrent laryngeal nerve is in a closer relationship with it. Lesions of superior glands dissect deep into the trachea-esophageal groove and maybe occasionally hidden. The recurrent laryngeal nerve is closely followed towards its entry to the larynx and is safeguarded. The thyroid lobe is rotated medially exposing the trachea-esophageal grove. The first assistant on the opposite side holds the thyroid lobe but should be cautious to avoid bleeding from the surface. The major source of blood supply is from the inferior thyroid artery or the posterior anastomotic plexus. Careful dissection is necessary to avoid injury to the vascular plexus and troublesome bleeding. The terminal cervical portion of the recurrent laryngeal nerve is closely applied to superior parathyroid lesions (Figs. 21.23, 21.24, 21.25).

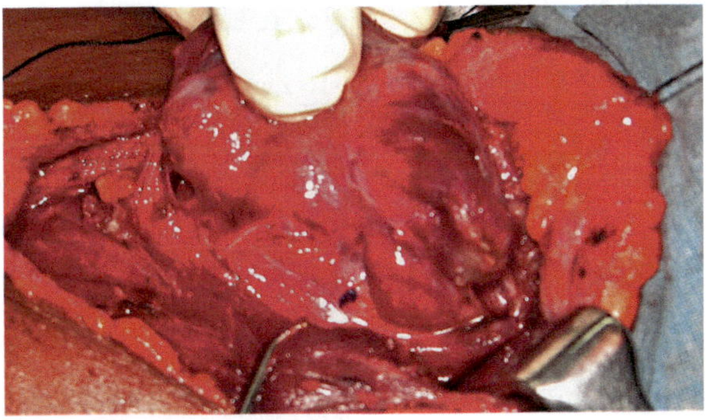

Fig. 21.23 Superior parathyroid adenoma close to the superior pole of thyroid

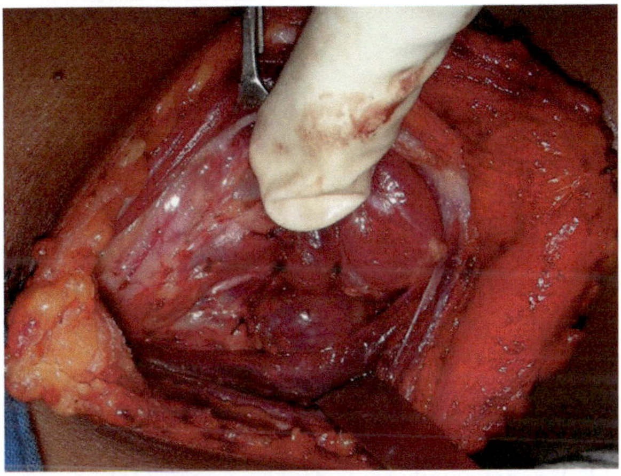

Fig. 21.24 Superior parathyroid lesion and RLN

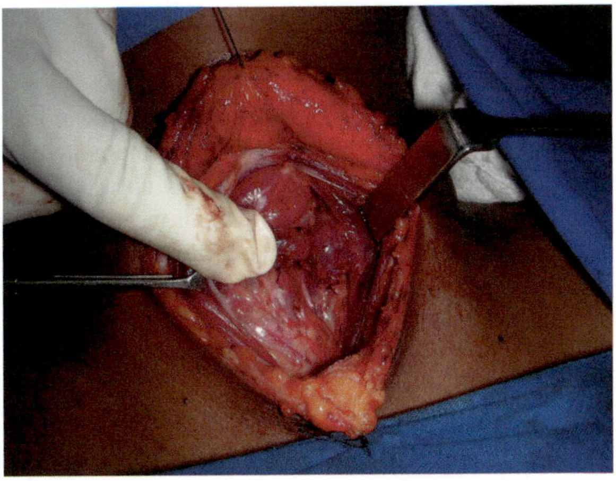

Fig. 21.25 The superior parathyroid adenoma delivered from tracheoesophageal grove

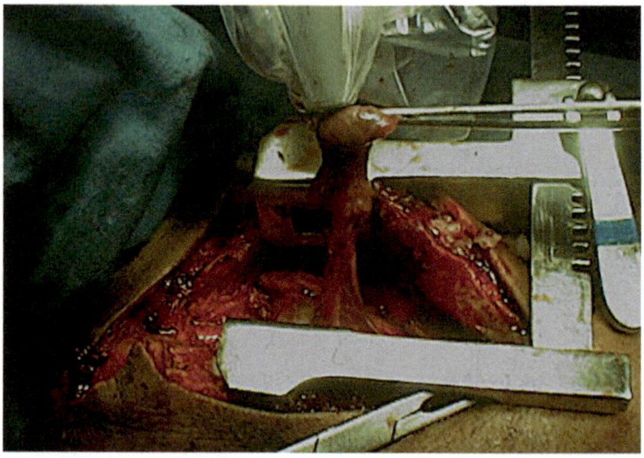

Fig. 21.26 Mediastinal parathyroid removed by sternotomy

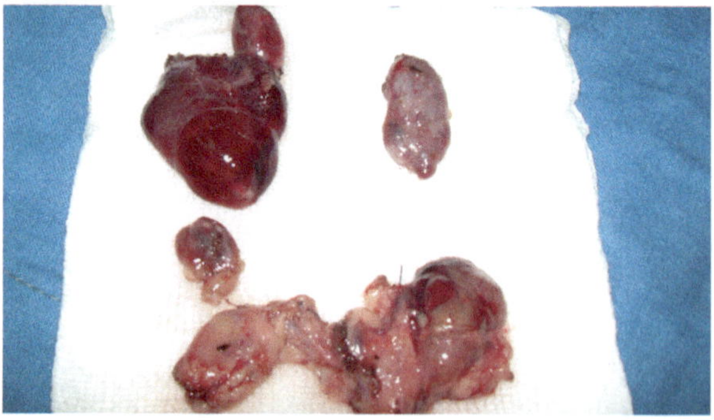

Fig. 21.27 Total parathyroidectomy with cervical thymectomy

13. Ectopic parathyroid adenomas are not infrequent. Mediastinum, carotid sheath retropharyngeal location are frequent locations for ectopic adenoma (Figs. 21.19 and 21.26).
14. When imaging fails to localize parathyroid lesion total neck exploration is the procedure of choice (Fig. 21.27). Systematic exposure of the normal position of the parathyroid is essential. The unusual locations such as carotid sheath, retro-esophageal areas, and finally mediastinum are explored. Intra-thyroid parathyroid lesions are usually identified by ultrasound.
15. The neck wound is closed without inserting a drain (Fig. 21.28). The divided strap muscles are approximated after ensuring hemostasis from the cut edges of the muscles. The cut edges are better approximated using horizontal mattress sutures using 3 "O" polylactic acid sutures.

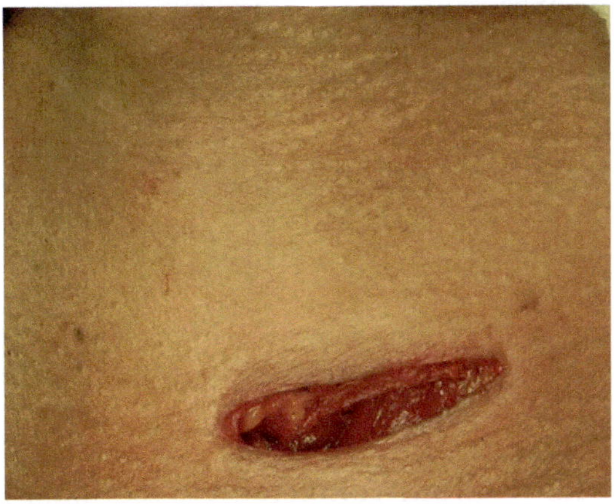

Fig. 21.28 The neck incision closed without a drain

21.2 Parathyroidectomy in Patients with Multiple Endocrine Neoplasia Type-1

Total parathyroidectomy with cervical thymectomy or sub-total parathyroidectomy with cervical thymectomy is the surgical option for Multiple Endocrine Neoplasia type-1. Bilateral neck exploration is mandatory. Prior imaging is not always supportive since only the most active lesion is lighted up in Scintigraphy. But preoperative and intra-operative ultrasounds are helpful to remove ectopic parathyroid lesions. Tracheo-esophageal grove is defined by mobilizing the thyroid lobe. The middle thyroid vein may have to be ligated for proper exposure. The recurrent laryngeal nerve is and traced down towards the mediastinum. The inferior parathyroid is lesion is very commonly encountered at this position. Traditionally all parathyroid glands are removed and a bit of relatively normal-looking gland removed for frozen section biopsy. After confirmation, the residual portion is minced and implanted in the non-dominant forearm muscle (preferably in brachioradialis). An alternate approach is to preserve in situ part of the relatively normal-looking gland (subtotal parathyroidectomy) (Fig. 21.29).

The thymus and inferior parathyroid glands share the embryological site of origin from the third pharyngeal pouch. The thymus occasionally harbors parathyroid remnants which gain significance in MEN. So cervical thymectomy is also performed along with parathyroidectomy. RLN is traced to the thoracic inlet on both sides and vestigial remnants of the thymus are removed. The inferior thyroid veins are secured carefully since cut ends may slip to the superior mediastinum causing troublesome bleeding.

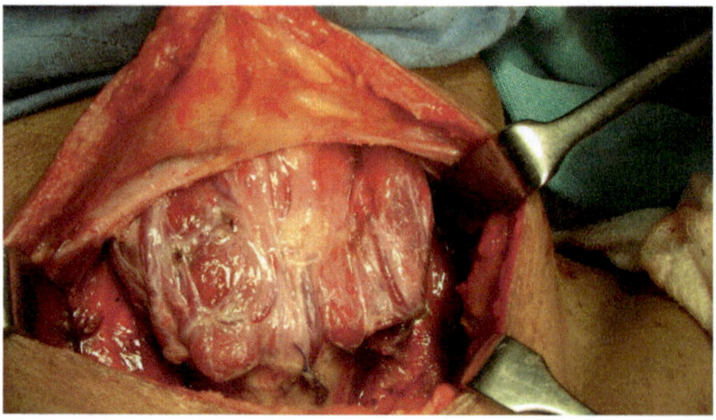

Fig. 21.29 Total neck exploration for multiple endocrine neoplasia

21.3 Parathyroidectomy: Secondary Hyperparathyroidism of Renal Origin

Hyperparathyroidism of renal origin is seen in end-stage renal disease patients on maintenance hemodialysis. Reversible changes in parathyroid glands, which are known as secondary hyperparathyroidism, are treated conservatively with diet regulations and medications such as vitamin D analogs and calcimimetics [3].

Parathyroidectomy is required in about 15% of patients after 10 years and 38% of patients after 20 years of ongoing dialysis therapy. Persistently elevated PTH values >800 pg/ml (>6 months) that remain nonresponsive to pharmacologic therapy are generally accepted as a criterion for parathyroidectomy [3, 4]. Persistent hypercalcemia or hyperphosphatemia (corrected serum calcium >10.2 mg/dl [>2.5 mmol/L] or phosphorus >5.5 mg/dl [>1.8 mmol/L]) despite patient compliance to diet and optimized medications is also considered indication for parathyroidectomy [5].

Four different approaches were reported for the surgical treatment of renal HPT [6]:

1. subtotal PTX with bilateral cervical thymectomy (3.5 parathyroid gland resection; subtotal PTX)
2. total PTX (TPTX) with auto-transplantation and bilateral cervical thymectomy (TPTX + AT)
3. TPTX without auto-transplantation and without thymectomy
4. TPTX without auto-transplantation and with bilateral cervical thymectomy (TPTX + BCT)

The first three procedures aim to maintain a long-standing residual production of PTH, whereas the goal of the fourth option is the complete elimination of PTH production. However, most experts recommend subtotal PTX and TPTX+AT as standard procedures. Thymectomy is indicated in conditions like MEN-1 and secondary hyperparathyroidism since the frequency of intrathymic parathyroid glands or parathyroid cells rests varies considerably between 14.8% and 45.3% [7].

21.4 Tertiary Hyperparathyroidism

The secondary hyperparathyroidism is reversible and when the stimulus ceases to exist reverts to normality or near normality. But a few progress to an irreversible condition known as tertiary hyperparathyroidism. One or two glands become autonomous and are usually diagnosed after a renal transplant.

Tertiary hyperparathyroidism is an absolute indication for parathyroidectomy and the procedure is similar to parathyroid adenoma.

21.4.1 Parathyroid Auto-transplantation

Auto-transplantation of parathyroid tissue is done following total parathyroidectomy. The genetic mutation in MEN-1 predisposes growth of implanted parathyroid tissue also and so an easily accessible location is preferred. The usual site for auto-transplantation is brachioradialis muscle mass. After confirming parathyroid tissue by frozen section biopsy, the remaining tissue is sliced into small pieces of <1 mm size. The muscle is exposed and using a 2 cm incision in Langer's line. The deep fascia is opened muscle fiber is split using blunt dissection. Five to six pieces of parathyroid tissue is implanted in a single pouch and the deep fascia is closed with 3-0 polyglactic acid sutures. Three or four pouches are made in a single incision.

21.4.2 Intraoperative PTH Assay

Intraoperative PTH assay is a very valuable tool to assess the successful removal of parathyroid lesions. But the usefulness in patients with MEN-1 is not established. But authors observed 10 min decline of serum PTH level below the lower limit of reference ranges ensures completeness and indicates hypocalcemia during the postoperative period.

21.4.3 Intraoperative Localization

Most of the operation theatres are equipped with a facility for ultrasound scanners and are used when exploration fails to identify the lesion.

A handheld gamma probe is an adjunct used usually when MIBI imaging indicates positive results but is not supported by ultrasound imaging (Fig. 21.30). The gamma probe is also helpful adenoma is suspected in an ectopic location (Fig. 21.31).

21.4.4 Radio-guided Parathyroidectomy

The widely used radiotracer is TC^{99} MIBI which gets concentrated in both thyroid and hyperfunctioning parathyroid glands. The washout from the thyroid gland is

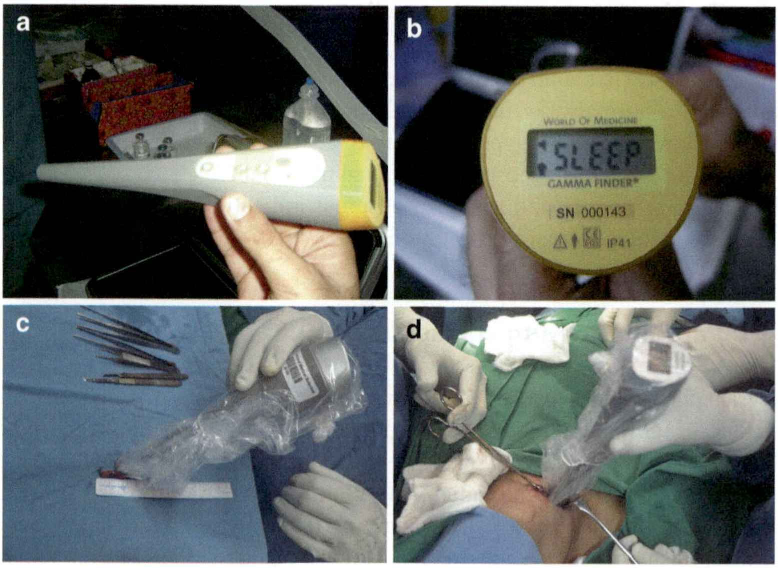

Fig. 21.30 The Gama probe used for intraoperative use

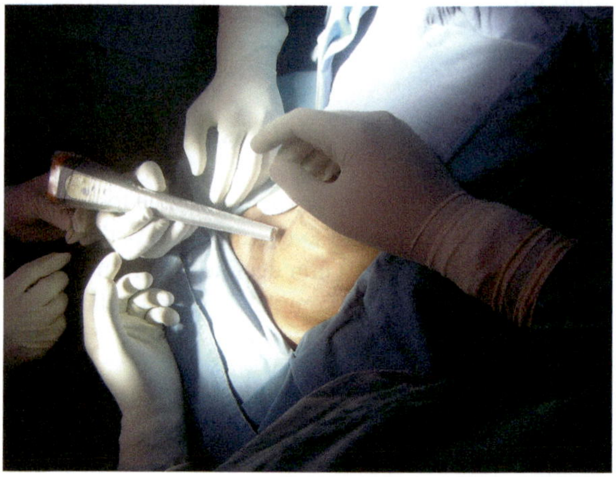

Fig. 21.31 Planning the incision with the Gama probe

quicker than parathyroid lesions and so 740 Mbq of MIBI is given as intravenous infusion 90–120 min before the start of the procedure.

It is advisable to do an intraoperative PTH assay to ensure completeness of removal of all hyperactive glands. The patient is anesthetized and positioned for a routine procedure. The incision is planned based on-site showing maximum count in the hand-held probe (Figs. 21.32 and 21.33).

A two to three-cm incision is made and dissected to expose the thyroid gland and proceed to mobilize the thyroid lobe. The probe is introduced to identify and locate the lesion. The activity measured over the hyperactive parathyroid gland is higher than thyroid and background activity. Targeted tissue demonstrates elevated count

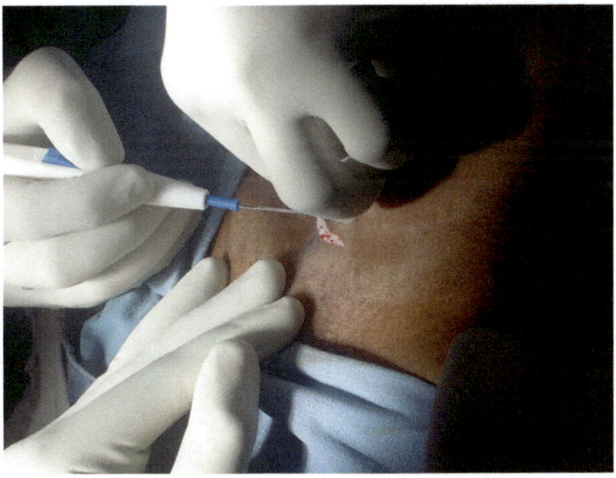

Fig. 21.32 Incision was made on the chosen site

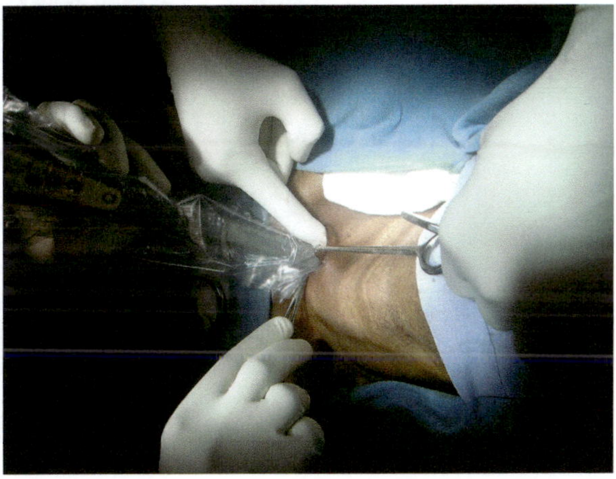

Fig. 21.33 Identifying the lesion with the probe

and an in vivo to background ratio >150% is strongly suggestive of parathyroid adenoma (Figs. 21.34, 21.35, and 21.36).

The excised tissue is reassessed with a probe away from the patient. Higher activity of more than 20% of background activity is suggestive of parathyroid adenoma (Fig. 21.37).

The probe is further inserted to search for additional incidental lesions. A second lesion may fail to light up when a more active lesion is in situ. The probe is used to search for a second hyperfunctioning gland while waiting for post-excision PTH values (Fig. 21.38).

Frozen section biopsy is not always necessary but helpful to confirm the diagnosis. But the result will not guarantee the completeness of excision when the multiglandular disease is suspected.

The incision is closed when the IOPTH assay is supportive of the complete removal of hyperfunctioning glands.

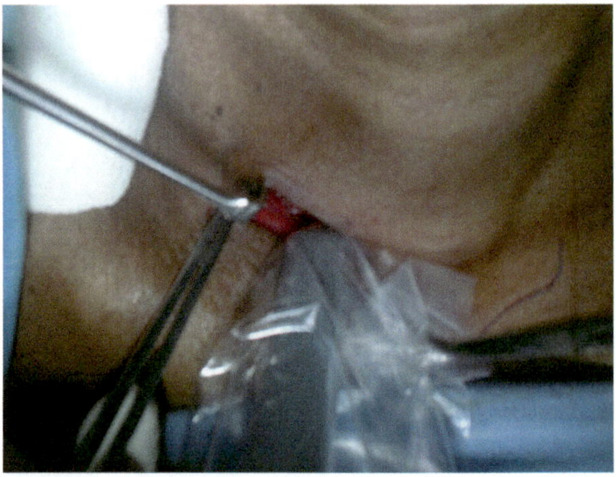

Fig. 21.34 The lesion identified with the probe

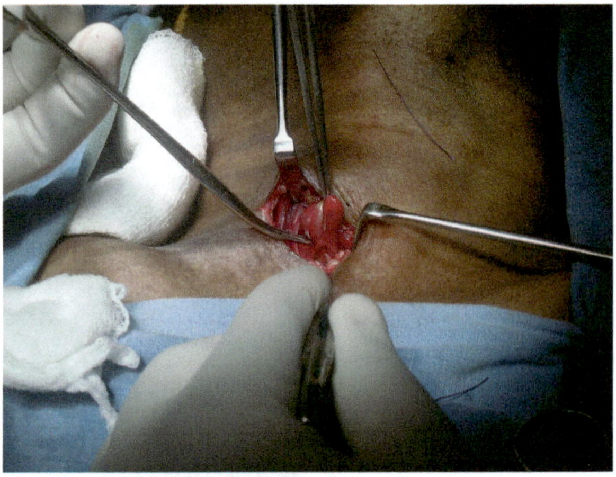

Fig. 21.35 The lesion being removed

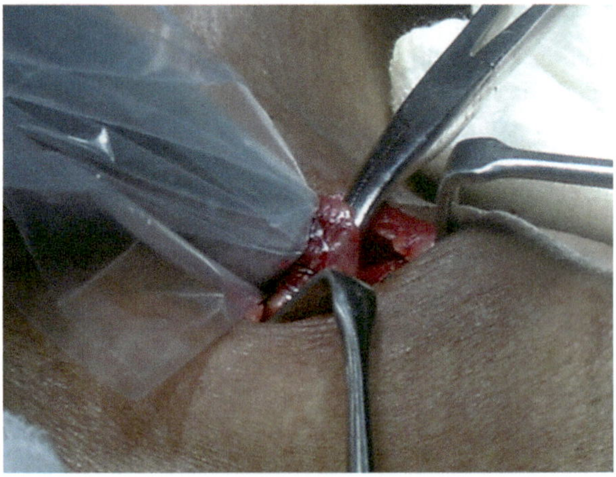

Fig. 21.36 Ensuring high radioactivity before removing the lesion

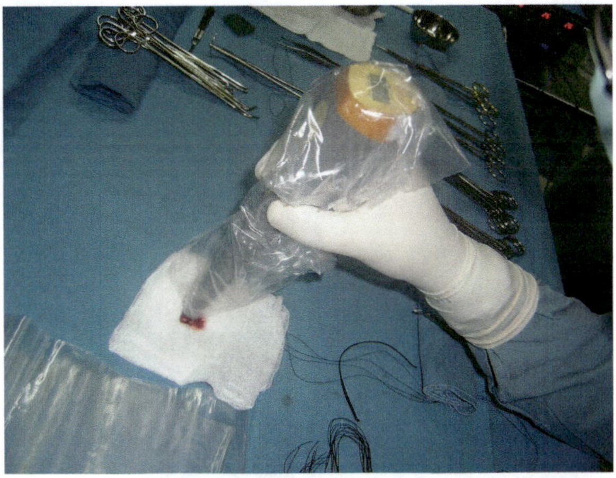

Fig. 21.37 Ensuring the high radioactivity on the removed lesion

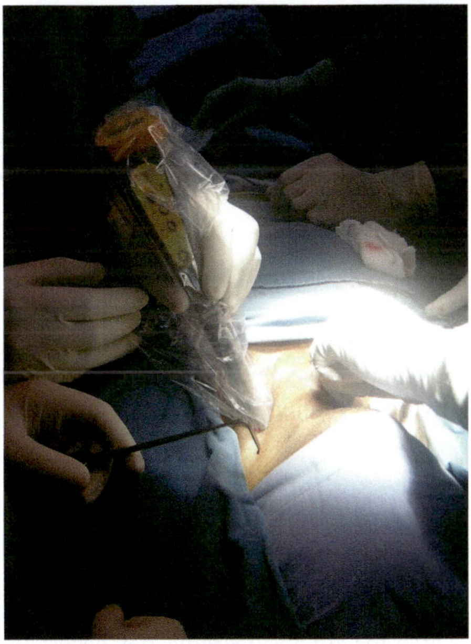

Fig. 21.38 Looking for residual lesion

21.4.5 Mini-parathyroidectomy

Focused parathyroidectomy can be done through an incision of 3–4 cm exactly over the lesion when concordance in two different imaging is available. The widely practiced protocol is to perform parathyroid scintigraphy followed by ultrasound for anatomical localization. Scintigraphy coupled with SPECT is a good alternative but hindered by a lack of universal availability. The procedure is done only when ably supported by intraoperative PTH assay to ensure completeness of the procedure.

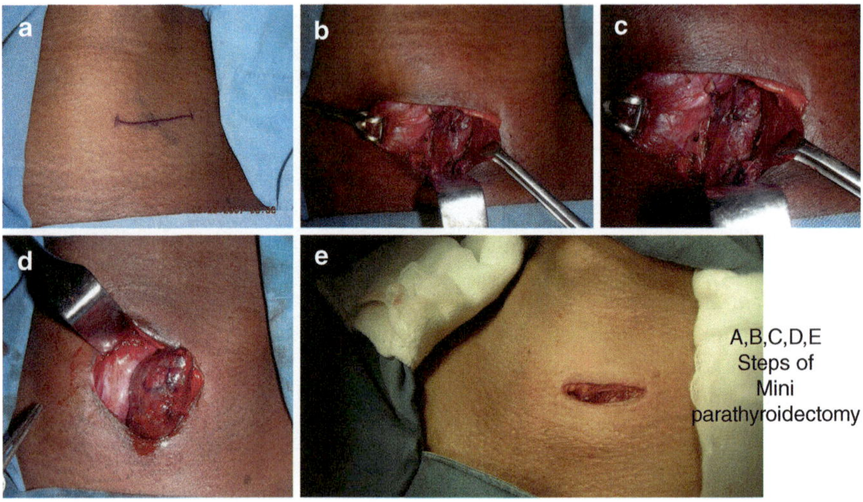

Fig. 21.39 Mini parathyroidectomy

The steps of the surgical procedure do not differ but need good precision in all movements (Fig. 21.39a–e). Parathyroid lesions in symptomatic diseases are generally larger in size and so require precise and gentle handling of tissues.

21.4.6 Postoperative Management

Many patients develop transient hypocalcemia following parathyroidectomy lasting for <4 days. Usually, this is corrected with oral calcium supplementation along with dihydroxy vitamin D.

21.4.7 Hungry Bone Syndrome

The term "hungry bone syndrome" (HBS) may be defined as prolonged hypocalcemia following parathyroidectomy. Serum corrected calcium remains below 8.4 mg/dl after the 4th postoperative day necessitating prolonged calcium supplementation. HBS is often encountered following parathyroidectomy for symptomatic PHPT. The prevalence among Indians seems to be high (21–87%) [8, 9].

HBS is related to the correction of dysregulation in bone formation and bone resorption induced by uncontrolled PTH secretion. Bone undergoes the cyclical process of osteoclastic bone resorption and remodeling with new bone formation by osteoblastic activity. PTH favors excess osteoclastic activity changing the rhythm in the cyclical activity. Successful parathyroidectomy curbs the osteoclastic activity and decreases the remodeling space. For the resultant gain in bone mass, more minerals such as calcium and magnesium are needed. This phenomenon is believed to cause a marked fall in serum calcium, phosphate, and magnesium levels which parallels the normalization of bone turnover.

Old age (>70), 25 OHD deficiency, preoperative high serum alkaline phosphatase levels (reflect the state of bone turnover), and evidence of bone disease such as osteitis fibrosa cystica are predictors of HBS. Large volume hyperactive parathyroid gland(s) are likely to be associated with HBS [10].

The hallmark of the condition is lasting hypocalcemia which may be mild (>7.62 mg/dl) or severe when serum total calcium level has gone below 7.62 mg/dl. The mild form is usually asymptomatic or with mild disturbances. Severe hypocalcemia is always symptomatic and may be associated with life-threatening manifestations.

Treatment of the HBS is aimed at replenishing the circulating calcium deficit, and in the longer term, at normalizing bone turnover leading to stable mineralization of the skeleton.

Mild hypocalcemia (>7.62 mg/dl (1.9 mmol/L). Mild hypocalcemia is treated with oral supplementation calcium and an active form of vitamin D. Two widely used calcium preparations are calcium carbonate (40% elemental calcium) and calcium citrate (21% elemental calcium). Calcium carbonate preparations (1000 g) twice daily before food and calcitriol 0.5 μg daily is recommended. The patient has to be reviewed in 4–7 days for monitoring the serum levels and the dose of supplementation is adjusted to achieve normalcy [11].

Acute severe symptomatic hypocalcemia (with seizure, laryngospasm, and tetany) is a medical emergency. The patient has to be monitored since rapid correction of hypocalcemia may cause hypotension and dysrhythmias. These patients require an intravenous infusion of calcium for quick response. One ampule (10 ml of 10% calcium gluconate) delivers 93 mg of elemental calcium. Intravenous rapid infusion of 10–20 ml of 10% calcium gluconate diluted in 100 ml of 5% dextrose is infused in 20 min. The rapid infusion may be repeated after 10–60 min when symptoms persist. After controlling symptoms, a patient is given a dose-adjusted maintenance infusion. Based on serum levels, calcium load in infusion is calculated (as 0.5–2 mg/kg/h). Quick intravenous infusion of calcium could result in arrhythmias and so patients should be carefully monitored. Oral supplementation is initiated after acute symptoms have subsided [11].

Infusion of calcium preparations containing >200 mg/100 ml through peripheral veins may cause vein wall damage and thrombophlebitis. Calcium infusion in large volume fluid causes fluid overload in patients with secondary hyperparathyroidism and in elderly patients with compromised cardiac function. Calcium phosphate deposition in the lungs, kidney, or other soft tissue may occur in patients receiving intravenous calcium especially in the presence of high serum phosphate levels.

Magnesium is an important electrolyte having a direct influence on various other electrolytes such as calcium and potassium. Normal serum magnesium level (1.46 and 2.68 mg/dl) is essential since it is the cofactor of many of the body's biochemical reactions. Decreased levels of magnesium impair the magnesium-dependent adenyl cyclase pathway and decrease the release of parathyroid hormone (PTH) which in turn decreases serum calcium. There is an increased uptake of magnesium for renewing bone following parathyroidectomy and so hypomagnesemia is a common accompaniment of hungry bone syndrome. Hypomagnesemia is found in 50–60% of intensive care patients. Hypomagnesemia is common following starvation and alcohol abuse and is associated with liver and kidney diseases [12].

Hypomagnesemia can induce or potentiate neuromuscular symptoms such as tremors, tetany (including positive Trousseau and Chvostek signs, muscle spasms, muscle cramps), and seizures. Cardiovascular manifestations include atrial and ventricular arrhythmias, cardiac ischemia, and ECG changes (widening of the QRS complex, peaked T waves, prolongation of the PR interval).

Correction of hypomagnesemia in a hemodynamically unstable patient is done as the emergent rapid infusion of 1–2 g of magnesium sulphate (in 100 ml of 5% dextrose) in about 15 min. In a symptomatic stable patient, an infusion is given slowly over 2 h. The serum magnesium level repletes quickly but the intracellular magnesium takes a longer duration to normalize. So, in patients with normal renal function supplementation is continued for 2 days more. But patients with compromised renal function (eGFR < 30 ml/min/1.73 m^2) dose adjustment is necessary to avoid hypermagnesemia [12].

Maintenance therapy may require oral administration of Mg^{2+} oxide (400 mg twice daily or three times daily)-400 mg (equivalent to 240 mg elemental magnesium) or magnesium gluconate 400 mg daily (270 mg elemental magnesium).

The doses of calcium and active vitamin D preparations required and the duration of treatment is guided by serum calcium and bone turnover marker levels. The duration may sometimes last in excess of 12 months after successful surgery. HBS patients require periodic evaluation of serum calcium, phosphate, and ALP levels for monitoring supplementation. Vitamin D deficiency is a common accompaniment and corrective supplementation is concurrently done.

Preoperative bisphosphonates and correction of vitamin D deficiency were thought to reduce the severity and duration of postoperative hypocalcemia. Further prospective studies are needed to optimize pre-and post-operative treatment strategies in patients with PHPT and skeletal manifestations at high risk for HBS [10].

Hungry bone syndrome following parathyroidectomy for secondary HPT is found in about 27% of patients. Risk factors included younger age, higher body weight, higher preoperative serum alkaline phosphatase, lower preoperative serum calcium level, and lower postoperative serum calcium. Calcium and magnesium correction is recommended as routine but requires careful adjustment to avoid fluid overload.

21.5 Complications

Complications following parathyroidectomy for PHPT are found in <2% of patients. Surgical site infection is rare and routine perioperative antibiotic therapy is not necessary. Other surgical site-related complications such as bleeding, hematoma, and seroma are found occasionally. Careful hemostasis should be ensured prior to wound closure.

Recurrent laryngeal nerve palsy is a potential complication but frequency shall be minimized by careful dissection. Occasional patients who have lesions with adhesions to soft tissue secondary to malignant infiltration or inflammation are candidates for nerve palsy.

The postoperative mortality (30 days) after parathyroidectomy for renal HPT is reported to range between 1% and 3% [13]. The single most preventable cause of postoperative death is hyperkalemia. Alkaline phosphatase is re-emerging as a better predictor of high/low bone turnover, and it has a more linear association with mortality in patients on dialysis than PTH [14].

References

1. Barletta G, De Feo ML, Del Bene R, Lazzeri C, Vecchiarino S, La Villa G, Brandi ML, Franchi F. Cardiovascular effects of parathyroid hormone: a study in healthy subjects and normotensive patients with mild primary hyperparathyroidism. J Clin Endocrinol Metab. 2000;85:1815–21.
2. Farahnak P, Ring M, Caidahl K, Farnebo L-O, Eriksson MJ, Nilsson I-L. Eur J Endocrinol. 2010;163(3):461–7.
3. Moe SM, Chertow GM, Coburn JW, Quarles LD, Goodman WG, Block GA, Drüeke TB, Cunningham J, Sherrard DJ, McCary LC, Olson KA, Turner SA, Martin KJ. Achieving NKF-K/DOQI bone metabolism and disease treatment goals with cinacalcet HCl. Kidney Int. 2005;67:760–71.
4. Lau WL, Obi Y, Kalantar-Zadeh K. Parathyroidectomy in the management of secondary hyperparathyroidism. CJASN. 2018;13(6):952–61. https://doi.org/10.2215/CJN.10390917.
5. Cheng SP, Liu CL, Chen HH, Lee JJ, Liu TP, Yang TL. Prolonged hospital stay after parathyroidectomy for secondary hyperparathyroidism. World J Surg. 2009;33(1):72–9.
6. Lorenz K, Bartsch DK, Sancho JJ, et al. Surgical management of secondary hyperparathyroidism in chronic kidney disease—a consensus report of the European Society of Endocrine Surgeons. Langenbecks Arch Surg. 2015;400:907–27. https://doi.org/10.1007/s00423-015-1344-5.
7. Pattou FN, Pellissier LC, Noël C, Wambergue F, Huglo DG, Proye CA. Supernumerary parathyroid glands: frequency and surgical significance in treatment of renal hyperparathyroidism. World J Surg. 2000;24(11):1330–4.
8. Agarwal G, Mishra SK, Kar DK, Singh AK, Arya V, Gupta SK, Mithal A. Recovery pattern of patients with osteitis fibrosa cystica in primary hyperparathyroidism after successful parathyroidectomy. Surgery. 2002;132:1075–83.
9. Bhansali A, Masoodi SR, Reddy KS, Behera A, das Radotra B, Mittal BR, Katariya RN, Dash RJ. Primary hyperparathyroidism in north India: a description of 52 cases. Ann Saudi Med. 2005;25:29–35.
10. Witteveen JE, van Thiel S, Romijn JA, Hamdy NA. Hungry bone syndrome: still a challenge in the post-operative management of primary hyperparathyroidism: a systematic review of the literature. Eur J Endocrinol. 2013;168(3):R45–53. https://doi.org/10.1530/EJE-12-0528.
11. Turner J, Gittoes N, Selby P, the Society for Endocrinology Clinical Committee. Society for endocrinology endocrine emergency in adult patients. Endocr Connect. 2016;5(5) https://doi.org/10.1530/EC-16-005.
12. Gragossian A, Bashir K, Friede R. Hypomagnesemia [Updated 2020 Sep 6]. In: StatPearls. Treasure Island, FL: StatPearls Publishing; 2020. Available https://www.ncbi.nlm.nih.gov/books/NBK500003.
13. Kestenbaum B, Andress DL, Schwartz SM, Gillen DL, Seliger SL, Jadav PR, et al. Survival following parathyroidectomy among United States dialysis patients. Kidney Int. 2004;66(5):2010–6.
14. Lau WL, Kalantar-Zadeh K, Kovesdy CP, Mehrotra R. Alkaline phosphatase: better than PTH as a marker of cardiovascular and bone disease? Hemodial Int. 2014;18(4):720–4.

GPSR Compliance

The European Union's (EU) General Product Safety Regulation (GPSR) is a set of rules that requires consumer products to be safe and our obligations to ensure this.

If you have any concerns about our products, you can contact us on ProductSafety@springernature.com

In case Publisher is established outside the EU, the EU authorized representative is:

Springer Nature Customer Service Center GmbH
Europaplatz 3
69115 Heidelberg, Germany

Batch number: 10732044

Printed by Printforce, the Netherlands

Pickens et al. provided a summary of the key components of an antibiotic stewardship program in the intensive care unit [18]. A better outcome in ICUs has been described by incorporating elements like antibiotic timeout, rapid diagnostics and laboratory testing, clinical pathways, computerized decision support, and infection control in addition to leadership, prospective audit, and feedback, which are from the general hospital setting.

8 Key Strategies for Antimicrobial Stewardship in the ICU

Patients with sepsis and septic shock in the ICU are at risk of antibiotic failure if incorrect antibiotics are administered. Antibiotic therapy in ICU patients can be divided into three phases. The initiation of empiric antibiotic therapy which should also be most the effective antibiotic regimen is based upon antibiogram and guidelines along with workup for sources of infection like imaging, cultures, and rapid molecular diagnostics. The second phase is the daily assessment of the severity of the illness and dosing and narrowing of antibiotics on the availability of the isolates and sensitivity of the organism. Microbiologists, infectious disease specialists, and critical care clinical pharmacist teams are important for the optimization of antibiotics. During these phases, the antibiotic dosing should be based on pharmacodynamics, pharmacokinetics, and organ dysfunction. De-escalation will lessen the selection pressure for resistance and cost-effectiveness and adverse effects of the broad-spectrum therapy. The final phase is the discontinuation of the antibiotics after an appropriate duration of therapy. During all these phases, cooperation and communication between the antimicrobial stewardship team members and data collection for future planning are invaluable [19, 20]. Figure 3 summarizes the phases of the therapy and the interventions in each phase of antimicrobial stewardship in the ICU.

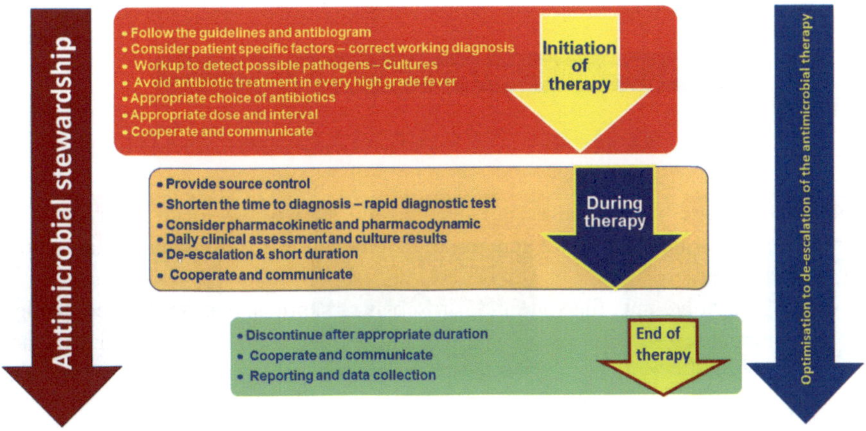

Fig. 3 Summary of the phases of the therapy and the interventions

9 Barriers to Antibiotic Stewardship in ICU

Life changes in the workplace are a key to improvement if it comes to quality improvement. However, the human tendency is to consider changes as challenges and a tendency to see them rather than as opportunities for learning and growth. As changes are costly, the cost comes in terms of time, finances, and changes in beliefs and behaviors to a new reality.

There is a reasonable amount of literature mentioning the issue of barriers, challenges, and difficulties in implementing antibiotic stewardship. These barriers or challenges depend on various factors like the status of the countries like developed, middle-income, or third-world countries; types of hospitals like primary care, general hospitals, or a tertiary care hospital; and facilities like outpatient, wards, or ICUs. There is a very meager amount of literature addressing this issue in ICUs.

Three main themes regarding antibiotic stewardship were identified from the interviews with pharmacists. Respondents described the importance of (1) a supportive organizational culture, (2) protected time for antibiotic stewardship, and (3) a cohesive organizational structure in the success of antibiotic stewardship programs. The three most common interventions of the AMS are shown in Fig. 4.

In a study by Alghamdi et al., they quizzed the antibiotic stewardship teams in the hospitals on certain questions [22]. They questioned them on issues like lack of adherence to guidelines, lack of administrative/management support, poor communication, lack of health information technology, need for education and training, shortage of ASP team members, poor communication, and physicians' fears and concerns, and their conclusion was "Several barriers to antibiotic stewardship adoption and implementation in hospitals were identified, including factors relating to the socio-political context of hospitals, organisational characteristics, and healthcare professionals' barriers."

In a systematic review of transnational research to identify commonly perceived barriers and facilitators in developed countries by Rzewuska M et al., in their analysis, they listed numerous barriers reported in the studies selected [23]. They classified them as environmental context and resources, goals, social influences,

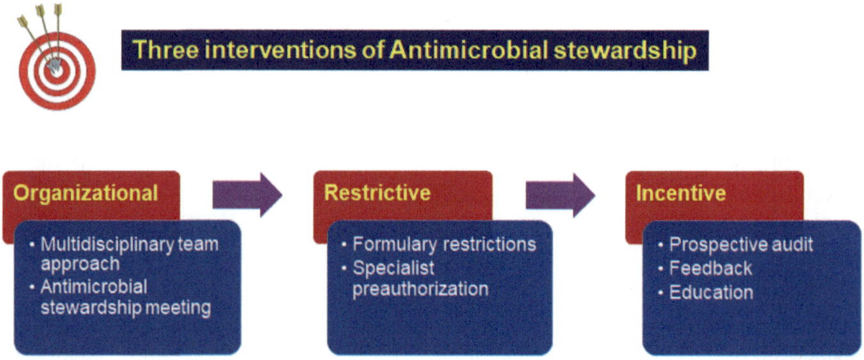

Fig. 4 Three most common interventions of the AMS [21]

behavioral regulation, knowledge, beliefs about consequences, social/professional role and identity, intentions, reinforcement, and skills. To mention a few, important barriers are the following:

- Lack of key personnel (e.g., infectious disease clinicians, pharmacy staff, microbiologists)
- Problems with data and information systems (e.g., inadequate information technology, lack of dedicated IT assistant, lack of good quality data, and resources to utilize it)
- Lack of financial resources and time
- Inadequate supply of laboratory provisions
- Limited antibiotic options available in settings with prevalent multidrug-resistant bacteria
- Poor communication, including interpersonal, within teams (e.g., inconsistency or conflict) and between private and public sectors
- Other higher-priority initiatives hindering the ASP
- The influence of local guidelines and clinical practice protocols and the lack of national and/or international standards required for a specific antibiotic stewardship strategy
- Lack of standards for measuring the performance of a specific antibiotic stewardship intervention
- Lack of knowledge about ASPs (e.g., due to poor education or inevitable loss of knowledge due to high staff turnover) and knowledge of current use of antibiotics
- Lack of willingness to change
- Medical professionals lacking relevant skills for a specific antimicrobial stewardship strategy (e.g., training in clinical microbiology)

10 Barriers Specific to ICU in Antimicrobial Stewardship

As mentioned in the hospital setup, all the barriers are descriptive in any ICU setup. Some specific barriers are peculiar to ICU and critical settings. The ICU physicians encounter patients with sepsis and septic shock in whom the exact source of infection is often unclear for the first 24–48 h which is the time for a workup to rule out the cause. This workup time is when empirical antibiotic plays an important role, and these empirical regimens must be necessarily broader. Inappropriate empirical antibiotic therapy at this stage may worsen outcomes, including mortality.

The most specific barriers that limit the success of antibiotic stewardship in the ICU are

1. Diagnostic uncertainty at the time of admission or stay in the ICU
2. Physicians fear factors and pressure to choose appropriate empiric therapy, fear of not adequately covering the causative pathogen
3. Underappreciation of the toxicity of antibiotics
4. The reluctance of ICU physicians to deescalate or stop antibiotics

These factors act synergistically and contribute to spiraling multiple broad-spectrum antibiotics [18].

11 Overcoming Obstacles to Implement Antimicrobial Stewardship Interventions in the ICU

11.1 Role of the Microbiology and Laboratory Antimicrobial Stewardship in ICUs

The clinical microbiology laboratory plays a significant role in promoting the appropriate use of antimicrobials. They work closely with doctors and infection management experts, compiling antibiogram data and periodically making decisions about diagnostic tests. Additionally, their selective reporting of susceptibility results can be a helpful tool in directing appropriate antibiotic treatment. Proper stewardship for diagnostic testing, such as ensuring correct usage of microbiologic assays, is also beneficial.

11.2 Role of Antibiogram in Antimicrobial Stewardship in ICUs

An antibiogram is a useful document that contains information about the susceptibility of bacterial isolates to antibiotics. It is compiled by microbiology laboratories over a specific period, usually a year. Clinicians rely on it to choose the best antibiotic therapy for their patients, and stewardship programs use it to develop custom therapeutic plans for their facilities and monitor resistance trends. The Clinical and Laboratory Standards Institute guidelines for antibiogram production are used in microbiology labs. Better empiric coverage for the treatment of infection is made possible by antibiogram. It also aids in selecting the best antibiotic combination till the organism has been identified and susceptibility information is available. Combination antibiograms are very useful when treating infections caused by multidrug-resistant organisms.

11.3 Reducing the Usage of High-Risk Medications

This strategy emphasizes the importance of appropriate antibiotic usage and helps reduce the risk of developing antimicrobial resistance. In this strategy, the antibiotics are prioritized according to the high- to low-priority necessity rating. The high-priority antibiotics are used only when there is an absolute necessity

[24]. In 2017, the WHO Expert Committee on *Selection and Use of Essential Medicines* developed the AWaRe classification of antibiotics. The abbreviation AWaRe stands for "**A**ccess, **Wa**tch, and **Re**serve." This was revised in 2021. These recommendations are based on the consequences of different antibiotics on antimicrobial resistance and support antibiotic stewardship efforts at local, national, and global levels. Figure 5 explains each category of recommendation of AWaRe. Figure 6 lists the medication according to the aware category as recommended by the WHO. The 2021 update of the AWaRe has a total of 258 antibiotics.

11.4 Preauthorization and Formulary Restriction

This practice of preauthorization of antibiotics is to restrict the use of antibiotics which are on the watch and reserve list. The physicians have to get authorization from the infectious disease specialist and/or infectious disease pharmacist. For ICUs, the practice is to allow prescription of these antibiotics for 72 h as this is the time for microbiology results to be available. Once organisms are isolated and sensitivity is available, the infectious disease specialist and/or infectious disease pharmacist has to approve to continue its use. This is called restrictive stewardship. This strategy may reduce the use of restricted antibiotics but may result in increased use of non-restricted antimicrobials [8].

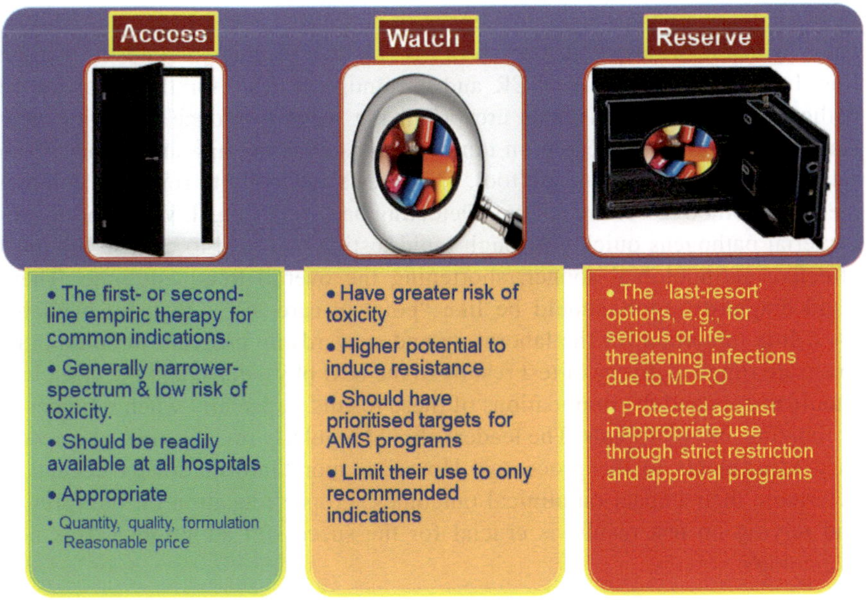

Fig. 5 Category of recommendation of the AWaRe by the WHO [25]

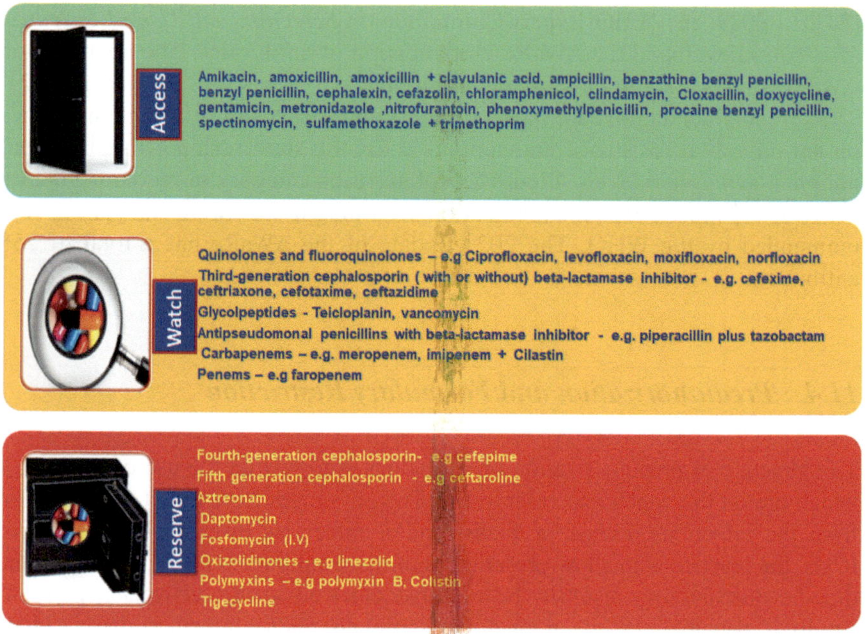

Fig. 6 List of the medication according to the AWaRe category as recommended by the WHO [25]

11.5 Role of Rapid Diagnostic Tools

In the ICU, one of the main challenges in antibiotic stewardship can be overcome with the widespread availability of rapid molecular diagnostic testing. Rapid diagnostic tools like PCR and chromogenic tests for the detection of pathogens and their resistance are promising ways to reduce antibiotic consumption. Extensive research on rapid diagnostic testing has shown that it is a reliable and precise diagnostic tool. These tests are likely to remain a complementary method to culture testing. Detecting bacterial, fungal, viral, and mycobacterial pathogens quickly through rapid testing can help to choose the most effective antimicrobial sooner, shortening the overall treatment time. These rapid diagnostic tools should be like "point-of-care tests" that are available 24/7. It is important for the laboratory and stewardship program to collaborate to ensure timely delivery of test results to the ICU physicians, and the clinicians should understand the implications of these results, especially when it comes to optimizing antimicrobials. The leaders should regularly review and make necessary adjustments to the processes. However, more information is needed to determine their impact on clinical outcomes. Proper education on interpreting and relying on test results is crucial for the successful adoption of this new technology.

11.6 Role of Serum Biomarkers

Procalcitonin, a biomarker for sepsis, can aid in determining whether to continue antibiotic medication. Serial procalcitonin (PCT) readings can shorten antibiotic treatments [26]. Serial PCT measurements are more helpful than a single test. On the contrary, escalation of empirical therapy with increasing PCT levels has been associated with unfavorable outcomes. Fungal markers are helpful tests for immunocompromised individuals at risk of invasive fungal illness to optimize the use of antifungals.

11.7 Role of Antibiotic Timeouts (ATO)

It is an important element of the antimicrobial stewardship program. It is a provider-led evaluation of the choice and ongoing need for antibiotics. In ICUs and critical care settings, empirical broad-spectrum antibiotics are started, considering the patients' critical situation. These empiric broad-spectrum antibiotics are not narrowed after the culture results become available. A well-structured antimicrobial timeout will remind the steward to reassess antibiotics' indication, spectrum, dose, and duration. The Centers for Disease Control and Prevention in the USA and the National Health Service in Great Britain have endorsed the antibiotic timeout strategy, encouraging ICU physicians to review antibiotic therapy on the third or fourth calendar day of treatment. An experimental trial by Thom KA et al. demonstrated that antibiotic timeouts at 48–72 h after empiric antibiotics decreased the use of inappropriate antibiotics but did not reduce overall antibiotic use [27].

In another study by Graber et al., electronic ATO implementation found a significant decrease in the two targeted antibiotics, piperacillin, and vancomycin. Antibiotic timeouts are crucial in reducing the incidence emergence of AMR organisms [28].

Electronic medical record systems will aid in tracking antibiotic timeouts. Antibiotic timeouts have a caveat the optimal timing has yet to be established.

11.8 Collaboration Between Infectious Diseases and ICU Physicians

Collaboration between the ASPs and intensivists makes a multi-professional approach toward determining the optimal antibiotics. ASPs and intensivists can optimize antibiotic therapy after mutual harmony. Rimawi RH et al. concluded in their study that "Active communication with an infectious disease practitioner can significantly reduce medical ICU antibiotic overuse by earlier modification or cessation of antibiotics without increasing mortality. This, in turn, can reduce healthcare costs, foster prodigious education, and strengthen relations between the subspecialties" [29].

11.9 Collaboration Between Infection Prevention and Control and Antimicrobial Stewardship Program

Prevention is better than cure. Prevention of healthcare-associated infections can result in decreased antibiotic use and decreased antibiotic resistance. Bacterial and fungal colonization is seen in the hands of healthcare workers. Preventive strategies like hand hygiene reduce the incidence of nosocomial infection. Two systems like antibiotic antimicrobial stewardship program and infection control need an intersection with similar goals. The antimicrobial stewardship program aims to reduce antibiotic usage and the emergence of antibiotic resistance, whereas infection control aims to reduce healthcare-associated infections. Both programs have strategies and frameworks and keep performance indicators. Antimicrobial stewardship programs are more effective when implemented alongside infection prevention and control measures. They collaborate with ASP leaders to synchronize the infection control programs and the ASP program. They offer information on infection surveillance and aid in data analysis and presentation for ASP. They liaise with the microbiologist in developing the antibiogram. Report HAI surveillance data to the ASP team about the outbreaks (for example, MDRO patterns over time). Additionally, they aid in developing, implementing, and evaluating policies and training front-line workers in ASP operations [30].

11.10 Computerized Decision Support System

A decision support system (DSS) is a computer program that helps organizations make decisions and judgments and take action. It is a highly effective tool that helps in implementing the ASP. A study by Evans et al. using the computerized decision support system found that patients received significantly fewer antibiotics and fewer doses of antibiotics [31].

A computerized decision support system helps with a variety of stewardship tasks, including de-escalation; antibiotic "timeouts," i.e., streamlining; scheduled antimicrobial reassessments; formulary restriction with preauthorization and prospective audit with intervention and feedback; and improved antimicrobial documentation (of indication and intended duration of therapy).

It offers a chance for prescribers to receive on-the-spot education using criteria, alerts, and access to instructions and instructional resources. Data gathering and auditing can be facilitated by the inclusion of forcing capabilities (such as requiring the justification for administering a restricted antibiotic prior to the execution of an order). Every prescriber in the institution receives consistent messaging.

11.11 *Infection Control*

Prevention is better than cure. Prevention of healthcare-associated infections can result in decreased antibiotic use and decreased antibiotic resistance. Bacterial and fungal colonization is seen in the hands of healthcare workers. Preventive strategies like hand hygiene reduce the incidence of nosocomial infection. IPs play a significant role in ASP activities. They collaborate with ASP leaders to synchronize the infection control programs and the ASP program. They offer information on infection surveillance and aid in data analysis and presentation for ASP. They liaise with the microbiologist in developing the antibiogram. Report HAI surveillance data to the ASP team about the outbreaks (for example, MDRO patterns over time). Additionally, they aid in developing, implementing, and evaluating policies and training front-line workers in ASP operations.

12 Impact and Outcome of the Antimicrobial Stewardship in ICU

Antimicrobial stewardship can have a significant impact on patients and numerous other outcomes. ASP should meet the primary goals to improve outcomes, prevent adverse events, and reduce antimicrobials. As any program's success is measured by its outcomes, the AMS looks into specific parameters like

- Emergence of antimicrobial resistance
- Reduction in the duration of therapy
- De-escalation of antibiotics based on microbiology data
- *C. difficile* infection rates
- Adverse event(s) related to antibiotic treatment
- Compliance with guidelines or treatment algorithm
- Clinical failure—broaden therapy, recurrence of infection
- Decrease in ICU length of stay
- Readmission rates
- Number of patients with infection due to multidrug-resistant organisms
- Mortality due to infection
- Excess days of therapy (i.e., unnecessary days of therapy avoided based on accepted targets and benchmarks)

There are enormous data and studies demonstrating the positive outcomes of implementing AMS in the intensive care unit (ICU) and critical care settings. The effect of AMS in ICU and critical care settings on the prevalence of infections caused by antibiotic-resistant bacteria and *C. difficile* infection was assessed in a systemic review and meta-analysis by Baur D. et al. It was determined that there had been a 51% decrease in the frequency of infections and colonization with multidrug-resistant gram-negative bacteria, a 48% decrease in the frequency of extended-spectrum B-lactamase-producing gram-negative bacteria, a 37% decrease in the frequency of MRSA infections, and a 32% decrease in the frequency of *C. difficile* infections [32].

The CDC's 2019 Antibiotic Resistance Threat Report has shown a reduction in mortality from 18% in 2013 to 28% in 2019 by AMR. The most impressive results were the 41% decline in VRE (vancomycin-resistant *Enterococcus*), 33% in carbapenem-resistant *Acinetobacter*, 29% in multidrug-resistant *Pseudomonas aeruginosa*, 21% in MRSA, and 50% in ESBL Enterobacteriaceae [33].

The study by Tamma et al. highlights the association between antibiotic use and adverse events in hospitalized patients, indicating the importance of AMS in critical care settings to minimize such occurrences [34].

AMS should never be implemented with the primary intention of cost reduction. Cost reduction may be a byproduct of AMS. The cost reduction can be direct from the reduction in the total number of unnecessary prescriptions or indirect effects like decreased length of stay and readmission. In the systematic literature review by Huebner et al. in 2019, there was strong evidence of direct cost reduction, but the indirect reduction results were vague [35].

One of the barriers to implementing the AMS and physicians' fear of de-escalating antibiotics was an increase in mortality. A systematic review by Lindsay et al. in 2019 concluded that AMS implementation can help reduce antibiotic use without negatively affecting mortality [36].

The restrictive approach leads to a decrease in the use of broad-spectrum antibiotics; the impact of restrictive antibiotic policy had a positive outcome by decreasing the emergence of resistance and ESBL Enterobacteriaceae [37].

One study by Le Terrier et al. showed that 53% of critically ill patients did not receive antibiotics after implementation of the ASP [38].

Feedback and prospective audits are methodologies used for assessing any program and using the findings to refine the process and improve the outcome. Feedback to the AMS team, especially the physicians prescribing antibiotics in the ICU, comes from the prospective audits. These strategies have also been proven to reduce antibiotic use and prevent resistance. The corroborative evidence was confirmed by the meta-analysis by DiazGranados et al. [39].

13 Conclusion

ICUs have a huge consumption of antibiotics because of the critically ill population, because ICUs are the centers for the emergence of multidrug-resistant microorganisms. Sometimes, there is inappropriate use of antibiotics due to uncertainty of diagnosis and ease of arability of antibiotics in ICUs. To prevent the overuse and misuse of antibiotics in ICUs, antimicrobial stewardship should be implemented, and ICU physicians should lead these programs. Antibiotic stewardship consists of a multiprofessional approach. The strategy the ICUs should adopt is to start empirical antibiotics according to the hospital antibiogram, optimizing antibiotics by following the five Ds. The PK/PD of critically ill patients has been considered in the dosing of antibiotics. Other strategies include rapid identification of the organisms, narrowing the spectrum of antibiotics, switching to monotherapy, and shortening antibiotic administration duration. Infection control and prevention is the cornerstone of every ASP these measures should combine to get the optimal outcome. The future of antimicrobial stewardship in ICUs can be improved by the core competency of critical care practitioners through educational interventions like training programs and continuing medical education, system innovations like computerized decision support, process indicator evaluation, and prospective audit and feedback to healthcare workers.

References

1. Antimicrobial Resistance Collaborators. Global burden of bacterial antimicrobial resistance in 2019: a systematic analysis. Lancet. 2022;399(10325):629–55.
2. O'Neill J. Tackling drug-resistant infections globally: final report and recommendations. London: HM Government and Wellcome Trust; 2016. Review on Antimicrobial Resistance, chaired by Jim O'Neill.
3. Strich JR, Palmore TN. Preventing transmission of multidrug-resistant pathogens in the intensive care unit. Infect Dis Clin N Am. 2017;31(3):535–50. https://doi.org/10.1016/j.idc.2017.05.010. Epub 2017 Jul 5.
4. WHO. Antimicrobial stewardship programmes in health-care facilities in low- and middle-income countries. A practical toolkit. Geneva: World Health Organization; 2019.
5. Barlam TF, et al. Implementing an antibiotic stewardship program: guidelines by the Infectious Diseases Society of America and the Society for Healthcare Epidemiology of America. Clin Infect Dis. 2016;62(10):e51–77.
6. Public Health Agency of Canada. Tackling antimicrobial resistance and antimicrobial use: a Pan-Canadian framework for action. 2015. https://www.canada.ca/en/healthcanada/services/publications/drugs-healthproducts/tackling-antimicrobial-resistance-use-pancanadian-framework-action.
7. Gould IM. Stewardship of antibiotic use and resistance surveillance: the international scene. J Hosp Infect. 1999;43:S253–60.
8. Dellit TH, Owens RC, McGowan JE Jr, Gerding DN, Weinstein RA, Burke JP, Huskins WC, Paterson DL, Fishman NO, Carpenter CF, Brennan PJ, Billeter M, Hooton TM; Infectious diseases society of america; society for healthcare epidemiology of america. infectious diseases society of america and the society for healthcare epidemiology of america guidelines for developing an institutional program to enhance antimicrobial stewardship. Clin Infect Dis. 2007;44(2):159–77. https://doi.org/10.1086/510393. Epub 2006 Dec 13. PMID: 17173212.
9. Society for Healthcare Epidemiology of America, Infectious Diseases Society of America and Pediatric Infectious Diseases Society. Policy statement on antimicrobial stewardship by the Society for Healthcare Epidemiology of America (SHEA), the Infectious Diseases Society of America (IDSA), and the Pediatric Infectious Diseases Society (PIDS). Infect Control Hosp Epidemiol. 2012;33(4):322–7.
10. World Health Organization. WHO guidelines on use of medically important antimicrobials in food-producing animals. Geneva: WHO Press; 2017.
11. Rhodes A, Evans LE, Alhazzani W, Levy MM, Antonelli M, Ferrer R, Kumar A, Sevransky JE, Sprung CL, Nunnally ME, Rochwerg B, Rubenfeld GD, Angus DC, Annane D, Beale RJ, Bellinghan GJ, Bernard GR, Chiche JD, Coopersmith C, De Backer DP, French CJ, Fujishima S, Gerlach H, Hidalgo JL, Hollenberg SM, Jones AE, Karnad DR, Kleinpell RM, Koh Y,

Lisboa TC, Machado FR, Marini JJ, Marshall JC, Mazuski JE, McIntyre LA, McLean AS, Mehta S, Moreno RP, Myburgh J, Navalesi P, Nishida O, Osborn TM, Perner A, Plunkett CM, Ranieri M, Schorr CA, Seckel MA, Seymour CW, Shieh L, Shukri KA, Simpson SQ, Singer M, Thompson BT, Townsend SR, Van der Poll T, Vincent JL, Wiersinga WJ, Zimmerman JL, Dellinger RP. Surviving sepsis campaign: international guidelines for management of sepsis and septic shock: 2016. Intensive Care Med. 2017;43(3):304–77. https://doi.org/10.1007/s00134-017-4683-6. Epub 2017 Jan 18. PMID: 28101605.

12. Ture Z, Güner R, Alp E. Antimicrobial stewardship in the intensive care unit. J Intensive Med. 2022;3:244. https://doi.org/10.1016/j.jointm.2022.10.001.
13. Ohl CA, Luther VP. Antimicrobial stewardship for inpatient facilities. J Hosp Med. 2011;6(Suppl 1):S4–S15.
14. Doron S, Davidson LE. Antimicrobial stewardship. Mayo Clin Proc. 2011;86(11):1113–23.
15. Improving antibiotic use in hospitals: a cornerstone approach to prevention of antibiotic resistance. Centers for Disease Control and Prevention. https://arpsp.cdc.gov/story/improving-antibiotic-use-hospitals.
16. Barlam TF, Cosgrove SE, Abbo LM, MacDougall C, Schuetz AN, Septimus EJ, et al. Executive summary: implementing an antibiotic stewardship program: guidelines by the Infectious Diseases Society of America and the Society for Healthcare Epidemiology of America. Clin Infect Dis. 2016;62:1197–202.
17. Schouten J, De Waele J. Antimicrobial stewardship in ICU. London: Academic Press; 2017. https://doi.org/10.1016/B978-0-12-810477-4.00016-7.
18. Pickens CI, Wunderink RG. Principles and practice of antibiotic stewardship in the ICU. Chest. 2019;156(1):163–71. https://doi.org/10.1016/j.chest.2019.01.013.
19. De Corte T, Van Hoecke S, De Waele J. Artificial intelligence in infection management in the ICU. Crit Care. 2022;26:79. https://doi.org/10.1186/s13054-022-03916-2.
20. Campion M, Scully G. Antibiotic use in the intensive care unit: optimization and de-escalation. J Intensive Care Med. 2018;33:647. https://doi.org/10.1177/0885066618762747.
21. Appaneal HJ, Luther MK, Timbrook TT, LaPlante KL, Dosa DM. Facilitators and barriers to antibiotic stewardship: a qualitative study of pharmacists' perspectives. Hosp Pharm. 2018;54:250. https://doi.org/10.1177/0018578718781916.
22. Alghamdi S, Atef-Shebl N, Aslanpour Z, Berrou I. Barriers to implementing antimicrobial stewardship programmes in three Saudi hospitals: evidence from a qualitative study. J f Glob Antimicrob Resist. 2019;18:284–90. https://doi.org/10.1016/j.jgar.2019.01.031.
23. Rzewuska M, Duncan EM, Francis JJ, Morris AM, Suh KN, Davey PG, Grimshaw JM, Ramsay CR. Barriers and facilitators to implementation of antibiotic stewardship programmes in hospitals in developed countries: insights from transnational studies. Front Sociol, vol. 5; 2020. p. 41.
24. Bankar NJ, Ugemuge S, Ambad RS, Hawale DV, Timilsina DR. Implementation of antimicrobial stewardship in the healthcare setting. Cureus. 2022;14(7):e26664.
25. https://www.google.com/url?sa=i&rct=j&q=&esrc=s&source=web&cd=&cad=rja&uact=8&ved=0CAIQw7AJahcKEwioy5_C25OAAxUAAAAAHQAAAAAQAw&url=https%3A%2F%2Fwww.who.int%2Fpublications-detail-redirect%2FWHO-MHP-HPS-EML-2022.02&psig=AOvVaw0O0STs77-2IOip_hsAh9Hh&ust=1689613658793897&opi=89978449
26. Prkno A, Wacker C, Brunkhorst FM, Schlattmann P. Procalcitonin-guided therapy in intensive care unit patients with severe sepsis and septic shock: a systematic review and meta-analysis. Crit Care. 2013;17(6):R291.
27. Thom KA, Tamma PD, Harris AD, Dzintars K, Morgan DJ, Li S, Pineles L, Srinivasan A, Avdic E, Cosgrove SE. Impact of a prescriber-driven antibiotic timeout on antibiotic use in hospitalized patients. Clin Infect Dis. 2019;68(9):1581–4. https://doi.org/10.1093/cid/ciy852. PMID: 30517592.
28. Graber CJ, Jones MM, Glassman PA, Weir C, Butler J, Nechodom K, Kay CL, Furman AE, Tran TT, Foltz C, Pollack LA, Samore MH, Goetz MB. Taking an antibiotic timeout: utilization and usability of a self-stewardship timeout program for renewal of vancomycin

and piperacillin-tazobactam. Hosp Pharm. 2015;50(11):1011–24. https://doi.org/10.1310/hpj5011-1011.
29. Rimawi RH, Mazer MA, Siraj DS, Gooch M, Cook PP. Impact of regular collaboration between infectious diseases and critical care practitioners on antimicrobial utilization and patient outcome. Crit Care Med. 2013;41(9):2099–107.
30. Assi M, Abbas S, Nori P, Doll M, Godbout E, Bearman G, Stevens MP. Infection prevention and antimicrobial stewardship program collaboration during the COVID-19 pandemic: a window of opportunity. Curr Infect Dis Rep. 2021;23(10):15.
31. Evans RS, Pestotnik SL, Classen DC. A computer-assisted management program for antibiotics and other antiinfective agents. N Engl J Med. 1998;338(4):232–8.
32. Baur D, Gladstone BP, Burkert F, et al. Effect of antibiotic stewardship on the incidence of infection and colonization with antibiotic-resistant bacteria and Clostridium difficile infection: a systematic review and meta-analysis. JAMA Intern Med. 2016;176(11):1639–48. https://doi.org/10.1001/jamainternmed.2016.5664.
33. Kadri SS. Key takeaways from the U.S. CDC's 2019 antibiotic resistance threats report for frontline providers. Crit Care Med. 2020;48(7):939–45.
34. Tamma PD, Avdic E, Li DX, Dzintars K, Cosgrove SE. Association of adverse events with antibiotic use in hospitalized patients. JAMA Intern Med. 2017;177(9):1308–15.
35. Huebner C, Flessa S, Huebner NO. The economic impact of antimicrobial stewardship programmes in hospitals: a systematic literature review. J Hosp Infect. 2019;102(4):369–76.
36. Lindsay PJ, Rohailla S, Taggart LR, Lightfoot D, Havey T, Daneman N, et al. Antimicrobial stewardship and intensive care unit mortality: a systematic review. Clin Infect Dis. 2019;68:748–56.
37. Schuts EC, Boyd A, Muller AE, Mouton JW, Prins JM. The effect of antibiotic restriction programs on prevalence of antimicrobial resistance: a systematic review and meta-analysis. Open Forum Infect Dis. 2021;8:070.
38. Le Terrier C, Vinetti M, Bonjean P, Richard R, Jarrige B, Pons B, et al. Impact of a restrictive antibiotic policy on the acquisition of extended spectrum beta-lactamase-producing Enterobacteriaceae in an endemic region: a before-and-after, propensity-matched cohort study in a Caribbean intensive care unit. Crit Care. 2021;25:261.
39. DiazGranados CA. Prospective audit for antimicrobial stewardship in intensive care: impact on resistance and clinical outcomes. Am J Infect Control. 2012;40:526–9.

Perioperative Sepsis and Septic Shock: An Update

Umm E Amara, Umme Nashrah, Abdelrahman Balal, Firdos Ummunnisa, Arshad Hussain Chanda, and Nissar Shaikh

Abstract Sepsis and septic shock were described by Hippocrates and had high morbidity and mortality, till Surviving Sepsis Guidelines were published in 2002. Although implementation of these protocols is low, nevertheless it effectively decreased the morbidity and mortality. Daily millions of surgeries are performed all over the world, and the rate of perioperative sepsis and septic shock is increasing. The anesthetist, acute care physician, surgeons, and intensivist should be aware of sepsis and its updates. Sepsis can occur at any age, but extremes of age remain a high risk for developing sepsis in the perioperative period. Emergency and trauma surgeries are at higher risk of developing sepsis in the perioperative period. The surge in proinflammatory cytokines, mitochondrial dysfunction, and disseminated intravascular coagulation are the main components of sepsis pathophysiology. Diagnosis of perioperative sepsis can be made with raised septic markers (C-reactive protein and procalcitonin) and with advanced hemodynamic monitoring. Early diagnosis and proper management are key to improved patient outcomes. The Sepsis Six is a good tool for the initial management of sepsis patients in the ward in the postoperative period. Delays in antimicrobial administration in these patients increase mortality. Proper hydration, organ-supportive care, correction of coagulopathy, hemodynamic support, and renal replacement therapy may be required in these patients. In patients with perioperative sepsis and septic shock, early and bold debridement and source control will have a better patient outcome.

U. E Amara
Apollo Institute of Medical Sciences, Hyderabad, Telangana, India

U. Nashrah
Deccan College of Medical Sciences, Hyderabad, Telangana, India

A. Balal · A. H. Chanda · N. Shaikh (✉)
Surgical Intensive Care Unit, Hamad Medical Corporation, Doha, Qatar
e-mail: achanda@hamad.qa; smaheboob@hamad.qa

F. Ummunnisa
Dr. Halima Al Tamimi, Obstetrics and Gynaecology Centre, Doha, Qatar

N. Shaikh, A. H. Chanda (eds.), *Applied Microbiology in Intensive Care Medicine*, https://doi.org/10.1007/978-981-97-4006-2_4

Keywords Acute care · Advanced hemodynamic monitoring · Antibiotics · Multiorgan dysfunction · Perioperative · Sepsis · Septic shock

1 Introduction

Hippocrates used the term sepsis to describe the disease consequence "dangerous biological decay that could potentially occur in the body." Sepsis affects more than a million patients annually, with mortality ranging from 28% to 50%. For sure, anesthetists and intensivists can play a vital role in resuscitation and continuity of care to optimize the patient's condition, as 40% of cardiac arrests in the perioperative period were attributed to sepsis and 77% of these patients die which results in high mortality [1].

In 1914, Hugo Schottmuller described sepsis as the state of microbial invasion of the blood circulation, causing signs and symptoms.

2 Definition

Sepsis was defined as a suspected or diagnosed source of infection and SIRS (systemic inflammatory response syndrome) in sepsis-1 and sepsis-2 definitions. The definition of SIRS was defined and includes organ dysfunction from hypotension and raised blood sugar levels to a disturbed level of consciousness. According to the sepsis-3 definition, "Sepsis is defined as life-threatening organ dysfunction caused by a dysregulated host response to infection." After the sepsis-3 definition, the terms SIRS and severe sepsis are no longer used, and septic shock in the previous definitions was hypotension with sepsis, not responding to fluid resuscitation. According to sepsis-3, "Septic shock is defined as a subset of sepsis in which underlying circulatory and cellular metabolism abnormalities are profound enough to substantially increase mortality."

The first consensus definition of sepsis was in 1992. It was a practical definition of sepsis aimed at early diagnosis and management. In 2012, there was an update in the definition of sepsis, with the inclusion of more organ dysfunction and widening the area of sepsis and thus noted for the severe sepsis definition [2].

The higher incidence of sepsis and septic shock in the 1990s was that all specialties used to treat sepsis differently with different approaches. One of the milestones in the management of sepsis and septic shock was the introduction of sepsis management guidelines for the first time; it guided all the physicians and surgeons to follow a set of directions for management. These guidelines were agreed upon by most of the critical care societies in the world, and these guidelines were published to be adopted and implemented by 2002.

Several publications from the literature showed improved sepsis outcomes, and the other specialties, particularly cardiology (early 1990s), also gave insight for formulating guidelines for sepsis. They proposed that whoever is treating myocardial infarction initially should administer aspirin and a statin, which showed a significant reduction in mortality of acute myocardial infarction [3].

The initial definition in 1992 relies more on or depends on SIRS (systemic inflammatory syndrome), and as we know by now, SIRS criteria are not specific.

The second sepsis definition added more parameters to SIRS, including organ dysfunction, and hence, there was no need for a severe sepsis definition; therefore, it was removed, whereas the most recent sepsis definition of SIRS has been removed.

To define organ dysfunction, the severity scores have been introduced: SOFA (Sequential Organ Failure Assessment) score or qSOFA (Quick Sequential Organ Failure Assessment) score. SOFA score of 2 or more defined organ dysfunction and qSOFA score include systolic blood pressure <90 mmHg, GCS (Glasgow Coma Scale) less than 15, and tachypnea (respiratory rate) more than 22 breaths per minute and can also serve as a quick guide for organ dysfunction and patients at risk. Hypotension (mean arterial pressure less than 65 mmHg), which necessitates the use of vasopressors to maintain hemodynamics, and high serum lactate of more than 2 mmol/L are now considered signs of septic shock [4].

3 Epidemiology

An increasing number of patients undergo surgical interventions each day, and postoperative sepsis is reported to be 11.6 cases per 1000 elective surgeries. Thus, the rate of postoperative sepsis is also increasing [5]. Cardiothoracic surgeries seem to have a higher incidence of sepsis (8.39%). Vogel et al. described esophageal, pancreatic, small bowel, and biliary surgeries as at higher risk for developing postoperative sepsis. Preoperatively, the patient may already be in sepsis and require surgical intervention for the source control as early as possible.

4 Risk Factors

The following are the risk factors for perioperative sepsis and septic shock:

1. Extremes of age, due to patients' frailty, are at higher risk.
2. Patients with comorbidities and diabetes mellitus.
3. Planned gastric and small bowel surgeries, pancreatic, esophageal, and biliary surgeries.
4. Male, race, gender, ethnicity, and socioeconomic status.
5. Hospital size and location.
6. Emergency and trauma surgeries, perforated viscus or musculoskeletal surgeries, and compound injuries [5].

5 Pathophysiology

Sepsis causes excessive production of nitric oxide and other vasodilating substances, thus depriving the metabolism of the regulation of the vascular tone, causing profound vasoplegia. With sepsis, there is frequently a decrease in endogenous vasopressin levels. This combination leads to a decrease in systemic vascular resistance, causing hypotension and septic shock [6, 7]. This vasodilatation affects both arterial and venous systems. Thus, the venous system becomes a larger blood reservoir, and no fluid resuscitation will cause a prolonged decrease in the cardiac preload, contributing to hypotension. Hypovolemia is exacerbated by endothelial damage, particularly glycocalyx causing loss of vascular integrity which causes leak and loss of fluid into the extracellular compartment contributing to hypotension and tissue edema. A decrease in myocardial contractility due to toxins leads to a combined reduction of preload, a decrease in myocardial contractility, and a decrease in systemic vascular resistance, causing a significant reduction in oxygen delivery to the body tissues. Initial reports of mitochondrial dysfunction in sepsis were reported from animal studies in the mid-1990s [8]. The altered mitochondrial morphological change with increased permeability, combined with the inhibition of electron transport chain enzymes by nitric oxide and the other reactive oxygen species, hormonal influences decrease in the mitochondrial activity and down-regulating mitochondrial protein expression, thus reducing energy supply [8].

The combined effect of circulatory failure, generation of lactic acidosis, hypoxia, cellular injury, and energy deprivation will lead to multiorgan dysfunction, and if these above injuries continue and without any improvement, it will lead to multiple organ failure. Details of multiorgan dysfunctions are described in Fig. 1.

6 Management

As aforementioned, sepsis is a syndrome of clinical signs and symptoms characterized by inflammation due to an infection. It progresses into septic shock with increased morbidity and mortality if not diagnosed. Sepsis-6 was developed by the UK Sepsis Trust [9]. It is a practical tool that is mainly useful for the initial management of patients with sepsis outside intensive care units. Sepsis-6 implementation was found to have a 50% reduction in mortality if performed within 1 h of diagnosis or suspected sepsis. It includes three items the patient should receive and three items we should take to monitor (Fig. 2). The ones to give to patients are oxygen administration to keep oxygen saturation (SpO_2) of ~94% (88–92% in COPD and CO_2 retention patients). Secondly, giving IV fluids and not exceeding 30 mL/kg. The third item is to give broad-spectrum antibiotics according to the hospital antibiogram. However, it is essential to take blood cultures before administering antibiotics. In the 3 take step, the first is to take blood for serum lactate, the second step is to monitor the urine output hourly and to keep a strict fluid balance, and the third

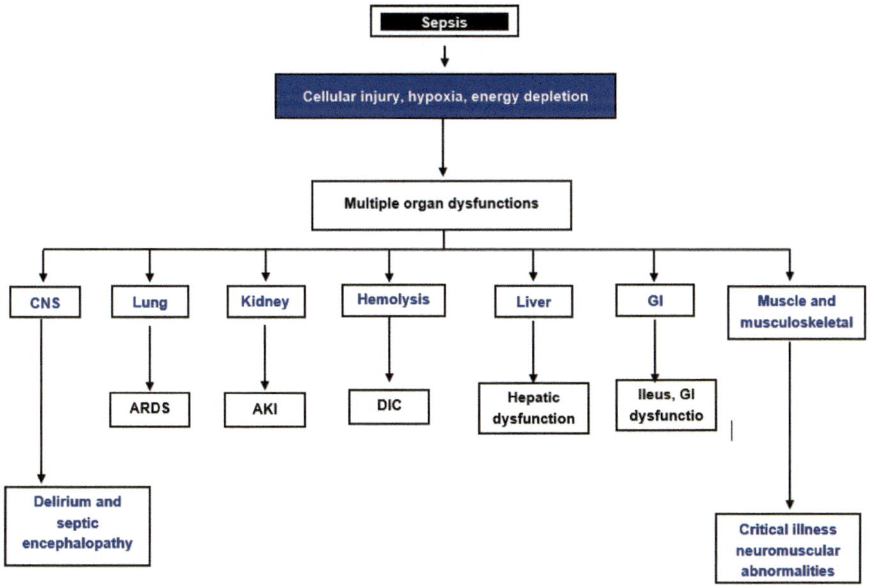

Fig. 1 Sepsis causing multiorgan dysfunction

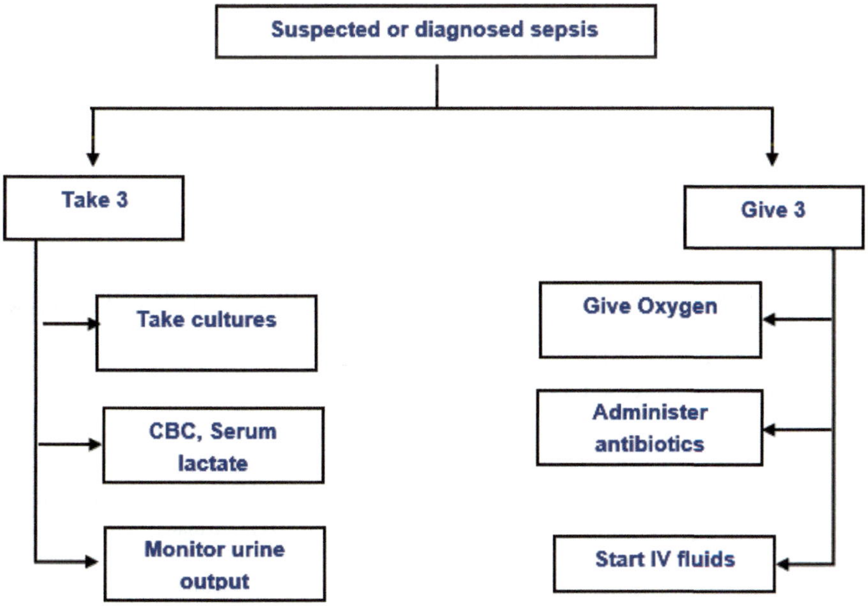

Fig. 2 The sepsis-6

and final step is to take blood cultures before administering antibiotics within 1 h. Call the critical care team if the patient is still hypotensive, has decreased conscious level or urine output, and is tachypneic or tachycardic and serum lactate is not trending downward or trending upward. These patients need intensive care therapy (Fig. 2: sepsis-6).

According to the Surviving Sepsis Campaign (SSC), there is no need for 3-h bundle implementations; a 1-h bundle is enough (Table 1).

6.1 Antimicrobial Administrations and Source Control

An essential component of the sepsis outcome is early and proper antibiotic administration. Inappropriate antibiotics increase mortality five-fold, and the initial 6-h delay in starting appropriate antimicrobial agents increases mortality by 7% per hour [10].

It is essential to take cultures before administering antibiotics to improve the yield of the cultures, narrow the spectrum once results are available, and prevent the development of resistance. Early source control is important to decrease bacterial load and for patient improvements. Soft tissue infections and acute abdomen with peritonitis are common diseases that require immediate surgical interventions. Early source control in these conditions reduces mortality significantly. Initial source control can be damage control, bedside, or by an interventional radiologist. Once the patient is stable, then they can be subjected to definitive intervention [11].

6.2 Hemodynamic Monitoring and Resuscitation

All these patients should have advanced hemodynamic monitoring with PiCCO®, LiDCO®, transesophageal Doppler, or Flow Trac targeting the optimization of preload with dynamic parameters. Large amounts of fluids should be avoided as it may overstretch myocytes and worsen pulmonary and interstitial edema leading to organ dysfunctions and increased mortality [12].

Table 1 SSC h-1 bundle (time 0 is the time of recognition of sepsis)

Measure serum lactate, and repeat if it is more than 2 mmol/L
Take cultures and start broad-spectrum antibiotics
Start rapid infusion of 30 mL/kg of crystalloid IV fluids
Start vasopressors if hypotension persists or keep MAP >65 mmHg

6.3 Vasopressors and Inotropes

In persistent hypotension (mean arterial pressure <65 mmHg) in spite of volume resuscitation, early induction of vasopressors reduces the incidence of pulmonary edema and arrhythmias [13]. It is reported that there is a 5.3% increase in mortality with each hour delay in starting vasopressors in patients with septic shock [14]. The vasopressor choice is noradrenaline due to its obvious potent vasoconstriction property. Dopamine is no longer used due to supraventricular arrhythmias and prolactin's immunosuppressive effects. The second vasopressor of choice is vasopressin, a potent vasopressor with action on vasopressin receptors and as a replacement therapy (relative deficiency of vasopressin in septic shock) [11].

It is suggested to start early vasopressin when the dose of noradrenaline reaches between 0.2 and 0.3 μg/kg/min. Adrenaline use was associated with metabolic side effects such as hyperglycemia and lactatemia. Angiotensin II is an upcoming vasopressor in refractory, a high cardiac output shock. It had the extra benefit of better renal system outcomes. The major concern with angiotensin II is the higher incidence of thromboembolism [11].

6.4 Perioperative Management

It is essential that patients should go for surgical intervention even though they are hemodynamically unstable and with high oxygen requirements. Carefully control the induction and maintenance of anesthesia to minimize the hemodynamic effects on the heart. It is recommended to use titrated doses of induction agents.

Etomidate and ketamine are the induction agents of choice that would maintain hemodynamic stability. One has to be careful as ketamine was also found to have an immunosuppressive effect on the neutrophils and cytokines in animal models [15].

Propofol has also shown to cause tumor necrosis factor inhibition including oxidative bursts and neutrophils. Etomidate is shown to cause adrenal suppression and is associated with increased mortality, and replacement therapy can be used after administering etomidate [11].

Analgesics are an essential component of anesthesia, and all opioids except remifentanil have hepatic metabolism and renal excretion. Hence, titrated doses of remifentanil are a better choice for these patients. The neuromuscular blocking agent (NMBA) is a particular concern. Aminosteroid NMBAs, particularly vecuronium and pancuronium, are metabolized in the liver and excreted in the urine, hence not suitable in patients with septic shock and organ dysfunction. The benzylisoquinoline neuromuscular blocking agents (atracurium and cisatracurium) are a drug of choice as they are metabolized by the Hoffman metabolism and by the liver. Succinylcholine is not recommended in these patients due to arrhythmias, which can be life-threatening.

6.5 Blood Transfusion

Source control surgical intervention may need blood and blood product transfusion; however, transfusion is not free from adverse effects such as transfusion-related lung injury or volume overload. Hence, transfusion should be restricted to packed RBC transfusion if hemoglobin is less than 7 g/dL and no cardiac comorbidities, hypoxia, or acute bleeding. Blood products should be directed by ROTEM or TEG and no platelet transfusion if it is above 20,000.

6.6 Glycemic Control

Hyperglycemia causes endothelial dysfunction and mitochondrial oxygen imbalance exacerbating infection and sepsis. Initially, in mid-2000, tight glycemic control was followed but found to have increased incidence of hypoglycemia, morbidity, and mortality. Currently, a liberal glycemic control is followed with target blood glucose of less than 180 g/dL [16].

6.7 Vitamins and Steroids

Vitamins were found to restore endothelial injury integrity, thus having beneficial effect in sepsis. Thiamine maintains the mitochondrial functions and improved lactic acid clearance. Combining these vitamins with hydrocortisone was found to decrease duration of shock and mortality.

7 Organ Support Therapy

7.1 Encephalopathy

The septic encephalopathy occurs due to neurotoxic ischemic inflammatory changes in septic patients manifested by delirium to coma and occurs in 30–70% of sepsis patients. Controlling sepsis is the primary treatment with daily spontaneous awakening trial (SAT) and spontaneous breathing trial (SBT).

7.2 *Lung Injury*

One of the frequent organ dysfunctions is lung injury. Intraoperative lung protective ventilation was found to reduce absolute mortality. Hence, in septic patients, perioperative lung protective ventilation and intermittent recruitment maneuvers in patients without lung pathology are recommended.

7.3 *Hepatic Dysfunction*

Hepatic dysfunction is seen in 34–46% of septic patients. It is 1.3 times more common in patients with chronic hepatic dysfunction or pathology. Hypoglycemia and hepatic dysfunction/failure with no lactate clearance are ominous signs. Care should be used when choosing medications to prevent hepatotoxicity. Avoiding prolonged hypotension and monitoring and managing coagulopathy are the main therapeutic approaches for these patients.

7.4 *Acute Kidney Injury*

Acute kidney injury (AKI) occurs in more than 50% of septic shock patients. Correction of hyperglycemia and use of steroids were found to decrease AKI. Avoiding sodium bicarbonate if the pH is more than 7.15 is advisable.

8 Conclusion

Sepsis and septic shock remained as potential life-threatening clinical entities in the perioperative period. Either the patient has sepsis when they first arrive and needs source control or they get it after surgery. Apart from emergency surgeries, planned gastric and small bowel surgeries and pancreatic, esophageal, and biliary surgeries are at higher risk for sepsis and septic shock. Pathophysiology of sepsis is over body reaction with proinflammatory cytokines surge, mitochondrial and multiple organ dysfunction. Sepsis and septic shock are diagnosed with the help of raised septic markers and advanced hemodynamic monitoring showing distributive shock. Initial resuscitation, earlier antibiotic administration, source control, and organ-supportive care are essential for improved patient outcomes.

References

1. Ben-Jacob TK, Sreedharan R, Nunnally ME, Chang BPM. Perioperative management of the septic patient. ASA Monitor. 2017;81:18–20.
2. Dellinger RP, et al. Surviving Sepsis Campaign: international guidelines for management of severe sepsis and septic shock. Crit Care Med. 2013;41:580–637.
3. Slade E, Tamber PS, Vincent J-L. The Surviving Sepsis Campaign: raising awareness to reduce mortality. Crit Care. 2003;7:1.
4. Singer M, et al. The third international consensus definitions for sepsis and septic shock (Sepsis-3). JAMA. 2016;315:801.
5. Vogel TR, Dombrovskiy VY, Carson JL, Graham AM, Lowry SF. Postoperative sepsis in the United States. Ann Surg. 2010;252:1065–71.
6. Landry DW, et al. Vasopressin deficiency contributes to the vasodilation of septic shock. Circulation. 1997;95:1122–5.
7. Kumar A. An alternate pathophysiologic paradigm of sepsis and septic shock: implications for optimizing antimicrobial therapy. Virulence. 2014;5:80–97.
8. Durand A, et al. Involvement of mitochondrial disorders in septic cardiomyopathy. Oxidative Med Cell Longev. 2017;2017:1–13.
9. Daniels R, Nutbeam T, McNamara G, Galvin C. The sepsis six and the severe sepsis resuscitation bundle: a prospective observational cohort study. Emerg Med J. 2011;28:507–12.
10. Kumar A, et al. Duration of hypotension before initiation of effective antimicrobial therapy is the critical determinant of survival in human septic shock*. Crit Care Med. 2006;34:1589–96.
11. Bughrara N, Diaz-Gomez JL, Pustavoitau A. Perioperative management of patients with sepsis and septic shock. Part II. Anesthesiol Clin. 2020;38:123–34.
12. Levy MM, et al. Mortality changes associated with mandated public reporting for sepsis. The results of the New York State Initiative. Am J Respir Crit Care Med. 2018;198:1406–12.
13. Bai X, et al. Early versus delayed administration of norepinephrine in patients with septic shock. Crit Care. 2014;18:532.
14. Varpula M, Tallgren M, Saukkonen K, Voipio-Pulkki L-M, Pettilä V. Hemodynamic variables related to outcome in septic shock. Intensive Care Med. 2005;31:1066–71.
15. Melamed R, Bar-Yosef S, Shakhar G, Shakhar K, Ben-Eliyahu S. Suppression of natural killer cell activity and promotion of tumor metastasis by ketamine, thiopental, and halothane, but not by propofol: mediating mechanisms and prophylactic measures. Anesth Analg. 2003;97(5):1331–9. https://doi.org/10.1213/01.ANE.0000082995.44040.07.
16. Griesdale DEG, et al. Intensive insulin therapy and mortality among critically ill patients: a meta-analysis including NICE-SUGAR study data. Can Med Assoc J. 2009;180:821–7.

Urosepsis and Septic Shock: A Simple Infection Progressing to Complex One

Umm E Amara, Umme Nashrah, Abdulrahman Balal, Nissar Shaikh, Arshad Hussain Chanda, and Khalid Mohammed Ahmed Al-Jalham

Abstract The systemic inflammatory response (SIRS) in the urinary tract, which includes the male genital organ (prostate), is known as urosepsis. One-fourth of all cases of sepsis are due to urosepsis. Sepsis is primarily caused by a microbial infection, although the host ultimately plays a major role in the development of the illness. The morbidity and mortality are high if it is not identified and treated promptly. All acute care physicians, anesthesiologists, intensivists, and urologists should be aware of this potentially fatal condition. The risk factors for urosepsis are impaired voiding, congenital obstruction of urinary flow, acquired obstruction, urinary tract interventions, metabolic disorder, and immunocompromised state. In community-acquired urosepsis, there are more UTIs with ESBL *E. coli*; 15% of urinary tract infections are caused by gram-positive bacteria, followed by *Proteus* species 15%, enterobacteria and *Klebsiella* 15%, and *Pseudomonas aeruginosa* in 5% and 15% of cases. The "CHAOS" theory is an excellent explanation of the interaction between the host's pro- and anti-inflammatory responses as well as the many stages or states of sepsis patients. Patients with urosepsis typically have a fever, widespread weakness, and flank pain. The patient will be tachycardic and tachypneic and could have borderline or low blood pressure. Leukocytosis and elevated sepsis indicators (CRP, procalcitonin) will be seen in a blood count. Emergency genitourinary tract ultrasonography allows for the quick identification of urosepsis. The common cause of urosepsis, blockages to the urinary system and urine flow, will be found with USG. As soon as urosepsis is suspected, start measuring urine output; take a blood

U. E Amara
Apollo Institute of Medical Sciences and Research, Hyderabad, Telangana, India

U. Nashrah
Deccan College of Medical Sciences, Hyderabad, Telangana, India

A. Balal · N. Shaikh (✉) · A. H. Chanda
Surgical Intensive Care Unit, Hamad Medical Corporation, Doha, Qatar
e-mail: smaheboob@hamad.qa

K. M. A. Al-Jalham
Medical and Executive Director, Ambulatory Care Centre, Hamad Medical Corporation, Doha, Qatar

N. Shaikh, A. H. Chanda (eds.), *Applied Microbiology in Intensive Care Medicine*, https://doi.org/10.1007/978-981-97-4006-2_5

and urine culture, CBC, and serum lactate; and start broad-spectrum antibiotics within an hour. Start giving oxygen and intravenous fluids, following the local hospital antibiogram.

For a better outcome, the source of urosepsis or septic shock must be controlled quickly. Percutaneous nephrostomy (PCN) and drainage of pus or infected urine are the two options. It typically has a high success rate and is performed with local anesthetic and USG guidance.

Keywords Urosepsis · Sepsis · Shock · Gram-negative bacteria · Antibiotics · Flank pain · Sepsis markers · Source control

1 Introduction

Urosepsis is the systematic inflammatory response (SIRS) caused by an infection in the urinary tract, including the male genital organ (prostate). Urosepsis accounts for ¼ of all sepsis cases. If urosepsis is not detected and managed early, it can quickly complicate into hypotension and hypoperfusion of the vital organs causing septic shock with higher morbidity and mortality [1].

2 Epidemiology

More recently, sepsis and urosepsis have been increasingly reported. A bulk of sepsis is caused by urosepsis. There is always precipitation or a risk factor for urosepsis. Around 78% of urosepsis occurs due to impaired urine flow due to obstruction. Patients may also develop urosepsis after instrumentation or intervention in the urinary tract. Ninety-two percent of these urosepsis patients need surgical intervention [2]. More recently, Shaikh et al. also described that most patients (60.9%) had obstructive uropathy, causing urosepsis. Traditionally described that females are at higher risk for urosepsis and septic shock, Shaikh et al. described equal distribution of urosepsis in both males and females [3].

3 Risk Factors

In patients with metabolic disease (diabetes mellitus), the more common risk is structural or functional abnormalities obstructing the urinary flow (Table 1).

Table 1 Risk factors for urosepsis and septic shock

Mechanism	Causes
Impaired voiding	I. Vesico-ureteral reflux II. Cystocele III. Neurogenic bladder
Congenital obstruction of urinary flow	I. Strictures II. Phimosis III. Ureterocele IV. Polycystic kidney
Acquired obstruction	I. Calculi II. Benign enlargement of prostate (BEP) III. Urinary tract tumors IV. Pregnancy V. Traumatic injury
Urinary tract interventions	I. Urinary catheters II. Ureteric stent III. Nephrostomy tube IV. Urological procedure
Metabolic disorder and immunocompromised state	I. Diabetes mellitus II. Azotemia III. Nephrocalcinosis

4 Classification of Urosepsis

According to the origin of urosepsis, it is classified into community-acquired or hospital-acquired urosepsis. Community-acquired urosepsis, also commonly called urosepsis, occurs when patients usually from the community present to the emergency department with signs and symptoms of sepsis, which is more frequent in females and has better outcomes, whereas hospital-acquired urosepsis occurs when patients are admitted to the hospital for other medical conditions and commonly get urosepsis due to urinary catheterization. This type of sepsis has significantly higher mortality [4].

5 Microbiology

The microbiology in community-acquired urosepsis (CAU) is gram-negative bacteria contributing 85% and gram-positive organisms contributing only 15%, whereas in hospital-acquired or catheter-associated urinary tract infections (CAUTIs), 66% of these infections are caused by gram-negative bacteria, and 21% are caused by gram-negative organisms, with increased frequency of ESBL (extended-spectrum B-lactamase) and multi-drug-resistant organisms in both gram-positive and gram-negative bacteria [4].

In our study of community-acquired urosepsis, the number of UTIs with *E. coli* (ESBL) is higher [3]. Other authors reported that 50% of urinary tract infections are caused by *E. coli*, *Proteus* species 15%, enterobacteria and *Klebsiella* 15%, and *Pseudomonas aeruginosa* in 5% and 15% caused by gram-positive bacteria 15% [5, 6].

6 Pathophysiology

Pan-European prevalence (PEP) and the pan-Euro-Asian prevalence (PEAP) found that 75 of these patients were found to have bacteremia in the hospital-acquired urosepsis [6], whereas the study showed that 21% of the total mixed community- and hospital-acquired urosepsis patients had bacteremia [7]. In our prospective urosepsis study, the rate of bacteremia in the community-acquired urosepsis was 71.5% [3].

6.1 Pathophysiology of Sepsis

The theory of "CHAOS" was initially proposed by Bone et al. (Fig. 1) and excellently describes the interaction between pro- and anti-inflammatory responses of the host with elucidated different phases or states of patients in sepsis [8, 9].

The underlying causative factor for the process of sepsis is the microbial pathogen, but eventually, the host is the one that drives the disease.

The bacterial cell wall ingredients or bacteria itself bind to the cellular co-receptor or receptor (CD14, Toll-like receptor TLR, TLR2 and 4, CD18, and selectin) to the macrophages, neutrophils, urothelial cells, and endothelial cells. These stimulate nuclear factor-B and protein kinase C, and these intracellular messengers, in turn, induce the cytokines IL1, IL6, IL8, TNF, and PAF production incorporating chemokines, prostaglandins, thromboxane, leukotrienes, and nitric oxide. The cytokine cell generates cooperative or antagonist action [10].

The initial activation of cellular immune responses to bacteria or bacterial products in the urogenital tract causes severe proinflammatory responses. The hemopoietic growth factor stimulates the formation of new neutrophils and the release of stored neutrophils. These neutrophils produce the bactericidal protease and oxygen radicals. B and T lymphocytes stimulate the antibody and cellular immune reaction, and the body's muscular proteins are utilized to generate the autoantibodies. The endothelial cell PAR and nitric oxygen production are triggered, leading to decreased vascular tone and hypotension. Again, these distorted endothelial cells result in loss of endothelial integrity, leading to capillary leakage and causing edema.

In most patients, they survive this proinflammatory course, and in a few patients, this proinflammatory phase continues into the counter-regulatory anti-inflammatory response syndrome, leading to an immunosuppression stage. In this stage, the macrophages and neutrophils undergo apoptosis, whereas the lymphocytes and dendrocytes undergo increased apoptosis. These increasing numbers of dysfunctional

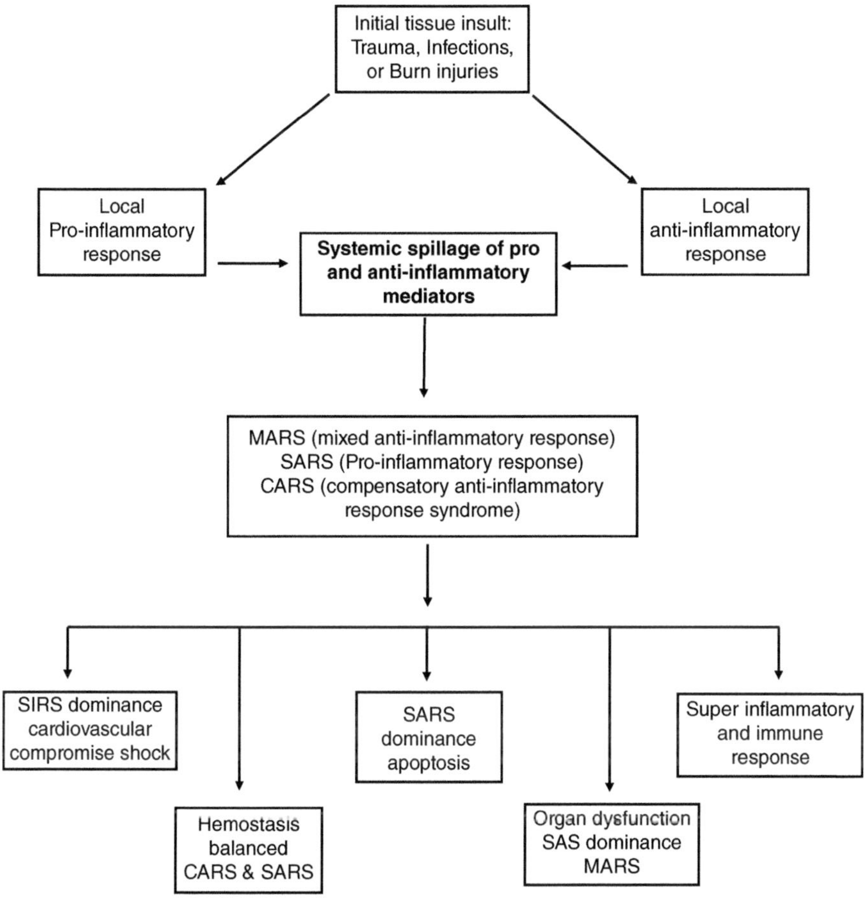

Fig. 1 Chaos theory in sepsis and septic shock

macrophages and neutrophil causes bystander damage to those cells, thus causing organ dysfunction.

In sepsis, apart from inflammation, the other systems involved are coagulation and autonomic and endocrine systems.

The complement system in sepsis is hyperactivated; this causes abnormalities of coagulation and fibrinolysis. The surface receptors on neutrophils and endothelial cells are upgraded and upregulated to increase mutual adhesions. Additionally, the plasminogen derivative inhibitor substances increased, and the protein pathway was inhibited, thus activating the coagulation system and complicating disseminated intravascular coagulation and thrombosis [11].

In autonomic dysfunction in sepsis patients, the sympathetic system modulates and innervates the lymphoid organs. The noradrenaline stimulates alpha-2 ($\alpha 2$) on macrophages, thus stimulating TNF alpha generation and contact with the B-receptor stimulation and inhibiting the release of proinflammatory cytokines. The proinflammatory mediators stimulate the hypothalamus and hypothalamic-pituitary-adrenal axis, thus inducing the release of cytotrophin and adrenocorticotrophic hormones, consequently causing a release of anti-inflammatory response by increasing the anti-inflammatory cytokines IL4–IL10 and supporting nuclear factor-κB (NF-κB) [12].

The parasympathetic system, through the vagal nerve and acetylcholine receptor, has the responsibility of sensing inflammation and informing the specific centers of the brain.

The release of acetylcholine reduces the proinflammatory cytokines, which are called the cholinergic anti-inflammatory pathway [13].

7 Diagnosis

7.1 Clinical Presentation

Urosepsis patients present with fever, generalized weakness, and flank pain. The patient will be tachypneic and tachycardic and may have borderline blood pressure or hypotension. A blood count will show leukocytosis and increased sepsis markers (CRP, procalcitonin). Most patients have urinary symptoms, from burning micturition to dysuria and sometimes urinary retention. If the patient is a stone-former, they may have a history of urosepsis.

Acute prostatitis can be ruled out by digital rectal examination if the prostate is not tender. In testicular palpation, testicular tenderness gives suspicion of epididymo-orchitis.

7.2 Imaging Studies

The rapid diagnosis of urosepsis is made by emergency ultrasonography of the genitourinary tract. USG will detect obstructions to the urinary tract and urine flow, which is the common etiology for urosepsis.

USG also detects urinary retention, prostatic enlargement, and abscess. In cases where the diagnosis is not confirmed with USG, CT is recommended. One must be careful of using contrast in patients with acute kidney injury due to urosepsis (USG showing renal calculi (Fig. 2)).

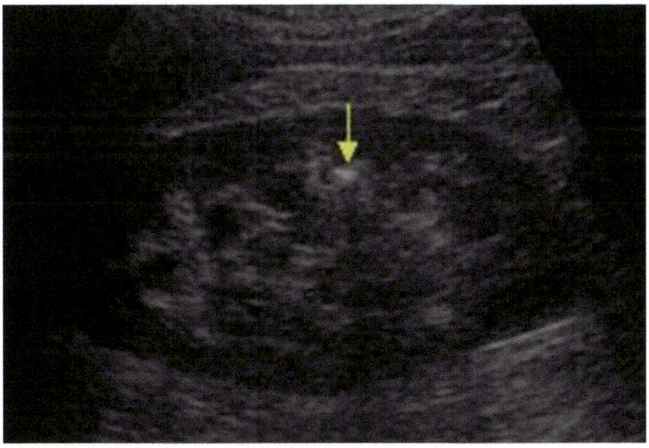

Fig. 2 Renal USG showing calculus

8 Management

As soon as we suspect urosepsis, within 1 h, take blood and urine culture, CBC, and serum lactate, start measuring urine output, give oxygen and intravenous fluids 30 m/kg bolus, and start broad-spectrum antibiotics or according to the local hospital antibiogram.

If the patient presents with septic shock, give oxygen or intubate, depending on respiratory status, and if fluid resuscitation is needed to build up the central venous pressure (CVP ~12 mmHg).

There should be advanced hemodynamic monitoring, and if MAP (mean arterial pressure) is lower than 65 mmHg, start vasopressor, preferably noradrenaline, early as it is a potent α agonist.

Giving early antibiotics is essential as each delay in antibiotic administration causes an increase in mortality by 10%. Admit these patients to intensive care therapy early to have appropriate resuscitation. If MAP is not maintained with 0.3 μg/kg/min of noradrenaline, they need to start vasopressin and hydrocortisone. Once culture results are released, we can select the appropriate antibiotics, thus decreasing the chances of antibiotic resistance [11].

8.1 Source Control

The early control of the source of urosepsis or septic shock is essential for a better outcome. There are two ways: percutaneous nephrostomy (PCN) and drainage of pus or infected urine. It is usually done under USG guidance and local anesthesia, with a high success rate.

The second method of source control and relief of urinary flow obstruction is by minimally invasive endoscopic insertion of stents. It is more used in patients with coagulopathy and hemodynamic instability, but there is no superiority of either treatment approach: PCN versus endoscopic stent [14].

In our prospective study, by source control within 24 h of presentation, the mortality was 2.6%, much lower than septic shock due to other etiologies [3]. Qiang et al. also reported a mortality in urosepsis of 6% [15].

9 Conclusion

Urosepsis is the medical term for the systemic inflammatory response (SIRS) in the urinary tract, which includes the genital organs including the prostate. Urosepsis is the cause of one-fourth of all sepsis cases. Although the host eventually has a significant impact on the progression of the illness, a microbial infection is the main cause of sepsis. If it is not discovered and treated promptly, there is a substantial risk of morbidity and mortality.

Impaired voiding, congenital obstruction of urine flow, acquired obstruction, urinary tract interventions, metabolic condition, and immunocompromised status are risk factors for urosepsis.

Typical symptoms of urosepsis are triad of fever, leukocytosis, and flank pain along with increased sepsis markers like CRP and procalcitonin. Using emergency genitourinary tract ultrasonography, urosepsis can be quickly diagnosed.

On suspicion of urosepsis follow the sepsis-6 bundle with specific stress on giving antibiotics within 1 h. All clinicians who treat acute patients should be aware of this potentially lethal syndrome.

References

1. Wagenlehner FME, Tandogdu Z, Bjerklund Johansen TE. An update on classification and management of urosepsis. Curr Opin Urol. 2017;27:133–7.
2. Hofmann W. [Urosepsis and uroseptic shock]. Z Urol Nephrol. 1990;83:317–24.
3. Shaikh N, Momin U, Atef Shible A, Al-Musalmani M, Ansari A. Community acquired urosepsis: a surgical intensive care experience. Qatar Med J. 2020;2020:8.
4. Urosepsis: flow is life. IntechOpen. https://www.intechopen.com/chapters/64483.
5. Wagenlehner FME, Weidner W, Naber KG. Pharmacokinetic characteristics of antimicrobials and optimal treatment of urosepsis. Clin Pharmacokinet. 2007;46:291–305.
6. Johansen TEB, et al. Hospital acquired urinary tract infections in urology departments: pathogens, susceptibility and use of antibiotics. Data from the PEP and PEAP-studies. Int J Antimicrob Agents. 2006;28(Suppl 1):S91–107.
7. Cardoso T, Ribeiro O, Costa-Pereira A, Carneiro A, A SACiUCI Study Group. Community-acquired and healthcare-related urosepsis: a multicenter prospective study. Crit Care. 2008;12:P8.
8. Bone RC, et al. Definitions for sepsis and organ failure and guidelines for the use of innovative therapies in sepsis. Chest. 1992;101:1644–55.
9. Wagenlehner FM, et al. Diagnosis and management for urosepsis: items in urosepsis. Int J Urol. 2013;20:963. https://doi.org/10.1111/iju.12200.

10. Russell JA. Management of Sepsis. N Engl J Med. 2006;355:1699–713.
11. Dellinger RP, et al. Surviving Sepsis Campaign guidelines for management of severe sepsis and septic shock. Crit Care Med. 2004;32:858–73.
12. John C, Buckingham J. Cytokines: regulation of the hypothalamo–pituitary–adrenocortical axis. Curr Opin Pharmacol. 2003;3:78–84.
13. Borovikova LV, et al. Vagus nerve stimulation attenuates the systemic inflammatory response to endotoxin. Nature. 2000;405:458–62.
14. Ramsey S, et al. Evidence-based drainage of infected hydronephrosis secondary to ureteric calculi. J Endourol. 2010;24:185–9.
15. Qiang X-H, Yu T-O, Li Y-N, Zhou L-X. Prognosis risk of urosepsis in critical care medicine: a prospective observational study. Biomed Res Int. 2016;2016:1–5.

Community-Acquired Pneumonia: Recent Advances

Jaweria Akram, Sayed Tarique Kazi, and Nissar Shaikh

Abstract An acute infection of the pulmonary parenchyma caused by microorganisms present in the community is known as community-acquired pneumonia (CAP). These infections are very dissimilar from the group of pathogens that can be contracted from people who are immunocompromised or who are hospitalized (nosocomial pneumonia). The pathogens that cause CAP can be classified as bacterial, viral, or fungal. We have listed all of the most prevalent creatures, and the list is thorough. In endemic regions, CAP can also be brought on by fungi, despite the fact that bacteria or viruses are the primary culprits. Despite significant improvements in detection methods, infections are only found in 38% of patients, 23% of which are composed primarily of one or more viruses, followed by 11% of bacteria and 3% of mixed bacteria. The infectious process of pneumonia begins with the entry of pathogenic organisms into the lung parenchyma. The choice of antibiotics for therapy depends on the patient's attitude, the severity of the disease, comorbidities, and risk factors for microorganisms with a high propensity for aggression. When selecting an antibiotic, local resistance rates should be taken into account.

Keywords Antibiotics · Bacteria · Community-acquired pneumonia · Fungus · Virus

1 Introduction

Community-acquired pneumonia (CAP) is an acute infection of pulmonary parenchyma caused by pathogens found in the community. These pathogens are distinctly different from the group of pathogens acquired in hospital settings (nosocomial

J. Akram · S. T. Kazi (✉)
Medical Intensive Care Unit, Hamad Medical Corporation, Doha, Qatar
e-mail: SKazi@hamad.qa

N. Shaikh
Surgical Intensive Care Unit, Hamad Medical Corporation, Doha, Qatar
e-mail: smaheboob@hamad.qa

N. Shaikh, A. H. Chanda (eds.), *Applied Microbiology in Intensive Care Medicine*, https://doi.org/10.1007/978-981-97-4006-2_6

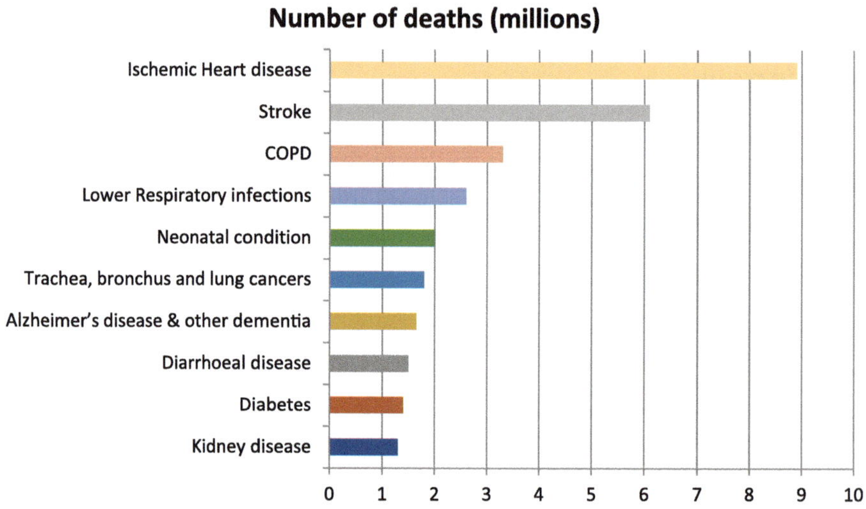

Fig. 1 Modified from WHO mortality data 2019 (https://www.who.int/news-room/fact-sheets/detail/the-top-10-causes-of-death)

pneumonia) or in immunocompromised patients. The term healthcare-associated pneumonia has now been abandoned in the new guidelines as it led to inappropriate use of antibiotics. CAP is a heterogeneous entity caused by varied pathogens and host responses. Hence, it can manifest itself from mild to severe life-threatening diseases and even death. It is one of the most common causes of visits to hospitals and inpatient admissions. In the USA, CAP is seen in over 4.5 million outpatient and emergency department visits [1]. Although the incidence of mortality has been decreasing over the last decade, it remains the fourth highest cause of mortality worldwide [2]. It still remains the leading cause of infectious disease–related deaths [3]. The introduction of pneumococcal conjugate vaccine (13 valent) and *Haemophilus* influenza type b (Hib) into national vaccination schedules is partly responsible for the reduction in rates and mortality of CAP. In this review, we will present the key differences between guidelines produced by the Infectious Diseases Society/American Thoracic Society (IDSA/ATS) with the European (Fig. 1).

2 Organism Responsible

The pathogens responsible for CAP can be grouped under bacterial, viral, or fungi. The list is exhaustive, and we have enumerated the most common organisms. Although CAP is usually caused by bacteria or viruses, it can also be caused by fungi in endemic areas. Despite major advances in detection techniques, pathogens are only detected in only 38% of patients. The majority of these include one or more viruses 23% followed by bacteria 11% and mixed in 3%. The most common pathogens are rhinovirus followed by influenza and *Streptococcus pneumoniae* [4] (Table 1).

Table 1 Microbiology of community-acquired pneumonia (CAP)

Bacteria	
Typical	*Streptococcus pneumoniae*
	Staphylococcus aureus
	Haemophilus influenzae
	Group A streptococci
	Moraxella catarrhalis
	Gram-negative *Kleibsella* and *E. coli*
Atypical	*Mycoplasma pneumoniae*
	Legionella spp.
	Chlamydia pneumoniae
	Coxiella burnetii
Virus	
	Influenza A and B viruses (various subtypes including H1N1)
	Parainfluenza viruses
	Rhinoviruses
	Coronaviruses (various subtypes including novel and SARS-CoV-2)
	Adenoviruses
	Respiratory syncytial virus
Fungi	
(Endemic areas)	*Histoplasma capsulatum*
	Blastomyces dermatitidis
	Coccidioides posadasii (valley fever)

3 Predisposing Factors

1. **Older Age**: The incidence of pneumonia steadily increases with age with the highest incidence in age above 85 years. Advanced age is not only associated with a higher incidence of CAP but also with more severe disease, greater need for hospitalization, and higher mortality [5].
2. **Smoking and Alcohol**: There is consistent evidence that smoking was associated with a higher incidence of pneumonia than non-smokers. Alcohol consumption of more than 41 g/day or a history of abuse has also been associated with increased risk [6].
3. **Comorbid Conditions**: Several comorbid conditions predispose patients to a higher risk of CAP. This includes obesity, diabetes, heart failure, stroke, and COPD. Among all the causes, the highest incidence is found in COPD patients [7].
4. **Environmental Factors**: Crowded living conditions, contact with young children at home, and pollutants in the environment also contribute to higher incidence. Then, there are specific factors like exposure to birds which can transmit *Chlamydia psittaci*. In old buildings, stagnant water systems or air conditioning can transmit *Legionella*.

4 Pathogenesis

The invasion of pathogenic organisms to lung parenchyma marks the beginning of the infectious process of pneumonia. The upper respiratory tract is known to contain diverse and multiple groups of organisms. In contrast, the lower respiratory tract including lung parenchyma was inherently considered to be sterile [8]. This assumption was based on culture-dependent techniques which have not changed since its discovery by Louis Pasteur in 1860. Now, with the advent of novel culture-independent, we know that the lungs contain diverse communities of microbes which form the lung microbiota [9]. These organisms are mainly anaerobes and similar to those found in the oral cavity, e.g., *Prevotella* and *Veillonella*. They cannot be cultured using the traditional culture techniques hence possibly the low rate of culture detection of 30% in pneumonia patients. There is a constant balance between these communities of microbes and the host response. If there is any disruption to this environment due to any of the above predisposing factors, pathogenic organisms are able to proliferate and cause pneumonia. The usual modes of transmission through which the patient acquires exogenous pathogens are [10]:

1. Inhalation: This is the usual mode of transmission for pathogens including bacteria and viruses. A particle size of 5 μm or less can transport up to 100 microorganisms bypassing the respiratory host defenses to reach the alveoli.
2. Aspiration: Upper airways and nasopharynx usually carry pathogenic organisms. The rate of nasal carriage of *S. pneumoniae* in healthy adults can vary from 5% to 30%. There is constant microaspiration of these organisms to the lower respiratory tract. But the normal lung defense mechanisms are able to protect against any invasion into the lungs. It is in conditions in which there is impaired cough reflex or mucociliary clearance that the organisms are able to cause infection.
3. Haematogenous: This mode of transmission is rare.
4. Direct spread from surrounding structures like lymph nodes.

5 Patterns of Involvement

1. Lobar Pneumonia: Classic pneumonia involving the entire lobe of the lung. The pathological process evolves in stages
   - Congestion (days 1–3), increased perfusion
   - Red hepatization (days 3–4), red and inflamed
   - Grey hepatization (days 5–7), grey liver–like appearance
   - Resolution (1–4 weeks), gradual aeration of the affected segment
2. Bronchopneumonia: It involves the bronchioles and adjacent alveoli. Patchy distribution involves the lower and middle lobes. It can lead to complications like necrotizing bronchopneumonia or pneumatocele formation if caused by virulent

organisms like *Staphylococcus*. This bronchopneumonia can be fatal as it causes tension pneumothorax due to rupture of the pneumatocele [11].

3. Interstitial Pneumonia: Usually involving the interstitial spaces seen in viral infections. It leads to patchy involvement of interstitial spaces.

6 Presentation

Diagnosis of pneumonia can be difficult in a busy emergency department as most symptoms are nonspecific. The most common presentation is fever, dyspnoea, and productive cough. It can sometimes be associated with pleuritic chest pain or a non-productive cough. Other generalized features such as malaise, fatigue, and chills are nonspecific. In severe cases, patients may present with hypoxia, confusion, or even loss of consciousness. Examination reveals rales and bronchial breathing over the area of consolidation. If the patient develops parapneumonic pleural effusion, it will be accompanied by loss of breath sounds and dullness to percussion on the affected area. As the diagnostic accuracy of signs and symptoms, American Thoracic Society and Infectious Diseases Society of America guidelines have recommended having radiographic evidence on chest XR in the definition of CAP.

7 Differential Diagnosis

There are other conditions which can mimic pneumonia (Table 2).

Table 2 Showing differential diagnosis of the CAP

Cardiorespiratory	Pulmonary edema
	Pulmonary embolism
	Pulmonary hemorrhage
Immune-mediated	Vasculitic diffuse alveolar hemorrhage
	Wegener's granulomatosis
	Cryptogenic organizing pneumonia
	Acute interstitial pneumonia
	Sarcoidosis
	Pulmonary alveolar proteinosis
	Systemic lupus erythematosus
	Polymyositis and dermatomyositis
	Acute and chronic eosinophilic pneumonia
Neoplastic	Lung cancer
	Endobronchial metastases
	Lymphoma
Others	Radiation pneumonitis

8 Investigations

The investigations carried out can be either for the establishment of diagnosis or to assess the severity of the disease, and they include the following:

8.1 Imaging

Chest radiographs are usually obtained for establishing the diagnosis of CAP, especially for all patients requiring hospital admission.

In very selected patients, further imaging with CT chest may be considered such as in immunocompromised patients, as in these patients, due to poor immune response, the chest radiograph may be negative.

The use of point-of-care lung ultrasound in the assessment of CAP has been demonstrated in many observational studies. The accuracy of these studies varies significantly with some studies demonstrating superiority in sensitivity of point-of-care lung ultrasound when compared to chest X-ray in the diagnosis of CAP [12–14].

8.2 Microbiology

Microbiology is not required in all cases of CAP. The severity of the disease, disposition of the patients, comorbidities, and epidemiological factors play a role in the decision to obtain microbiology. This may include obtaining respiratory Gram stain and culture, blood cultures, testing for *Legionella*, pneumococcal urine antigen, *Chlamydophila* antigen and/or PCR, mycoplasma pneumonia PCR, and testing for influenza.

8.3 Other Investigations

Other additional tests can be performed to aid the establishment of diagnosis or assess the severity of the disease, which may include

- Arterial blood gases.
- Urea and electrolytes for severity assessment.
- C-reactive protein as a baseline and to help the diagnosis.
- Full blood count.
- Liver function tests.
- Host transcriptional signatures in peripheral blood can be used as a diagnostic tool to help differentiate between bacterial and viral pneumonia, thus avoiding unnecessary antimicrobial use [15]. Details of investigations were done in CAP patients (Table 3).

Table 3 Showing details of investigations done on CAP patients

Investigations	American Thoracic Society and Infectious Diseases Society of America, 2019 [16]	British Thoracic Society guidelines, 2009 [17]	ERS guidelines, 2011 [18]
Chest X-ray	The guidelines were based on studies that used radiographic criteria for defining CAP along with clinical signs and symptoms	All patients admitted to the hospital should have chest X-rays For patients treated as outpatients, imaging is only required if there is any concern regarding the diagnosis, if there is a suboptimal response to treatment, or if the patient is at high risk for lung pathology such as cancer	In patients treated as outpatients, chest X-rays should only be considered when the diagnosis is in doubt All patients admitted to the hospital should have chest X-rays
CT chest	Not mentioned	There is no routine role of CT scans while investigating for CAP	Not mentioned
US chest	Not mentioned	Not mentioned	Not mentioned
1. Microbiology			
Purulent sputum examination	Should not be performed in patients with mild pneumonia who are treated as outpatients For patients treated as inpatients, it can be obtained in patients with severe CAP or in those who are being empirically treated for MRSA or *P. aeruginosa*, have a history of MRSA or *P. aeruginosa* infection, or were hospitalized and received parenteral antibiotics, whether during the hospitalization event or not, in the last 90 days	Should not be performed in patients treated as outpatients, except if they fail to respond to treatment To be performed in patients with moderate CAP without prior antibiotic use and those with severe CAP	Should not be performed in patients with mild pneumonia who are treated as outpatients In patients treated as inpatients, gram stain and culture should be performed when can be obtained and processed in a timely manner
Bronchoalveolar lavage	Not mentioned	Can be considered in patients with high-severity CAP	It is preferred in case of non-resolving pneumonia

(continued)

Table 3 (continued)

Investigations	American Thoracic Society and Infectious Diseases Society of America, 2019 [16]	British Thoracic Society guidelines, 2009 [17]	ERS guidelines, 2011 [18]
Bronchoscopic sampling of the lower respiratory tract	Not mentioned	Can be considered in patients with high-severity CAP	Can be considered in intubated patients and selected non-intubated patients, if gas exchange status allows
Transthoracic needle aspiration		Can be considered in patients with severe CAP if the facility is skilled in this technique	Consider only for severely ill patients with a focal infiltrate in whom less invasive measures were not diagnostic
Blood cultures	Recommended in patients with severe CAP or in those who are being empirically treated for MRSA or *P. aeruginosa*, were previously infected with MRSA or *P. aeruginosa*, or were hospitalized and received parenteral antibiotics, whether during the hospitalization event or not, in the last 90 days	Recommended for all patients with moderate and high-severity CAP	Should be obtained in all patients with CAP who require hospitalization
Testing for *Legionella*	Recommended in adults with CAP with epidemiological factors, such as association with a *Legionella* outbreak, recent travel, or in adults with severe CAP It can be performed by testing *Legionella* urinary antigen and by sending for lower respiratory tract secretions or *Legionella* nucleic acid amplification testing	Recommended for all patients with high-severity CAP, for other patients with specific risk factors, and for all patients with CAP during outbreaks *Legionella* urinary antigen with culture of respiratory secretions should be sent	Recommend to perform urinary antigen detection for patients admitted to the hospital due to disease severity or when there is clinical or epidemiological suspicion

(continued)

Table 3 (continued)

Investigations	American Thoracic Society and Infectious Diseases Society of America, 2019 [16]	British Thoracic Society guidelines, 2009 [17]	ERS guidelines, 2011 [18]
Pneumococcal urine antigen tests	Recommended for patients with severe CAP	Recommended for all patients with moderate or high severity	Recommend performing urinary antigen in patients admitted to the hospital due to the severity of illness Pneumococcal antigen from pleural fluid sample should also be obtained in case thoracocentesis is performed in the presence of parapneumonic effusion
Chlamydophila antigen and/or PCR detection tests	Not mentioned	Recommended for patients with high-severity CAP or if there is a strong suspicion of psittacosis *Chlamydophila* antigen and/or PCR detection tests should be obtained	Recommended only to test for the diagnosis of atypical pneumonia Testing with a combination of IgM Antibody and PCR are considered appropriate approaches
Testing for influenza	Recommended during times of high influenza activity Rapid influenza molecular assay is preferred over a rapid influenza diagnostic test	Recommend paired serology tests for patients with high-severity CAP with no other microbiological diagnosis and who fail to improve and/or when there are epidemiological risk factors	Recommend considering performing molecular tests during the winter season
Testing for SARS-CoV-2	Not mentioned	Not mentioned	Not mentioned
Test for TB	Not mentioned	Recommend testing of sputum for Mycobacterium tuberculosis in patients with a persistent productive cough, along with malaise, weight loss or night sweats, or risk factors for tuberculosis	Not mentioned

(continued)

Table 3 (continued)

Investigations	American Thoracic Society and Infectious Diseases Society of America, 2019 [16]	British Thoracic Society guidelines, 2009 [17]	ERS guidelines, 2011 [18]
Arterial blood gases	Required for assessment for ICU admission	Recommended for all patients requiring admission	Required for assessment for ICU admission
Urea and electrolytes	Not mentioned	Recommended for all patients requiring admission	Not mentioned
C-reactive protein	Not mentioned	Recommended for all patients requiring admission	Level of CRP can be used to make the diagnosis of CAP. CRP level >100 mg/L makes pneumonia likely, and CRP <20 mg/L makes CAP less likely
Full blood count	Not mentioned	Recommended for all patients requiring admission	Not mentioned
Liver function tests	Not mentioned	Recommended for all patients requiring admission	Not mentioned
Procalcitonin	Recommend against using initial serum procalcitonin to affect the decision of empiric antibiotic therapy as this test alone cannot differentiate between viral and bacterial pathogens in the case of CAP	Not mentioned	Insufficient studies to make a recommendation

9 Patient Disposition

Clinical judgment, along with validated clinical prediction scores such as the Pneumonia Severity Index (PSI) and the CURB-65, can be used to decide patient disposition. PSI is preferred over CURB-65, as it identifies low risk better and is a better predictor of mortality [19].

9.1 Ambulatory Care

Patients with PSI scores of I to II and CURB-65 scores of 0 can be considered for outpatient treatment.

9.2 Hospital Admission

Hospital admission is warranted for patients who are hypoxemic on room air or have substantial decline from baseline or with signs or symptoms of sepsis or quickly immediately progressive disease or are suspected to have infection with aggressive pathogens. In addition to this, patients who are unable to take medications orally, have cognitive or functional disability or social issues impairing compliance, or are unable to seek medical help in case of clinical worsening should also be admitted to the hospital.

Along with the clinical judgment, patients with PSI scores of ≥III and CURB-65 scores ≥1 should also be hospitalized.

9.3 Intensive Care Unit (ICU) Admission

ICU admission is warranted for patients with septic shock or respiratory failure necessitating mechanical ventilation.

Other than the above-mentioned patients, clinical judgment along with three or more minor criteria from IDSA/ATS 2007 criteria can be used (Table 4).

A detailed comparison of different societal guidelines for CAP patient disposition is given in Table 5.

Table 4 Showing major and minor criteria for diagnosis of severe CAP

Minor criteria	Major criteria
Respiratory rate ≥30 breaths/min	Septic shock with need for vasopressors
PaO_2/FiO_2 ratio ≤250	Respiratory failure requiring mechanical ventilation
Multilobar infiltrates	
Confusion/disorientation	
Uremia (blood urea nitrogen level ≥20 mg/dL)	
Leukopenia[a] (white blood cell count <4000 cells/μL)	
Thrombocytopenia (platelet count <100,000/μL)	
Hypothermia (core temperature <36 °C)	
Hypotension requiring aggressive fluid resuscitation	

[a] Due to infection alone (i.e., not chemotherapy-induced). 2007 Infectious Diseases Society of America/American Thoracic Society criteria for defining severe community-acquired pneumonia [20]

Table 5 Comparison between the American Thoracic Society and Infectious Diseases Society of America (2019), British Thoracic Society guidelines (2009), and ERS guidelines (2011) for the patient disposition of patients with CAP

	American Thoracic Society and Infectious Diseases Society of America, 2019 [15]	British Thoracic Society guidelines, 2009 [16]	ERS guidelines, 2011 [17]
Ambulatory care	Patients with PSI scores of I to II and CURB-65 scores of 0 can be considered for outpatient treatment	Patients who have a CURB-65 score of 0 do not normally require hospitalization	Patients who have a CURB-65 score of 0 do not normally require hospitalization
Hospital admission	Recommend admitting patients to the hospital as per clinical judgment, along with patients with PSI scores of ≥III and CURB-65 scores ≥1	Patients who have a CURB-65 score of 1 or 2 should be referred to the hospital Patients who have a CURB-65 score of 3 or more need urgent hospital admission	Patients who have a CURB-65 score of 1 or more should be considered for hospitalization
Intensive care unit (ICU) admission	Recommend direct admission to an ICU for patients with hypotension requiring vasopressors or respiratory failure requiring mechanical ventilation Also, recommend admitting patients to ICU using the IDSA/ATS 2007 minor severity criteria together with clinical judgment	Recommend considering admitting patients to ICU with CURB-65 scores of 4 and 5	Recommend to admit patients to ICU with signs and symptoms reflecting acute respiratory failure, severe sepsis, or septic shock and radiographic extension of infiltrates, as well as severely decompensated comorbidities

10 Treatment

The choice of antibiotics for the treatment depends upon patient disposition, severity of the disease, comorbidities, and risk factors for aggressive pathogens. Local resistance rates should be considered when choosing a particular antibiotic. Details regarding the choice of antibiotics are mentioned in Table 4.

10.1 Duration of Antibiotics

For patients who are improving clinically, antibiotic therapy can be stopped after a minimum of a total of 5–7 days [21]. For patients with severe pneumonia with suspected or confirmed aggressive pathogens such as MRSA or gram-negative pathogens or with complications, the duration can be extended to 14–21 days depending on clinical response. In Table 6, details of CAP management according to different societies are mentioned.

Further difference in the treatment of CAP patients is mentioned in Table 7.

Table 6 Treatment of CAP from different societies

Treatment	American Thoracic Society and Infectious Diseases Society of America, 2019 [16]	British Thoracic Society guidelines, 2009 [17]	ERS guidelines, 2011 [18]
– Outpatient	See below management depends on the presence or absence of comorbidities and risk factors	Monotherapy with amoxicillin Doxycycline or clarithromycin is appropriate as an alternative	Recommend monotherapy with amoxicillin or tetracycline Macrolides are good alternatives in countries with low pneumococcal macrolide resistance In case of clinically relevant bacterial resistance rates against all first-choice agents, treatment with levofloxacin or moxifloxacin may be considered
No comorbidities or risk factors for MRSA or *Pseudomonas aeruginosa* (Risk factors are prior respiratory isolation of MRSA or *P. aeruginosa* or recent hospitalization and received parenteral antibiotics in the last 90 days)	Monotherapy with amoxicillin, doxycycline, or macrolide given that local pneumococcal resistance is <25%		
With comorbidities (Chronic heart, lung, liver, or renal disease, diabetes mellitus, alcoholism malignancy, or asplenia)	Combination therapy with amoxicillin/clavulanate or cephalosporin and macrolide or doxycycline Monotherapy with respiratory fluoroquinolone		
– Inpatient treatment			

(continued)

Table 6 (continued)

Treatment	American Thoracic Society and Infectious Diseases Society of America, 2019 [16]	British Thoracic Society guidelines, 2009 [17]	ERS guidelines, 2011 [18]
Non-severe CAP	**Standard regimen**: β-Lactam + macrolide or respiratory fluroquinolone **Prior respiratory isolation of MRSA** MRSA coverage should be added and get cultures/nasal PCR to guide de-escalation or continuation of the therapy **Prior respiratory isolation of *Pseudomonas aeruginosa***: *P. aeruginosa* coverage should be added and obtain cultures to allow de-escalation or confirmation of the need for continued therapy **Recent hospitalization and parenteral antibiotics and locally validated risk factors for MRSA** Recommend adding MRSA coverage only once cultures or rapid nasal PCR come back positive **Recent hospitalization and parenteral antibiotics and locally validated risk factors for *P. aeruginosa*** Recommend adding *P. aeruginosa* coverage only once cultures or rapid nasal PCR comes back positive	**Low-severity CAP:** In patients with low-severity CAP who require hospital admission for other reasons, such as unstable co-morbidities, oral amoxicillin is preferred If oral therapy is contraindicated, parenteral options are intravenous amoxicillin, benzylpenicillin, or clarithromycin **Moderate CAP:** Combination therapy with oral therapy with amoxicillin and a macrolide is recommended for patients requiring hospital admission If oral therapy is contraindicated, recommended parenteral choices include intravenous amoxicillin or benzylpenicillin, together with clarithromycin If the patient does not tolerate penicillins or macrolides, oral doxycycline is the recommended alternative Oral levofloxacin and oral moxifloxacin are other alternative choices When oral therapy is contraindicated in those intolerants of penicillins, recommended parenteral choices include levofloxacin monotherapy or a second- or third-generation cephalosporin together with clarithromycin	**Non-severe CAP that requires hospitalization without risk factors for *P. aeruginosa* or MRSA** – Combination therapy: Beta-lactam plus—macrolide – Monotherapy Non-antipseudomonal cephalosporin, respiratory fluoroquinolone, or ertapenem **CAP that requires hospitalization with risk factors for MRSA** – Combination therapy Teicoplanin ± rifampin – Monotherapy vancomycin, linezolid, or clindamycin

Treatment	American Thoracic Society and Infectious Diseases Society of America, 2019 [16]	British Thoracic Society guidelines, 2009 [17]	ERS guidelines, 2011 [18]
Severe CAP	**Standard regimen**: β-Lactam + macrolide or β-lactam + fluroquinolone **Prior respiratory isolation of MRSA** Add MRSA coverage and obtain cultures/ nasal PCR to allow de-escalation or continuation of the therapy **Prior respiratory isolation of *Pseudomonas aeruginosa***: Add *P. aeruginosa* coverage and obtain cultures to allow de-escalation or continuation of the therapy **Recent hospitalization and parenteral antibiotics and locally validated risk factors for MRSA:** Add MRSA coverage and obtain nasal PCR and cultures to allow de-escalation or confirmation of the therapy **Recent hospitalization and parenteral antibiotics and locally validated risk factors for *P. aeruginosa*** Add coverage for *P. aeruginosa* and obtain cultures to allow de-escalation or confirmation of the therapy	**Severe CAP** Intravenous combination of a broad-spectrum b-lactamase stable antibiotic with a macrolide is recommended If patients are allergic to penicillin, second- or third-generation cephalosporin can be used, together with clarithromycin	**Severe CAP that requires hospitalization without risk factors for *P. aeruginosa* or MRSA** – Combination therapy Non-antipseudomonal cephalosporins plus macrolide – Monotherapy Respiratory fluoroquinolone **Severe CAP that requires hospitalization with risk factors for *P. aeruginosa*** – Combination therapy Antipseudomonal cephalosporin or acyl ureidopenicillin/beta-lactamase inhibitor or carbapenem plus Ciprofloxacin or macrolide + aminoglycoside

Table 7 Further management of CAP as per different societies

	American Thoracic Society and Infectious Diseases Society of America, 2019	British Thoracic Society guidelines, 2009	ERS guidelines, 2011 [17]
Change from parenteral to oral antibiotics	Not mentioned	Recommend changing patients from parenteral to oral antibiotics as soon as clinical improvement occurs with normalization of temperature, given that there is no contraindication to the oral route	Recommend to guide this decision according to the resolution of the most prominent clinical features at admission
Use of corticosteroids	Use of corticosteroids in adults with no severe CAP and severe influenza pneumonia is not recommended Recommend steroids for patients with CAP and refractory septic shock as per the Surviving Sepsis Campaign recommendations	Routine use of steroids is not recommended in the treatment of high-severity CAP	Steroids are not recommended in the treatment of pneumonia
Anti-influenza treatment	Recommend starting anti-influenza treatment in patients with CAP who test positive for influenza, independent of duration of illness before diagnosis	Recommend anti-influenza treatment for patients with CAP who test positive for influenza, independent of duration of illness before diagnosis	Recommend empirical anti-viral therapy in only high-risk patients who have typical influenza symptoms
Complications of CAP	Not mentioned	– For patients with parapneumonic effusion, early thoracocentesis should be performed – Patients with empyema or pleural fluid with pH < 7.2 should have early and effective pleural fluid drainage – In the presence of lung abscess, unusual respiratory pathogens including anaerobes, *S. aureus*, Gram-negative enteric bacilli, and *S. milleri* should be considered. This needs a lengthy course of antibiotic therapy of up to 6 weeks depending on clinical response and may also need surgical intervention	– Thoracente sis should be performed in hospitalized patients with CAP when a significant pleural effusion is present

Duration of antibiotics	For patients who are improving clinically with antibiotic treatment for CAP, the antibiotic therapy can be stopped after a total of 5 days	Recommend 7 days of appropriate antibiotics for patients treated as outpatients or those with mild to moderate CAP requiring inpatient admission Recommend 7–10 days of therapy in those with high-severity microbiologically undefined pneumonia The duration may be extended to 14 or 21 days according to clinical response	Recommend duration of treatment to up to a maximum of 8 days in a responding patient Biomarkers, particularly PCT, may guide shorter treatment duration
Vaccination	Not mentioned	Pneumococcal and influenza vaccination should be performed as per the guidelines	Pneumococcal and influenza vaccination should be performed as per the guidelines
Smoking cessation	Not mentioned	Smoking cessation advice should be recommended to all patients who are active smokers according to smoking cessation guidelines issued by the health education authority	Not mentioned

Table 8 Comparison between the American Thoracic Society and Infectious Diseases Society of America (2019), British Thoracic Society guidelines (2009), and ERS guidelines (2011) for the follow-up of CAP

	American Thoracic Society and Infectious Diseases Society of America, 2019 [16]	British Thoracic Society guidelines, 2009 [17]	ERS guidelines, 2011 [18]
Follow-up	In adults with CAP whose symptoms have resolved within 5–7 days, no follow-up chest imaging is required	Follow-up clinical review should be planned for all patients around 6 weeks Chest radiographs should be performed after 6 weeks for patients who have persistent symptoms or signs or who are at high risk of malignancy, such as smokers and those aged 50 years	Not mentioned

11 Follow-Up

Patients with CAP should have a follow-up after almost 6 weeks, and those with persistent symptoms or signs or at higher risk of malignancy should have follow-up chest imaging at the same time (Table 8).

12 Conclusions

CAP remains a significant cause of morbidity, mortality, and health costs globally, despite improvements in its management over the last decade. There has been a significant increase in the incidence of CAP in Europe over the past decade, probably because of the increasing age of the population and the greater number of comorbidities. It is hoped that advances in molecular diagnostic tests will help us detect the causative pathogens in pneumonia, especially antibiotic-resistant pathogens. Preventive strategies for CAP, including the surveillance of trends in microbial etiology, may help improve vaccination policies in this specific population.

References

1. National Ambulatory Medical Care Survey (NAMCS) and National Hospital Ambulatory Medical Care Survey (NHAMCS) 2009–2010. https://www.cdc.gov/nchs/data/ahcd/combined_tables/2009-2010_combined_web_table01.pdf.
2. Xu J, Murphy SL, Kochanek KD, Bastian BA. Deaths: final data for 2013. Natl Vital Stat Rep. 2016;64:1.
3. WHO mortality data 2019. https://www.who.int/news-room/fact-sheets/detail/the-top-10-causes-of-death.

4. Jain S, Self WH, Wunderink RG, et al. Community-acquired pneumonia requiring hospitalization among U.S. adults. N Engl J Med. 2015;373(5):415–27. https://doi.org/10.1056/NEJMoa1500245. Epub 2015 Jul 14.
5. Ramirez JA, Wiemken TL, Peyrani P, et al. Adults hospitalized with pneumonia in the United States: incidence, epidemiology, and mortality. Clin Infect Dis. 2017;65(11):1806–12. https://doi.org/10.1093/cid/cix647.
6. Teepe J, Grigoryan L, Verheij TJ. Determinants of community-acquired pneumonia in children and young adults in primary care. Eur Respir J. 2010;35:1113–7.
7. Almirall J, Bolibar I, Serra-Prat M, et al. New evidence of risk factors for community-acquired pneumonia: a population-based study. Eur Respir J. 2008;31:1274–84.
8. Dickson RP, Erb-Downward JR, Huffnagle GB, et al. Towards an ecology of the lung: new conceptual models of pulmonary microbiology and pneumonia pathogenesis. Lancet Respir Med. 2014;2(3):238–46. https://doi.org/10.1016/S2213-2600(14)70028-1.
9. Dickson RP, Erb-Downward JR, Huffnagle GB. The role of the bacterial microbiome in lung disease. Expert Rev Respir Med. 2013;7:245–57. [PubMed: 23734647].
10. Woodhead M, Blasi F, Ewig S, Garau J, Huchon G, Ieven M, et al. Guidelines for the management of adult lower respiratory tract infections—full version. Clin Microbiol Infect. 2011;17(Suppl 6):E1–59. https://doi.org/10.1111/j.1469-0691.2011.03672.x.
11. Maheboob SN, Louon AM, Ahmed AH. Necrotizing fasciitis complicated by multiple pneumoceles. Saudi Med J. 2003;24(12):1394–6.
12. Liu XL, Lian R, Tao YK, et al. Lung ultrasonography: an effective way to diagnose community-acquired pneumonia. Emerg Med J. 2015;32:433–8.
13. Cortellaro F, Colombo S, Coen D, et al. Lung ultrasound is an accurate diagnostic tool for the diagnosis of pneumonia in the emergency department. Emerg Med J. 2012;29:19–23.
14. Amatya Y, Rupp J, Russell FM, et al. Diagnostic use of lung ultrasound compared to chest radiograph for suspected pneumonia in a resource-limited setting. Int J Emerg Med. 2018;11:8.
15. Siljan WW, Sivakumaran D, Ritz C, Jenum S, Ottenhoff TH, Ulvestad E, Holter JC, Heggelund L, Grewal HM. Host transcriptional signatures predict etiology in community-acquired pneumonia: potential antibiotic stewardship tools. Biomark Insights. 2022;17:11772719221099130. https://doi.org/10.1177/11772719221099130. PMID: 35693251; PMCID: PMC9174553.
16. Metlay JP, et al. Diagnosis and treatment of adults with community-acquired pneumonia. An official clinical practice guideline of the American Thoracic Society and Infectious Diseases Society of America. Am J Respir Crit Care Med. 2019;200(7):e45–67. https://www.atsjournals.org/doi/full/10.1164/rccm.201908-1581ST.
17. Lim WS, Baudouin SV, George RC, et al. BTS guidelines for the management of community acquired pneumonia in adults: update 2009. Thorax. 2009;64:iii1–55.
18. Woodhead M, Blasi F, Ewig S, Garau J, Huchon G, Ieven M, Ortqvist A, Schaberg T, Torres A, van der Heijden G, Read R, Verheij TJ, Joint Taskforce of the European Respiratory Society and European Society for Clinical Microbiology and Infectious Diseases. Guidelines for the management of adult lower respiratory tract infections—full version. Clin Microbiol Infect. 2011;17 Suppl 6(Suppl 6):E1–59. https://doi.org/10.1111/j.1469-0691.2011.03672.x. PMID: 21951385; PMCID: PMC7128977.
19. Aujesky D, Auble TE, Yealy DM, Stone RA, Obrosky DS, Meehan TP, et al. Prospective comparison of three validated prediction rules for prognosis in community-acquired pneumonia. Am J Med. 2005;118:384–92.
20. Mandell LA, Wunderink RG, Anzueto A, Bartlett JG, Campbell GD, Dean NC, Dowell SF, File TM Jr, Musher DM, Niederman MS, Torres A, Whitney CG. Infectious Diseases Society of America/American Thoracic Society consensus guidelines on the management of community-acquired pneumonia in adults. Clin Infect Dis. 2007;44(Suppl_2):S27–72. https://doi.org/10.1086/511159.
21. Tansarli GS, Mylonakis E. Systematic review and meta-analysis of the efficacy of short-course antibiotic treatments for community-acquired pneumonia in adults. Antimicrob Agents Chemother. 2018;62:e00635-1.

Septic Abdomen: An Update on Its Microbiology and Management

Abdelrahman Balal, Umm E Amara, Umme Nashrah, Syed Muhammad Ali, Maliha Thapur, and Nissar Shaikh

Abstract Gastrointestinal infections present complex challenges requiring meticulous understanding and innovative strategies for effective diagnosis and management. This comprehensive review amalgamates insights from four distinct yet interconnected topics such as peritonitis, acute cholangitis, acute calcular cholecystitis, and pyogenic liver abscess to offer a holistic perspective on evolving paradigms in gastrointestinal infection management.

Peritonitis, arising from peritoneal inflammation, requires a timely multi-tiered approach. Primary peritonitis necessitates timely antimicrobial intervention, while secondary peritonitis stemming from gastrointestinal disruptions necessitates surgical intervention, resuscitation, and targeted antimicrobial treatment. Tertiary peritonitis demands a comprehensive multidisciplinary approach, encompassing intensive care, surgical source control, and drainage.

Acute cholangitis, a consequence of biliary obstruction, mandates prompt diagnosis, antibiotic therapy, and biliary drainage. Acute calcular cholecystitis involves gallstone-related inflammation, with laparoscopic cholecystectomy as the gold standard. Balancing antibiotic therapy with timely intervention optimizes outcomes. For pyogenic liver abscess, early diagnosis is challenging yet imperative. Diverse patho-

A. Balal (✉) · N. Shaikh
Surgical Intensive Care Unit, Hamad Medical Corporation, Doha, Qatar
e-mail: ABalal@hamad.qa

U. E Amara
Apollo Medical College, Apollo Institute of Medical Sciences, Hyderabad, Telangana, India

U. Nashrah
Deccan College of Medical Sciences, Hyderabad, Telangana, India

S. M. Ali
Department of Surgery, Hamad Medical Corporation, Doha, Qatar

M. Thapur
Department of Infectious Disease, Hamad Medical Corporation, Doha, Qatar

N. Shaikh, A. H. Chanda (eds.), *Applied Microbiology in Intensive Care Medicine*, https://doi.org/10.1007/978-981-97-4006-2_7

gens demand tailored antibiotic regimens, with percutaneous drainage emerging as a cornerstone.

This comprehensive review navigates gastrointestinal infections, presenting a unified strategy for early diagnosis and multidimensional management. Insights from each condition form a comprehensive strategy, highlighting timely intervention, tailored therapies, and multidisciplinary approaches to improve patient outcomes.

Keywords Gastrointestinal infections · Peritonitis · Acute cholangitis · Acute calcular cholecystitis · Pyogenic liver abscess · Multidisciplinary approach

1 Introduction

Intrabdominal infections are the second most common source of sepsis and are associated with high morbidity and mortality. Intra-abdominal sepsis encompasses a range of complex and potentially life-threatening conditions that arise from infection and inflammation within the abdominal cavity; due to multiple and polymicrobial sources, abdominal sepsis progresses rapidly and seriously. The effective treatment of intra-abdominal infection consists of early recognition of the disease, control of the source, appropriate antibiotic therapy, and stabilization in the intensive care setting with a proper surgical approach.

Primary peritonitis, once prevalent in the pre-antibiotic era, remains a distinct entity characterized by infection of the peritoneal cavity unrelated to intra-abdominal or gastrointestinal causes. Despite a decline in incidence due to the advancement in antibiotic therapies, it continues to affect specific demographics, particularly females, the elderly, and individuals with ascites. Various sources, ranging from hematologic to genital, can lead to this condition, reminding us of the diverse origins of abdominal infections.

Acute cholangitis, on the other hand, originates from bacterial invasion of the biliary system, often as a result of bile duct obstruction. The spectrum of its clinical presentation varies from mild and self-limiting to severe and life-threatening, necessitating prompt intervention. Choledocholithiasis is the primary cause of obstructions, accompanied by factors such as malignancy, benign strictures, and biliary procedures, which can all contribute to the onset of acute cholangitis.

In the context of acute calcular cholecystitis, gallstone-related inflammation of the gallbladder wall spurred by gallstone-induced cystic duct obstruction stands as a prevalent challenge. Although most cases stem from gallstone presence, a minority are attributed to acalculous cholecystitis. Factors such as family history, genetics, ethnicity, gender, and age contribute to gallstone formation. With a notable percentage of untreated patients progressing to acute cholecystitis, understanding the intersection of biliary colic, choledocholithiasis, and associated complications becomes paramount.

Finally, hepatic abscesses emerge as critical entities within the spectrum of intra-abdominal sepsis, often life-threatening infections resulting from microbial invasion of the liver parenchyma through various routes, including the bile ducts, bloodstream, and contiguous spread. The causative agents encompass a spectrum of bacteria, parasites, and rarely fungi. Geographical differences influence the prevalence of specific pathogens, highlighting the multifaceted nature of liver abscesses. Although historically linked to appendicitis, etiology has shifted to biliary tract diseases, emphasizing the dynamic nature of infectious pathways. The evolving understanding of hepatic abscesses highlights the pivotal roles of accurate diagnosis and timely intervention, especially in cases with subtle and non-specific clinical presentations.

2 Peritonitis: We Should Be Vigilant in the Initial Management

Peritonitis is an inflammation and infection of the peritoneum. Peritonitis is frequently divided into primary and secondary peritonitis. Primary peritonitis (a forgotten entity) is the infection of the peritoneal cavity not directly caused by intra-abdominal or gastrointestinal pathologies [1].

It was more frequent in the pre-antibiotic era. In the post-antibiotic era, the incidence has decreased drastically. It is more common in females, increasing age and ascites. The presence of ascites due to any etiology increases the risk for primary peritonitis. Hematologic, lymphatic, gastrointestinal, genital/fallopian tubes are the possible cause of infections [2].

2.1 *Primary Peritonitis*

Patients with primary peritonitis may present with abdominal pain and tenderness, fever, vomiting, diarrhea or constipation, and dysuria. The cultures usually grow a single organism, and commonly found organisms in ascitic fluids are *Streptococcus pneumoniae*, *Staphylococcus aureus*, *E. coli*, and *Klebsiella pneumoniae*. In rare cases of primary peritonitis, cultures could be negative. Primary peritonitis should be diagnosed from history, physical examination, imaging studies, and laboratory and microbiological results. Management of primary peritonitis is supportive care and systemic broad-spectrum antimicrobial therapy. Few patients will need peritoneal lavage if they are not improving in the first 24–48 h. Primary or spontaneous peritonitis occurs when microbes, particularly bacteria, seed the peritoneum.

Spontaneous or primary bacterial peritonitis needs the presence of media for bacterial growth in the peritoneum. The presence of ascitic fluid, in general, or even remanent of the peritoneal dialysis fluids, is a suitable medium for bacterial growth in the peritoneum. Around 25% of all bacterial infections in certain patients are due to primary spontaneous peritonitis, and peritoneal dialysis patients every 2 years suffer at least once from primary peritonitis [3].

The initial treatment of primary peritonitis is to start broad-spectrum antibiotics and titrate to a narrow spectrum once the cultures and sensitivities are available. In cases of sever primary peritonitis, removal of the peritoneal dialysis catheter, in addition to antimicrobial therapy, is often necessary, due to the high risk of infection seeding and subsequent development catheter-related peritonitis [3].

2.2 *Secondary Peritonitis*

Secondary peritonitis is peritoneal inflammation and infection due to physical or functional disruption of the functional integrity of the gastrointestinal tract. Secondary peritonitis is usually due to perforation of the bowel, ischemic bowel, volvulus, and traumatic injuries with blood in the peritoneal cavity. Secondary peritonitis remains a potential threat to the lives of all human age groups, races, and socioeconomic status. It affects both developed and developing countries equally. Secondary peritonitis is a major etiology of loss of life, livelihood, and resources. It causes devastating economic and psychological burdens.

Secondary peritonitis accounts for 1% of all emergency admissions and seems a common cause of the occurrence of sepsis in the world [4]. Secondary peritonitis also surprisingly affects astronauts beyond the Earth. There are challenges in secondary peritonitis affecting astronauts in space. These individuals are immunocompromised and have a physiological de-adaptation of the cardiovascular system, multidrug-resistant bacteria and microbes from space, and limited support and surgical resources [5].

2.2.1 Epidemiology

The research by Anya et al. found that secondary peritonitis is diagnosed in 9.3 per 1,000 hospital admissions. Furthermore, approximately 11% of these patients had severe sepsis, while up to 74% had single-organ dysfunction, and 20% developed multiorgan dysfunction [6].

In a large European multi-center study, most patients had community-acquired infections (79%), and 21% had nosocomial infections. The majority of nosocomial infections recovered in the postoperative period, and 99% of them were due to anastomotic leakage [7].

2.2.2 Pathophysiology of Secondary Peritonitis

Peritonitis and intra-abdominal infections are the second most frequent cause of sepsis. Diffuse infection of the peritoneum secondary to the loss of integrity of the gastrointestinal tract leads to sepsis and septic shock with multiorgan dysfunction, thus increasing morbidity and mortality.

As bacterial resistance continues to rise, the growth of antimicrobial-resistant microorganisms within the peritoneum significantly impacts the prognosis of abdominal sepsis. Notably, the incidence of ESBL-producing (Extended-spectrum beta-lactamase) E. coli tripled between 2000 and 2008. Similarly, Klebsiella pneumoniae resistance has risen by 20%, and pseudomonal infections have emerged as an independent risk factor for increased mortality [8, 9].

2.2.3 Inflammatory Cytokines

The mismatch of the dysregulated immune response is the key to multiorgan dysfunction. Toll-like receptors (TLRs) are pattern recognition receptors expressed on immune and endothelial cells, which are the key to the inflammatory response. The protein kinase cascades are activated within the cells, resulting in the degeneration of pro- and anti-inflammatory cytokines. The contaminated peritonitis provides an ongoing source for PAMPs (pathogen-associated molecular patterns), and the damaged abdominal viscera produces DAMPs (damage-associated molecular patterns). The net result is providing continuous pro-inflammatory cytokines for systemic inflammatory response causing fever and vascular leak [10].

Secondary peritonitis triggers the release of abundant proinflammatory cytokines within the ascitic fluids. This inflammatory ascitic fluids then translocate into the systemic circulation through the lymphatic system. Therefore, draining the infected ascitic fluids may prevent and reduce the systemic spillover [11].

2.2.4 Secondary Peritonitis and Intra-abdominal Hypertension

The secondary peritonitis and subsequent fluid resuscitation lead to raised intra-abdominal pressure, causing intra-abdominal hypertension (IAH) (Table 1). The humoral effects of raised IAH cause reduced blood flow to the bowel mucosa and increased mucosal barrier permeability leading to irreversible mitochondrial damage and necrosis of gut mucosa with increasing translocation of bacteria into the systemic circulation [12].

Table 1 Anastomotic leakage as per systemic sequences

Colorectal	40%
Gastrointestinal	32%
Biliary	50%
Pancreatic	11%
Urinary	2%

2.2.5 Diagnosis

Clinical Signs and Symptoms

Although the parietal and visceral peritoneum have a common embryological origin, it has different innervation and manifestations. The irritation of the parietal peritoneum generates a sharp, constant, and localized pain associated with rigidity and guarding of overlying muscles. The affected visceral peritoneum causes colicky, paroxysmal pain.

The initial evaluation of abdominal pain must be focused and rapid, with a goal to assess the severity of the patient's condition, rapid triage to resuscitations, and immediate surgery and imaging studies or conservative therapy. Observations should concentrate on overall patient's general condition, vital signs, timing, location, and characteristics of abdominal pain.

The examination should identify rigidity, rebound tenderness, and guarding all over the abdomen, indicating generalized peritonitis or peritoneal signs limited to one or more abdominal quadrants indicating localized peritonitis.

The laboratory parameters will have a special role in diagnosing acute abdomen, pancreatitis, appendicitis, and cholecystitis.

An elevated white blood cell count, known as leukocytosis, suggests the presence of a bacterial infection. Additionally, increased serum lactate levels indicate tissue hypoperfusion associated with sepsis and an acute abdomen with mesenteric ischemia and bowel involvement. Although elevated serum lactate is a common finding, it has limited diagnostic specificity (36–53%). Moreover, markedly elevated pancreatic amylase and lipase levels, more than three times the normal range, are associated with pancreatic pathology. High levels of C-reactive protein and procalcitonin further indicate bacterial infections, postoperative infections in patients with acute abdomen, and the potential need for surgical interventions [13].

Imaging Studies Ultrasonography helps diagnose hepatobiliary and urinary tract acute pathology. Enhanced CT contrast (computed tomography) has become essential in the modern workup for peritonitis. In stable patients with acute abdomen, a CT reported by a senior radiologist gives a correct diagnosis in more than 90% of the cases. It is helpful in decisions regarding operative versus conservative treatment and whether percutaneous drainage would be helpful [14].

2.2.6 Management the Secondary Peritonitis

The cornerstone of management is surgical intervention. Still, it is equally essential to resuscitate these patients as per Surviving Sepsis Campaign Guidelines, as most of them are in sepsis and septic shock.

The advantages of early administration of antibiotics in bacteremia patients are known to be beneficial. In patients with secondary peritonitis, it is crucial to administer broad-spectrum antibiotics, depending on the expected bacteria, and then narrow the spectrum as cultures and susceptibility are available, as it will reduce bacterial resistance. Each 30-min delay in administering antibiotics in patients with secondary peritonitis increases the mortality rate with an odd ratio of 1.02 [15].

Antifungal prophylaxis should be considered in patients with secondary peritonitis, particularly in the setting of nosocomial infections, high gastrointestinal perforations, prolonged antibiotic treatment, and immunocompromised states [16].

The surgical interventions should be targeted to control and prevent ongoing peritoneal infections and source control. Sometimes, it is better to keep the abdomen open or unable to close it. Relaparotomies with a temporary closure of the abdomen are the current practice [17].

2.2.7 The Outcome of Secondary Peritonitis

The overall mortality for peritonitis is 6%; it will increase to 34% if peritonitis gets complicated into sepsis shock. The risk factors apart from sepsis and septic shock for mortality in these patients are elderly age; diabetes; premorbid organ dysfunction, mainly cardiovascular, renal, liver, and neurological disorders; non-appendicular abdominal infections; delay interventions for more than 24 h; and extent of peritonitis [6].

Patients with four quadrants of peritonitis have a very poor prognosis (mortality of 36%). Fecal peritonitis is also a poor prognostic indicator, with a mortality of 38%. Diabetes is an independent risk for poor outcomes in secondary peritonitis, with a mortality rate of 50%, compared to the standard of 19% [18].

3 Tertiary Peritonitis (TP)

Although exploratory laparotomy and surgical infective source control are the standard care for secondary peritonitis, in several patients, the intraperitoneal infections and peritonitis persist or recur for more than 48 h following surgical source control; it is termed tertiary peritonitis. The etiological factors for tertiary peritonitis are mentioned in Table 2.

Table 2 Etiology of tertiary peritonitis include

Appendicitis
Diverticulitis
Perforated ulcer
Necrosis of the bowel
Pancreatitis
Postoperative peritonitis

3.1 Risk Factors for Tertiary Peritonitis

There are various risk factors for tertiary peritonitis, such as elderly patients, female gender, organ dysfunctions and failure, malignancy, malnutrition, delay in presentation of more than 24 h, diffuse and generalized peritonitis, fecal peritonitis, repeated surgical procedures, and fungal and nosocomial infections [19].

3.2 Microbiology of Tertiary Peritonitis

TP is associated with gram-negative and enteric microbes: *Enterobacter* (up to 35%), *Enterococcus* (up to 20%), *Actinobacter*, *Citrobacter*, *Pseudomonas*, and fungal species (up to 30%) [19].

3.3 Diagnosis

The relaparotomy is the most commonly used method to diagnose and treat tertiary peritonitis. Along with peritonitis, patients are frequently in sepsis and septic shock.

The pro-inflammatory markers, leukocytosis, leukopenia, raised C-reactive protein, and pro-calcitonin will help diagnose TP. The peritoneal fluid consistency and microbiology also help in diagnosis. The patients frequently have organ dysfunctions; hence, the severity of the disease can be calculated by various severity scores from SOFA to SAPS score. The Mannheim Peritonitis Index (MPI) has a higher potential for the diagnoses and assessment of tertiary peritonitis severity [20]. In these patients, CT abdomen may show collections and other changes.

3.4 Management

Patients with tertiary peritonitis are frequently managed in surgical intensive care settings by multidisciplinary teams and physicians, including intensivists, surgeons, infectious disease specialists, clinical pharmacists, dietitians, physiotherapists, and critical care nurses.

Early diagnosis and effective and immediate therapeutic interventions should be started in tertiary peritonitis patients as most of these patients are in sepsis and septic shock with multidrug-resistant bacteria. Appropriate antibiotic and antifungal coverage along with organ-supportive resuscitation is essential in the intensive care therapy settings. Abdominal source control is a cornerstone in the management of TP patients. Surgical interventions are required to eliminate the abdominal source of infection and reduce the peritoneal contaminants [21]. A critical point in treating tertiary peritonitis is avoiding planned relaparotomies, as it did not show any beneficial effects compared to laparotomy [22].

In contrast, the relaparotomies reduce the percutaneous drainage and length of ICU stay in TP patients [22]. Percutaneous image-guided collection drainage in patients with tertiary peritonitis is increasingly used and helps to drain infected collections and contaminants and decrease the microbiological load. The peritoneal drainage fluid helps identify microbes and narrows the spectrum of antibiotic treatment.

3.5 Mortality and Outcome

The mortality of tertiary peritonitis ranges from 30% to 60%. Early diagnosis and earlier immediate interventions are essential for better outcomes [23]. Advanced age, delayed interventions, malignancy, cardiovascular disease, hepatic disorder, end-stage renal disease, higher severity scores, and multidrug-resistant bacterial infections are independent risk factors for increased mortality in patients with tertiary peritonitis [24].

4 Navigating Acute Cholangitis: Unveiling Pathways to Diagnosis and Dynamic Management

4.1 Introduction

Acute cholangitis is a clinical condition that results from bacterial infection of the biliary system, most frequently secondary to partial or total obstruction of the bile duct or hepatic ducts. Clinical manifestations may be mild and self-limiting to severe and life-threatening, requiring urgent intervention [25]. Choledocholithiasis is the primary etiology of biliary obstruction. In addition, malignancy, benign strictures, and biliary procedures may also result in acute cholangitis.

4.2 Pathophysiology

The pathophysiology of cholangitis includes biliary obstruction, increased intraluminal pressure, and contaminated bile [26]. Human bile is sterile under normal circumstances, and the biliary system is typically protected from infection by anatomical barriers (such as the sphincter of Oddi and tight connections between hepatocytes), physical mechanisms (such as bile flow and biliary mucus), chemical components (such as bile salts), and immunological defense (Kupffer cells and immunoglobulin) [27].

Additionally, the sphincter of Oddi prevents bacterial migration from the duodenum into the biliary system. In the case of biliary obstruction, bile becomes stagnant in the biliary system, which causes an increase in intraductal pressure, Kupffer cell malfunction, a widening of the tight junction between cholangiocytes, and a decrease in IgA production [27].

"Choledochal pressure" plays an essential role in the pathogenesis of acute cholangitis. The intraductal biliary pressure typically ranges between 7 and 14 cm H_2O. In partial or complete biliary obstruction, the intraductal pressure can rise to 20–30 cm H_2O, resulting in cholangiovenous reflux of infected bile, which in turn causes systemic bacteremia and endotoxemia [26]. In addition, the widespread release of inflammatory mediators such as tumor necrosis factor (TNF), soluble TNF receptors, interleukin (IL)-1, IL-6, and IL-10 results in severe hemodynamic impairment [28].

Sepsis is common in the case of suppurative biliary infections. The mortality rate has fallen to 5–15% despite antibiotics and endoscopic retrograde cholangiopancreatography (ERCP) interventions, which in the past had substantial morbidity of 80–90% mortality [26].

4.3 Clinical Presentation

Clinical presentation of acute bacterial cholangitis ranges from mild illness to septic shock. The majority of patients have underlying biliary pathology, often in the form of gallstones, malignant obstruction, or recent bile duct instrumentation.

Over 90% of patients typically have a fever, whereas 60% and 10% have jaundice and abdominal pain. Charcot's triad (fever, abdominal pain, and jaundice) may appear in 50–90% of patients. Reynolds' pentad, which affects less than 14% of cholangitis patients, is characterized by altered mental state and hypotension in those with Charcot's triad [29].

4.4 Microbiology

The most frequently found pathogens isolated in acute cholangitis are coliform organisms. These include *Escherichia coli* (25–50%), *Klebsiella* species (15–20%), *Enterococcus* species (10–20%), and *Enterobacter* species (5–10%). Anaerobic bacteria, such as *Bacteroides fragilis* and *Clostridium perfringens*, may sometimes cause acute cholangitis, especially in patients with previous biliary surgery and the elderly population [30]. Bile cultures are positive in 80–100% of the cases and between 30% and 87% involve isolating multiple organisms [29, 31]. Bacteremia, found in 21–83% of patients with cholangitis, frequently included the same bacteria as in bile cultures. *E. coli* (59%), *Klebsiella* spp. (16%), *P. aeruginosa* (5%), and *Enterococcus* spp. (4%) were the most frequently isolated pathogens from the blood culture [29].

4.5 Management

Acute cholangitis is a medical emergency. Once biliary obstruction with bacterial infection is suspected, a set of diagnostic and therapeutic modalities should be implemented. The exact sequence of events depends on the degree of diagnostic uncertainty, the severity of the illness, and existing medical comorbidities. However, every patient management includes decisions about four interventions: antibiotic therapy, diagnostic imaging, biliary drainage, and the role of surgery.

As cholangitis is due to infection and obstruction of the biliary system, both aspects must be addressed. Patients with acute cholangitis require resuscitation with IV fluids, correction of electrolytes, coagulopathy, and prompt treatment with broad-spectrum empiric antibiotics after blood cultures have been obtained. Antibiotic treatment is given to control sepsis and local inflammation. Biliary drainage, by ERCP or percutaneous transhepatic cholangiography, is often required for a successful clinical outcome.

4.6 Pharmacological Therapy

The initial empiric antibiotic therapy should be broad-spectrum, targeting the typical pathogens such as gram-negative aerobic enteric organisms (*E. coli*, *Klebsiella spp.*, *Enterobacter spp.*), gram-positive species (*Enterococcus* and *Streptococcus spp.*), and anaerobes (*Bacteroides fragilis*, *Clostridium perfringens*). The choice of antibiotics depends on multiple factors, including the patient's renal function, hepatic function, drug allergies, comorbidities, hospital-acquired (multiple or resistant organisms like *Pseudomonas*, vancomycin-resistant *Enterococcus* (VRE), carbapenem-resistant Enterobacteriaceae (CRE), or methicillin-resistant *Staphylococcus aureus* (MRSA)) or community-acquired infection (single agents like *E. coli*, *Klebsiella*, or *Enterococcus*), and also on the severity of cholangitis [28].

The initial choice should be piperacillin/tazobactam, ticarcillin/clavulanate, ceftriaxone plus metronidazole, or ampicillin/sulbactam. If the patient is allergic to penicillin, ciprofloxacin plus metronidazole, carbapenems, or gentamicin plus metronidazole are alternatives [32]. Although it is recommended to use antibiotics that are excreted into bile, a biliary obstruction may result in reduced excretion. After successful drainage, it takes around 24 h for antibiotics to reach the biliary system [32].

The antibiotics should be further evaluated and adjusted according to the susceptibility patterns of cultured microorganisms. The antibiotics should be continued for 7–10 days [33]. However, 3 days may be sufficient in patients with adequate biliary drainage.

In patients with mild to moderate cholangitis, cephalosporin, like cefoxitin, provides broad-spectrum gram-positive and gram-negative activity; however, anaerobic coverage is poor. Fluoroquinolones such as ciprofloxacin and levofloxacin also offer broad-spectrum coverage for gram-positive and gram-negative aerobes, including *Pseudomonas* spp., but lack anaerobic coverage. Although moxifloxacin, a newer generation of fluoroquinolone, has better anaerobic activity, it is less effective against *Pseudomonas* species [31].

Fluoroquinolones can be switched from intravenous to oral regimens once patients can tolerate orals because of their similar bioavailability [31]. Even though ampicillin and gentamicin are widely used, disadvantages include limited anaerobic coverage, frequent microbial resistance, and nephrotoxicity [32].

Empirical antimicrobials should provide pseudomonal and enterococcal coverage in critically ill patients with severe cholangitis. Piperacillin/tazobactam and carbapenems such as imipenem or meropenem have excellent broad-spectrum activity in critically ill patients. Although the need for empirical *Enterococcus* coverage is still debated, the antibiotic regimen should include enterococcal coverage if the organism is isolated in the blood or bile [31, 32, 34].

4.7 Biliary Drainage

In addition to antibiotics, successful management of acute cholangitis often requires biliary drainage, either through a transhepatic approach or by performing sphincterotomy and biliary stenting during ERCP [35]. Many patients respond well to antibiotic therapy and resuscitation, which makes biliary drainage a semi-elective procedure performed within the subsequent 72 hours.

However, some patients, especially those with severe cholangitis, fail to respond to antibiotic therapy or experience clinical deterioration within 12–24 hours, necessitating urgent biliary decompression. Biliary drainage remains the mainstay treatment for managing cholangitis, as antibiotic therapy alone is often ineffective in the presence of biliary obstruction, which impairs the excretion of antibiotics into the biliary system.

Endoscopic retrograde cholangiopancreatography (ERCP) is now the preferred technique for biliary drainage. Although the optimal timing of biliary drainage has not been evaluated in randomized trials, patients who are hypotensive at the time of presentation are highly likely to require urgent biliary decompression. Urgent biliary decompression should be performed in patients presenting with evidence of sepsis, jaundice, potential biliary obstruction, and cholangitis, as well as those who fail to improve with antibiotic treatment [26, 30]. The procedure is usually performed in the endoscopy suite with fluoroscopy under moderate sedation, supplemental oxygen, and appropriate cardiopulmonary monitoring. In the setting of hemodynamic instability, the procedure may be performed in the intensive care unit.

Although there are several well-known complications associated with ERCP, such as pancreatitis, bleeding, and perforation, ERCP is successful in over 98% of patients, is considerably safer, and has many advantages over surgical or percutaneous drainage techniques [36].

Sphincterotomy and stone extraction should be performed with or without a trans-papillary biliary stent placement in patients with choledocholithiasis. If a patient has coagulopathy or is taking anti-platelet medication, there is occasionally a higher risk of bleeding after biliary sphincterotomy. In those circumstances, biliary stent placement is possible without sphincterotomy. The placement of a trans-papillary biliary stent should provide adequate drainage in the event of biliary stricture. The old stent should be removed and replaced with a new one if the existing stent becomes obstructed due to the development of bacterial biofilm and bile sludge [37]. Percutaneous transhepatic biliary drainage is an alternative when ERCP expertise is not available or technically not possible. Surgical intervention is appropriate only for those patients who are unable to undergo or fail ERCP and PTBD [31].

5 Acute Cholecystitis: From Pathophysiology to Precision Management

5.1 Introduction

Acute calcular cholecystitis (ACC) is an acute inflammatory disease of the gallbladder wall caused by gallstone obstruction of the cystic duct in approximately 90–95% of people diagnosed with the disease [38]. Less frequently, about 5–10% of patients with acute cholecystitis have acalculous cholecystitis, in which the gallbladder becomes acutely inflamed without gallstones. Acute calcular cholecystitis represents between 3% and 10% of hospitalizations for abdominal pain [39]. In developed countries, the prevalence of gallstones among adults ranges between 10% and 15%. Family history, genetic predisposition, ethnic origin, female sex, and the patients' age are all independent factors contributing to gallstone formation [40].

Biliary colic affects 1–4% of all patients with cholelithiasis annually. Although most gallstones are asymptomatic, acute cholecystitis occurs in 10–20% of untreated patients. Most patients with acute cholecystitis have a history of biliary colic. However, some have had no prior history of biliary symptoms. Additional flare-ups of pain or inflammation are frequent following an initial attack of acute cholecystitis. A small percentage of individuals may also have choledocholithiasis, cholangitis, or gallstone pancreatitis in addition to acute cholecystitis [41].

Due to a high likelihood of hospital readmission for the same symptoms, pancreatitis, or cholecystitis within a discrete time frame, surgical treatment accompanied by antibiotic therapy remains the standard once a patient experiences symptoms or complications associated with gallstones [38, 42].

5.2 Pathophysiology

More than 90% of cases of acute cholecystitis are associated with cholelithiasis (acute calculous cholecystitis). The key factors in pathogenesis appear to be obstruction of the cystic duct caused by gallstone or impacted biliary sludge at the gallbladder neck resulting in biliary stasis, which triggers an inflammatory cascade causing inflammation, ischemia, and necrosis of the gallbladder wall. The gallbladder becomes enlarged and tense as a result of inflammation. It may also develop wall thickening and exudate of pericholecystic fluid. The inflammation is usually initially sterile, but most patients eventually develop secondary infections with enterococci, Enterobacteriaceae, or anaerobes [43].

Necrosis and gangrene might develop on the gallbladder's wall causing gangrenous cholecystitis. The gallbladder's wall or lumen may develop gas because of bacterial superinfection with gas-forming organisms (emphysematous cholecystitis). The gallbladder may rupture without proper treatment, leading to a liver abscess or generalized peritonitis.

5.3 Diagnosis

Acute cholecystitis (AC) is diagnosed based on symptoms and signs of inflammation localized to the right upper quadrant. The pain is typically severe in the epigastrium or right upper quadrant. It tends to present postprandially or at night, commonly radiating to the back. Patients also frequently experience nausea and vomiting. Typically, acute cholecystitis starts with a biliary colic attack, often in a patient who has previously experienced attacks. The persistent pain in the right upper quadrant and Murphy's sign (arrest of inspiration during palpation of the right upper quadrant due to pain) should be used to distinguish acute cholecystitis from biliary colic [41].

On physical examination, tenderness and guarding in the right upper quadrant associated with localized peritonitis are present in 95.7% of patients and are common findings [44]. The right upper quadrant Murphy's sign is pathognomonic of acute cholecystitis, with a sensitivity of 62% and a specificity of 96% for acute cholecystitis [45].

In patients with severe cholecystitis, mild jaundice (serum bilirubin concentrations of 60 mol/L) may occur in patients with acute cholecystitis as a result of edema and inflammation surrounding the biliary tract as well as direct pressure from the distended gallbladder obstructing the biliary tree. However, frank jaundice is uncommon, and the presence of bilirubin concentrations of more than 60 mol/L should raise suspicion of concomitant choledocholithiasis (a gallstone in the common bile duct) or Mirrizi's syndrome (obstruction of the common bile duct or common hepatic duct by external compression from an impacted gallstone in Hartmann's pouch) [39, 46].

Patients typically have leukocytosis with left-shift and immature bands. The C-reactive protein (CRP) level is frequently elevated [41]. However, no clinical finding or laboratory test is sufficient to establish or exclude cholecystitis without further testing. A complete blood count and a serum chemistry panel, which includes a liver function test, a bilirubin level, and a serum lipase level (to rule out acute pancreatitis in the differential diagnosis), are recommended laboratory tests in the workup for acute cholecystitis.

Sometimes, acute cholecystitis can result in systemic sepsis and organ failure, usually in gangrenous or emphysematous cholecystitis. Delays in diagnosis are common in elderly patients because physical examinations and laboratory results may be normal, and the only symptoms may be a change in mental status or a decrease in food intake [41].

Abdominal ultrasonography is the initial imaging modality to diagnose acute cholecystitis due to its relatively low cost, easy accessibility, quick examination duration, and lack of ionizing radiation. It detects cholelithiasis in about 98% of patients. Common radiological findings in acute cholecystitis are thickened gallbladder with pericholecystic fluid (fluid around the gallbladder), gallbladder distention, and ultrasonographic Murphy's sign [39].

Gallbladder distension (41%) and gallbladder wall thickening (59%) were among the acute cholecystitis CT findings in acute cholecystitis. Other CT findings included sub-serosal edema (31%), pericholecystic fluid collection, pericholecystic fat density, and significant gallbladder bile attenuation (24%) [47]. Therefore, no single CT feature can be used to diagnose acute cholecystitis, and due to its risk of ionizing radiation exposure, CT is usually indicated when sonography is non-diagnostic or patients have confusing symptoms or to rule out other differential diagnoses [48].

The "Tokyo Guidelines" classify acute calcular cholecystitis into three categories: severe forms (Grade 3), moderate forms (Grade 2), and mild forms (Grade 1). The severe forms (Grade 3) correspond to acute cholecystitis with associated organ failure (neurologic, hematologic, hepatic, respiratory, renal, or cardiac). While moderate forms (Grade 2) correspond to acute cholecystitis with symptoms of marked inflammation (disease duration >72 h, leukocytosis >18,000/mm^3, palpable mass or right subcostal guarding, or symptoms of biliary peritonitis or gangrenous acute cholecystitis), mild forms (Grade 1) are acute cholecystitis and do not meet the criteria for Grades 2 or 3 [49, 50].

Table 3 summarizes the diagnostic criteria for acute cholecystitis according to the Tokyo Guidelines [51].

5.4 Microbiology

Enteric bacteria (*Escherichia coli*, *Klebsiella*, *Enterococcus*, and anaerobic bacteria) are the most common causative microorganisms in acute cholecystitis. The significance of blood culture results is relatively low because they are rarely positive. Additionally, positive blood cultures do not seem to alter the choice of antibiotics or

Table 3 Diagnostic criteria for acute cholecystitis based on 2018 Tokyo Guidelines

Local signs of inflammation • Murphy sign • Right upper quadrant mass, pain, or tenderness
Systemic signs of inflammation • Fever • Elevated C-reactive protein level • Elevated white blood cell count
Imaging findings characteristic of acute cholecystitis • Suspected diagnosis: One local sign plus one systemic sign • Definite diagnosis: One local sign plus one systemic sign plus one image finding characteristic of acute cholecystitis

Adapted from Yokoe et al. [51]

the duration of treatment [38]. However, a positive bile culture is associated with developing a severe form of cholecystitis. Thus, especially in severe cases, bile samples should be obtained for culture and susceptibility testing during surgical, laparoscopic, or percutaneous intervention. When necessary, a sample of the gallbladder wall should be taken separately for culture and histopathology [52]. Grande et al. reported that the presence of infected bile did not affect the likelihood of surgical site infection in patients having surgery for acute or chronic cholecystitis [53].

5.5 *Treatment*

Most patients with acute cholecystitis respond to conservative initial management; as the gallstone disimpacts and returns to the gall bladder, the cystic duct might empty. However, if the gallstone impaction is persistent, complications such as advanced cholecystitis (gangrenous cholecystitis or gall bladder empyema) or perforation may occur. In most cases, the inflammatory process can be reduced by taking immediate action to rest the gallbladder. Patients should be fasted and given appropriate analgesics and intravenous fluids for rehydration.

5.6 *Antibiotic Therapy*

Acute cholecystitis is an inflammatory process at the beginning; however, a subsequent bacterial infection of the gallbladder can arise in up to 20% of patients due to continuous bile stasis due to cystic duct occlusion by calculus and edema [39]. Therapy with appropriate antimicrobial coverage is an essential component in the management of patients with acute cholecystitis. Antibiotics are recommended in complicated cholecystitis and are considered the primary therapy in patients undergoing delayed surgery or observation.

According to the Infectious Diseases Society of America's recommendations, antimicrobial therapy should be started if an infection is suspected based on laboratory and clinical findings (more than 12,500 white blood cells per cubic millimeter or a temperature of more than 38 °C) and radiographic evidence (e.g., air in the gallbladder or gallbladder wall) [54]. Additionally, antibiotics are advised for routine use in elderly, diabetic, or immunocompromised patients as well as for prophylaxis in cholecystectomy patients to prevent septic complications even when an infection is not suspected [41].

Antibiotic therapy should include coverage against microorganisms in the Enterobacteriaceae family. Gram-negative aerobes like *Escherichia coli* and *Klebsiella pneumonia* as well as anaerobes, mainly *Bacteroides fragilis*, are the most frequently isolated microorganisms in biliary infections.

The pathogenicity of enterococci in biliary tract infections remains unclear, and specific coverage against these microorganisms is not routinely advised for community-acquired biliary infections. However, enterococcal infection should always be assumed and treated in specific immunosuppressed patients, such as those who have undergone hepatic transplantation [47]. Empiric selection is essential in such cases since antibiotic therapy has a significant prognostic influence, particularly in severe sepsis. Inadequate initial antimicrobial therapy is an independent predictor of mortality [47].

Empiric treatment should involve antibiotics with good biliary distribution and potent activity against the most common bacteria involved with acute cholecystitis, as shown in Table 4.

The choice of antibiotics depends on multiple factors, including individual patient characteristics, documented multidrug-resistant bacterial colonization, pharmacokinetics, pharmacodynamics, adverse side effects of treatment, the results of antibiotic sensitivity testing, and the severity of the disease. When choosing the initial empiric antibiotic regimen, local bacteriologic susceptibility data should be considered. Moreover, the Gram stain of intraoperative samples can also guide antibiotic selection (Table 5).

Penicillins are routinely used to treat biliary infections. Aminopenicillins, like amoxicillin, are excreted in the bile unchanged. Amoxicillin is three times more concentrated in bile than plasma in patients with normal biliary function.

Table 4 Antibiotics commonly used to treat biliary tract infections and their biliary penetration ability [47]; Antibiotics Bile/Serum Concentration Ratio (ABSCR)

Good penetration efficiency (ABSCR ≥ 1)	Low penetration efficiency (ABSCR < 1)
Piperacillin/tazobactam (4.8)	Ceftriaxone (0.75)
Tigecycline (>10)	Cefotaxime (0.23)
Amoxicillin/clavulanate (1.1)	Meropenem (0.38)
Ciprofloxacin (>5)	Ceftazidime (0.18)
Ampicillin/sulbactam (2.4)	Vancomycin (0.41)
Cefepime (2.04)	Amikacin (0.54)
Levofloxacin (1.6)	Gentamicin (0.30)
Penicillin "G" (> 5)	
Imipenem (1.01)	

Table 5 Antimicrobial regimens suggested for acute calculous cholecystitis [47]

Community-acquired	Healthcare-associated
1. **Beta-lactam/beta-lactamase inhibitor combination-based regimens** Amoxicillin/clavulanate (in stable patients) Ticarcillin/clavulanate (in stable patients) Piperacillin/tazobactam (in unstable patients) 2. **Cephalosporins-based regimens** Ceftriaxone + metronidazole (in stable patients) Cefepime + metronidazole (in stable patients) Ceftazidime + metronidazole (in stable patients) 3. **Carbapenem-based regimens** Ertapenem (in stable patients) Imipenem/cilastatin (only in unstable patients) Meropenem (only in unstable patients) 4. **Fluoroquinolone-based regimens** (in case of allergy to beta-lactam) Ciprofloxacin + metronidazole (only in stable patients) Levofloxacin + metronidazole (only in stable patients) Moxifloxacin (only in stable patients) 5. **Glycylcycline-based regimen** Tigecycline (in stable patients if risk factors for ESBLs)	Tigecycline + piperacillin/tazobactam (in stable patients) Imipenem/cilastatin ± teicoplanin (only in unstable patients) Meropenem ± teicoplanin (only in unstable patients)

Modified from Ansaloni et al. [47]

Fluoroquinolones also have high bioavailability. For instance, ciprofloxacin has a biliary concentration of 28–45 times higher than its plasma concentration [38, 39].

For patients with mild or moderate acute cholecystitis and no risk factors for resistance, ciprofloxacin and metronidazole may be an alternative to amoxicillin/clavulanic acid.

The combination of piperacillin and tazobactam is an effective alternative. As the concentration of piperacillin is highest in bile (25% in active form, 60 times higher concentration in the bile than in the plasma), however, tazobactam only achieves an effective concentration in the bile during the first 3 h following administration due to its distinct pharmacokinetics [38]. The glycylcyclines, like tigecycline, have a wide range of activity and high availability in bile and the gallbladder's wall [49].

The prevalence of ESBL and carbapenemase-producing bacteria in non-urinary tract isolates should be considered when choosing empirical antimicrobial therapy. Carbapenems, piperacillin/tazobactam, and aminoglycosides are recommended treatments for ESBL Enterobacteriaceae. Carbapenems, piperacillin/tazobactam, and aminoglycosides are empirically recommended for *Pseudomonas* if the prevalence of ceftazidime resistance is greater than 20% until culture and susceptibility testing findings are available [49].

Since most aminoglycosides, carbapenems, cephalosporins, and penicillins are eliminated via the kidneys, their dosage should be adjusted in patients with impaired renal function. Ceftriaxone dosage does not need to be changed in patients with renal insufficiency but should be in patients with hepatic insufficiency. Aminoglycosides are "concentration-dependent" antibiotics; in most cases, they should be given as a single daily dose. While the peak concentration and bactericidal activity are correlated, toxicity is related to the residual or trough level. Therefore, peak and trough serum aminoglycoside concentrations should be routinely monitored to optimize its use.

Despite the limited data, the 2018 Tokyo recommendations advise administering antibiotics before and during operative intervention for patients with uncomplicated cholecystitis [49].

5.7 *Surgical Treatment*

Cholecystectomy is the standard course of treatment for acute calculous cholecystitis. Laparoscopic cholecystectomy is the standard of care in the USA. Laparoscopic cholecystectomy, performed within 3 days of diagnosis, is the first-line therapy for acute cholecystitis. Cholecystectomy significantly reduces the initial inoculum, optimizes antibiotics' action and distribution, and improves the course of treatment.

Percutaneous cholecystostomy tube (PCT) is a minimally invasive procedure which can be an alternative to cholecystectomy in patients with acute cholecystitis and significant perioperative risks, such as elderly patients with multiple medical comorbidities and critically ill patients at the time of diagnosis. The percutaneous drainage catheter is placed under image guidance into the gallbladder lumen [55]. PCT decompresses the gallbladder while draining infection and can improve cholecystitis, allowing additional time to stabilize the patient before cholecystectomy [56]. Although there is evidence that PCT usage has increased over the past 20 years, the ideal patient group to receive PCT is still unclear [57].

6 Emerging Strategies in Pyogenic Liver Abscess: Unraveling Pathways from Diagnosis to Dynamic Management

6.1 *Introduction*

The term hepatic abscess "HA" refers to a life-threatening suppurating infection caused by the invasion and growth of microorganisms within the healthy or diseased liver parenchyma. Microbial invasion of the liver parenchyma can occur through the bile ducts, bloodstream (hematogenic, typically portal), or contiguous spread, mainly through the gallbladder bed [58]. The causes of HA might be bacterial, parasitic, mixed (pyogenic superinfection of parasitic abscess), or rarely fungal [58].

Entamoeba histolytica is the most common pathogen that leads to amoebic illnesses. While the pyogenic abscesses are typically polymicrobial, some organisms, such as *E. coli*, *Klebsiella*, *Streptococcus*, and *Staphylococcus*, are seen more frequently. Anaerobes are also commonly found in pyogenic abscesses [59].

In the USA, the incidence of pyogenic liver abscess (PLA) is 2.3 per 100,000, predominantly in older men. Diabetes and cancer are considered to be risk factors for the development of pyogenic liver abscesses [60].

There is a variation in the casual frequency based on geographical regions. In Southeast Asia and Africa, amoebic contamination is the most frequent cause, while in Western countries, 80% of liver abscesses are bacterial. They may develop due to intra-abdominal biliary infections that contaminate the biliary tract simultaneously or as a result of non-biliary infections seeding the portal system causing hepatic artery/portal vein bacteremia. Either way, they are caused by bile duct contamination [58].

Historically, appendicitis was the leading cause of liver abscesses. However, this number has dropped to fewer than 10% since improved methods of diagnosis and treatment of appendicitis have become available. Nowadays, the majority of pyogenic liver abscesses are due to biliary tract disease, specifically biliary stones, strictures, malignancy, and congenital anomalies [59]. However, up to 55% of patients with PLA have no clear risk factors, and these cases are called cryptogenic [61].

Early diagnosis and treatment are crucial. Diagnosing hepatic abscesses may be challenging, as the clinical presentation is usually subtle and non-specific (abdominal pain, fever, nausea, and vomiting). A high index of suspicion is the cornerstone of preventing misdiagnosis and improving prognosis [62]. Early diagnosis and treatment are crucial in managing these patients because the presentation may be subtle and not specific [62].

6.2 Microbiology

Many different types of bacteria have been identified as potential causes of pyogenic liver abscess (PLA). Still, there is a lack of published data on the associations of bacterial species with clinical presentation, radiological findings, and prognosis [63].

Regarding causative pathogens, *Escherichia coli*, Enterobacteriaceae, anaerobes, and other gastrointestinal flora have been identified as the most frequent bacteria causing pyogenic liver abscesses. In the past 20 years, *Klebsiella pneumoniae* has become more prevalent, accounting for 50% and 88% of PLA in the Asian population, with increasing incidence in South Africa, Europe, and the USA [61, 64].

Klebsiella pneumoniae–related hepatic abscess is unique in that it can cause distant "septic metastases" (ocular or neurologic in 12% of cases versus 0% for other bacterial HAs), and it is rarely associated with intra-abdominal disease (less than 1% of cases) [65].

In addition to bacteria, *Entamoeba histolytica*, a protozoan parasite, is also a well-known non-bacterial cause of liver abscess. It affects the liver by causing amebic colitis, seeding the portal system, migrating to the liver, and causing an amebic liver abscess. In Europe, incidences of amoebic liver abscess (ALA) are uncommon and are typically imported, although, in some highly endemic countries, ALA can be more common than (PLA) [66].

6.3 Treatment

Untreated pyogenic liver abscesses are almost uniformly fatal. During the pre-antibiotic era, the management of pyogenic liver abscesses was exclusively surgical. However, the treatment of this disease has changed significantly. In recent decades, combined antibiotic therapy and percutaneous drainage have become the cornerstones of treatment, along with treatment of the underlying disease or cause of the infection, significantly improving patients' prognosis, with a mortality rate falling from 70% to 6.31%. Early indicators of successful treatment include the absence of fever, complete resolution of pain, and return to normal levels of leukocytosis and CRP. Reversal of imaging abnormalities typically takes place somewhat later [58, 67].

Sonographic or CT-guided needle percutaneous drainage is currently the first-line treatment of hepatic abscess. Needle aspiration also allows the identification of the causative organism.

In selected patients, antibiotics can be used as the sole therapy, particularly in those with multiple small liver abscesses less than 3–5 cm, with favorable outcomes reported in carefully selected patients. In one study, antibiotic therapy alone had a 100% success rate for unilocular HA 3 cm [68].

Needle aspiration and catheter aspiration are both effective in large hepatic abscesses larger than 3 cm in diameter, with equivalent morbidity and mortality and duration of hospital stay [69].

Surgical drainage is rarely indicated when interventional radiology is readily available. However, surgical drainage may be indicated if percutaneous treatment for hepatic abscess fails, for large abscesses >5 cm, multilocular HA, ruptured abscess, or peritonitis [68, 70].

6.4 Pharmacological Therapy

Empirical therapy with broad-spectrum antibiotics directed against the common causative organisms, i.e., antibiotics, should cover Enterobacteriaceae, anaerobes, streptococci, enterococci, and *Entamoeba histolytica*. Such antibiotic regimens include third-generation cephalosporins plus metronidazole, beta-lactam + beta-lactamase inhibitor plus metronidazole, or synthetic penicillin plus aminoglycosides and metronidazole. Alternatively, in case of penicillin allergy, fluoroquinolones or carbapenems can be used instead for cephalosporins or penicillin in case of allergy or unavailability [58, 59]. When choosing the initial empiric antibiotic regimen, local bacteriologic susceptibility data should be considered.

There is little formal data regarding the optimum duration of antibiotic therapy. Still, most units use a regimen of 2-week parenteral treatment, followed by a more prolonged course (4–6 weeks) of oral antibiotics, commonly with amoxicillin and sulbactam or a combination of fluoroquinolone (e.g., ciprofloxacin) and metronidazole.

The choice of antibiotic therapy is modified according to the susceptibility patterns of cultured microorganisms, if any. The patient's clinical response and repeated imaging studies determine the duration of antibiotic therapy and the need for further intervention.

6.5 *Interventional Radiology*

Once the patient has been stabilized, percutaneous catheter drainage of the abscess should be performed. Success with percutaneous catheter drainage with antimicrobial therapy is achieved in 70–90% of cases. The drainage catheter is usually left in place until the drainage is reduced to a minimum, usually 1–2 weeks after placement [71–73].

Repeated needle and catheter aspiration is effective in large hepatic abscesses larger than 3 cm in diameter, with equivalent morbidity, mortality, and hospital stay [69]. However, aspiration without placement of an indwelling catheter may require careful follow-up and often repeat imaging procedures comparing the response to therapy.

Percutaneous aspiration should not be delayed because of concern that the space-occupying lesion is an echinococcal cyst. Although it is commonly believed that accidental aspiration of an echinococcal cyst carries a significant risk of anaphylaxis, available data do not support this belief [74]. Complications of catheter aspiration may include bleeding, catheter tract infection, and biliary-cutaneous fistula.

6.6 *Endoscopic Therapy*

Hepatic abscesses may be classified as communicating or non-communicating with the biliary tree [75]. Suppose bile is present in a hepatic abscess. In that case, one must look for biliary tract obstruction, such as choledocholithiasis or, less frequently, tumoral stricture or bile duct disease, which is often unrecognized and undiagnosed (i.e., sclerosing or ischemic cholangitis).

When cholangitis is complicated by hepatic abscess, treatment of the primary underlying problem (e.g., ERCP for choledocholithiasis or via endoscopic sphincterotomy) is critical for preventing recurrences since persistent obstruction will result in failure of treatment of the hepatic abscess [74, 76].

6.7 *Surgery*

Surgical drainage is reserved for patients with complicated abscesses (e.g., multilocular or ruptured abscesses), incomplete or difficult percutaneous drainage, inadequate response to medical therapy, those requiring surgical intervention as primary treatment for other disease processes, or those with recurrent disease [71, 74]. Laparoscopic drainage, guided by laparoscopic ultrasound, is also effective and is associated with reduced morbidity [74]. Early intervention (drainage) is warranted in the vast majority of cases.

7 Conclusion

In conclusion, the intricate landscape of intra-abdominal infections, including peritonitis, acute cholangitis, acute calcular cholecystitis, and pyogenic liver abscess, demands meticulous understanding and dynamic management. Timely diagnosis, guided by clinical assessment and microbiological insights, forms the cornerstone of effective treatment. The review illuminates the critical significance of timely intervention in these conditions. From the early commencement of broad-spectrum antimicrobial therapy in peritonitis to the prompt diagnosis and multidimensional approaches in acute cholangitis and acute calcular cholecystitis, each case highlights the significance of well-timed strategies.

Moreover, the review recognizes the evolving microbial landscape, underscoring the necessity of targeted antibiotic regimens guided by local susceptibility patterns. The emergence of percutaneous drainage as a cornerstone intervention for pyogenic liver abscesses further showcases the evolution of management approaches.

In essence, this narrative review bridges the knowledge gap by presenting a cohesive perspective on gastrointestinal infections. By highlighting the evolving paradigms in diagnosis and management, medical professionals are empowered to optimize patient outcomes through early intervention, precise therapies, and a collaborative multidisciplinary approach. As the field of gastrointestinal infection management continues to evolve, the insights shared within this narrative review will undoubtedly contribute to advancing patient care and enhancing clinical outcomes.

References

1. Wilcox CM, Dismukes WE. Spontaneous bacterial peritonitis. A review of pathogenesis, diagnosis, and treatment. Medicine (Baltimore). 1987;66(6):447–56.
2. West KW. Primary peritonitis. St. Louis, MO: Mosby; 1988. p. 1345–8.
3. Salzer WL. Peritoneal dialysis-related peritonitis: challenges and solutions. Int J Nephrol Renov Dis. 2018;11:173–86.
4. Murray CJL, Vos T, Lozano R, Naghavi M, Flaxman AD, Michaud C, et al. Disability-adjusted life years (DALYs) for 291 diseases and injuries in 21 regions, 1990–2010: a systematic analysis for the Global Burden of Disease Study 2010. Lancet. 2012;380(9859):2197–223.

5. Ball CG, Kirkpatrick AW, Williams DR, Jones JA, Polk JD, Vanderploeg JM, et al. Prophylactic surgery prior to extended-duration space flight: is the benefit worth the risk? Can J Surg J Can Chir. 2012;55(2):125–31.
6. Anaya DA, Nathens AB. Risk factors for severe sepsis in secondary peritonitis. Surg Infect. 2003;4(4):355–62.
7. Sartelli M, Catena F, Ansaloni L, Leppaniemi A, Taviloglu K, van Goor H, et al. Complicated intra-abdominal infections in Europe: a comprehensive review of the CIAO study. World J Emerg Surg WJES. 2012;7(1):36.
8. Sartelli M, Bassetti M, Martin-Loeches I, editors. Abdominal sepsis: a multidisciplinary approach. Cham: Springer; 2018. [cited 2023 Jul 29]. (Hot Topics in Acute Care Surgery and Trauma). http://link.springer.com/10.1007/978-3-319-59704-1.
9. Vincent JL, Sakr Y, Sprung CL, Ranieri VM, Reinhart K, Gerlach H, et al. Sepsis in European intensive care units: results of the SOAP study*. Crit Care Med. 2006;34(2):344–53.
10. Johnson D, Mayers I. Multiple organ dysfunction syndrome: a narrative review. Can J Anaesth J Can Anesth. 2001;48(5):502–9.
11. Cavriani G, Domingos HV, Soares AL, Trezena AG, Ligeiro-Oliveira AP, Oliveira-Filho RM, et al. Lymphatic system as a path underlying the spread of lung and gut injury after intestinal ischemia/reperfusion in rats. Shock. 2005;23(4):330–6.
12. Roberts DJ, Ball CG, Kirkpatrick AW. Increased pressure within the abdominal compartment: intra-abdominal hypertension and the abdominal compartment syndrome. Curr Opin Crit Care. 2016;22(2):174–85.
13. Ross JT, Matthay MA, Harris HW. Secondary peritonitis: principles of diagnosis and intervention. BMJ. 2018;361:k1407.
14. Weir-McCall J, Shaw A, Arya A, Knight A, Howlett DC. The use of pre-operative computed tomography in the assessment of the acute abdomen. Ann R Coll Surg Engl. 2012;94(2):102–7.
15. Barie PS, Hydo LJ, Shou J, Larone DH, Eachempati SR. Influence of antibiotic therapy on mortality of critical surgical illness caused or complicated by infection. Surg Infect. 2005;6(1):41–54.
16. Blot SI, Vandewoude KH, De Waele JJ. Candida peritonitis. Curr Opin Crit Care. 2007;13(2):195–9.
17. Van Ruler O, Boermeester MA. Surgical treatment of secondary peritonitis: a continuing problem. Chirurg. 2017;88(S1):1–6.
18. Koperna T, Schulz F. Prognosis and treatment of peritonitis. Do we need new scoring systems? Arch Surg. 1996;131(2):180–6.
19. Mishra S, Tiwary S, Mishra M, Gupta S. An introduction of tertiary peritonitis. J Emerg Trauma Shock. 2014;7(2):121.
20. Billing A, Fröhlich D, Schildberg FW. Prediction of outcome using the Mannheim peritonitis index in 2003 patients. Peritonitis Study Group. Br J Surg. 1994;81(2):209–13.
21. Hudspeth AS. Radical surgical debridement in the treatment of advanced generalized bacterial peritonitis. Arch Surg. 1975;110(10):1233.
22. van Ruler O, Mahler CW, Boer KR, Reuland EA, Gooszen HG, Opmeer BC, et al. Comparison of on-demand vs planned re-laparotomy strategy in patients with severe peritonitis: a randomized trial. JAMA. 2007;298(8):865–72.
23. Marques HS, Araújo GRL, da Silva FAF, de Brito BB, Versiani PVD, Caires JS, et al. Tertiary peritonitis: a disease that should not be ignored. World J Clin Cases. 2021;9(10):2160–9.
24. Chromik AM, Meiser A, Hölling J, Sülberg D, Daigeler A, Meurer K, et al. Identification of patients at risk for development of tertiary peritonitis on a surgical intensive care unit. J Gastrointest Surg. 2009;13(7):1358–67.
25. Attasaranya S, Fogel EL, Lehman GA. Choledocholithiasis, ascending cholangitis, and gallstone pancreatitis. Med Clin North Am. 2008;92(4):925–60.
26. Buyukasik K, Toros AB, Bektas H, Ari A, Deniz MM. Diagnostic and therapeutic value of ERCP in acute cholangitis. ISRN Gastroenterol. 2013;2013:1–4.

27. Sung JY, Costerton JW, Shaffer EA. Defense system in the biliary tract against bacterial infection. Dig Dis Sci. 1992;37(5):689–96.
28. Ahmed M. Acute cholangitis—an update. World J Gastrointest Pathophysiol. 2018;9(1):1–7.
29. Hanau LH, Steigbigel NH. Acute (ascending) cholangitis. Infect Dis Clin N Am. 2000;14(3):521–46.
30. Kinney TP. Management of ascending cholangitis. Gastrointest Endosc Clin N Am. 2007;17(2):289–306.
31. Jain MK, Jain R. Acute bacterial cholangitis. Curr Treat Options Gastroenterol. 2006;9(2):113–21.
32. Qureshi WA. Approach to the patient who has suspected acute bacterial cholangitis. Gastroenterol Clin N Am. 2006;35(2):409–23.
33. Van Den Hazel SJ, Speelman P, Tytgat GNJ, Dankert J, Van Leeuwen DJ. Role of antibiotics in the treatment and prevention of acute and recurrent cholangitis. Clin Infect Dis. 1994;19(2):279–86.
34. Nichols RL, Muzik AC. Enterococcal infections in surgical patients: the mystery continues. Clin Infect Dis. 1992;15(1):72–6.
35. Melzer M, Toner R, Lacey S, Bettany E, Rait G. Biliary tract infection and bacteraemia: presentation, structural abnormalities, causative organisms and clinical outcomes. Postgrad Med J. 2007;83(986):773–6.
36. NIH state-of-the-science statement on endoscopic retrograde cholangiopancreatography (ERCP) for diagnosis and therapy. NIH Consens State Sci Statements. 2002;19(1):1–26.
37. Donelli G, Guaglianone E, Di Rosa R, Fiocca F, Basoli A. Plastic biliary stent occlusion: factors involved and possible preventive approaches. Clin Med Res. 2007;5(1):53–60.
38. Fuks D, Cossé C, Régimbeau JM. Antibiotic therapy in acute calculous cholecystitis. J Visc Surg. 2013;150(1):3–8.
39. Gallaher JR, Charles A. Acute cholecystitis: a review. JAMA. 2022;327(10):965.
40. Everhart JE, Khare M, Hill M, Maurer KR. Prevalence and ethnic differences in gallbladder disease in the United States. Gastroenterology. 1999;117(3):632–9.
41. Strasberg SM. Clinical practice. Acute calculous cholecystitis. N Engl J Med. 2008;358(26):2804–11.
42. Schuld J, Glanemann M. Acute cholecystitis. Visc Med. 2015;31(3):163–5.
43. Claesson B, Holmlund D, Mätzsch T. Biliary microflora in acute cholecystitis and the clinical implications. Acta Chir Scand. 1984;150(3):229–37.
44. Kanafi Vahed L, Khedmat L. Frequency of symptoms associated with gallstone disease: a hospital-based cross sectional study. Eur J Transl Myol. 2018;28(2). [cited 2023 Aug 6]. https://www.pagepressjournals.org/index.php/bam/article/view/7412.
45. Jain A, Mehta N, Secko M, Schechter J, Papanagnou D, Pandya S, et al. History, physical examination, laboratory testing, and emergency department ultrasonography for the diagnosis of acute cholecystitis. Acad Emerg Med. 2017;24(3):281–97.
46. Indar AA. Acute cholecystitis. BMJ. 2002;325(7365):639–43.
47. Ansaloni L, Pisano M, Coccolini F, Peitzmann AB, Fingerhut A, Catena F, et al. 2016 WSES guidelines on acute calculous cholecystitis. World J Emerg Surg. 2016;11(1):25.
48. Marincek B. Nontraumatic abdominal emergencies: acute abdominal pain: diagnostic strategies. Eur Radiol. 2002;12(9):2136–50.
49. Gomi H, Solomkin JS, Schlossberg D, Okamoto K, Takada T, Strasberg SM, et al. Tokyo guidelines 2018: antimicrobial therapy for acute cholangitis and cholecystitis. J Hepatobiliary Pancreat Sci. 2018;25(1):3–16.
50. Okamoto K, Suzuki K, Takada T, Strasberg SM, Asbun HJ, Endo I, et al. Tokyo guidelines 2018: flowchart for the management of acute cholecystitis. J Hepatobiliary Pancreat Sci. 2018;25(1):55–72.
51. Yokoe M, Hata J, Takada T, Strasberg SM, Asbun HJ, Wakabayashi G, et al. Tokyo guidelines 2018: diagnostic criteria and severity grading of acute cholecystitis (with videos). J Hepatobiliary Pancreat Sci. 2018;25(1):41–54.

52. Yoshida M, Takada T, Kawarada Y, Tanaka A, Nimura Y, Gomi H, et al. Antimicrobial therapy for acute cholecystitis: Tokyo guidelines. J Hepatobiliary Pancreat Surg. 2007;14(1):83–90.
53. Grande M, Torquati A, Farinon AM. Wound infection after cholecystectomy. Correlation between bacteria in bile and wound infection after operation on the gallbladder for acute and chronic gallstone disease. Eur J Surg Acta Chir. 1992;158(2):109–12.
54. Solomkin JS, Mazuski JE, Baron EJ, Sawyer RG, Nathens AB, DiPiro JT, et al. Guidelines for the selection of anti-infective agents for complicated intra-abdominal infections. Clin Infect Dis. 2003;37(8):997–1005.
55. Smith TJ, Manske JG, Mathiason MA, Kallies KJ, Kothari SN. Changing trends and outcomes in the use of percutaneous cholecystostomy tubes for acute cholecystitis. Ann Surg. 2013;257(6):1112–5.
56. Akhan O, Akıncı D, Özmen MN. Percutaneous cholecystostomy. Eur J Radiol. 2002;43(3):229–36.
57. Dimou FM, Adhikari D, Mehta HB, Riall TS. Outcomes in older patients with grade III cholecystitis and cholecystostomy tube placement: a propensity score analysis. J Am Coll Surg. 2017;224(4):502–11.e1.
58. Lardière-Deguelte S, Ragot E, Amroun K, Piardi T, Dokmak S, Bruno O, et al. Hepatic abscess: diagnosis and management. J Visc Surg. 2015;152(4):231–43.
59. Liver abscess—StatPearls—NCBI Bookshelf. [cited 2023 Jan 2]. https://www.ncbi.nlm.nih.gov/books/NBK538230/.
60. Khim G, Em S, Mo S, Townell N. Liver abscess: diagnostic and management issues found in the low resource setting. Br Med Bull. 2019;132(1):45–52.
61. Serraino C, Elia C, Bracco C, Rinaldi G, Pomero F, Silvestri A, et al. Characteristics and management of pyogenic liver abscess: a European experience. Medicine (Baltimore). 2018;97(19):e0628.
62. Ruiz-Hernández JJ, León-Mazorra M, Conde-Martel A, Marchena-Gómez J, Hemmersbach-Miller M, Betancor-León P. Pyogenic liver abscesses: mortality-related factors. Eur J Gastroenterol Hepatol. 2007;19(10):853–8.
63. Neill L, Edwards F, Collin SM, Harrington D, Wakerley D, Rao GG, et al. Clinical characteristics and treatment outcomes in a cohort of patients with pyogenic and amoebic liver abscess. BMC Infect Dis. 2019;19(1):490.
64. Tsai FC, Huang YT, Chang LY, Wang JT. Pyogenic liver abscess as endemic disease, Taiwan. Emerg Infect Dis. 2008;14(10):1592–600.
65. Wang J, Liu Y, Lee SS, Yen M, Chen YS, Wang JH, Wann S, et al. Primary liver abscess due to *Klebsiella pneumoniae* in Taiwan. Clin Infect Dis. 1998;26(6):1434–8.
66. Menon A, Kizhakkekarammal P, Rao G. Amoebic vs pyogenic liver abscesses: a comparative study in a tertiary care hospital. J Acad Clin Microbiol. 2015;17(2):89.
67. Su YJ, Lai YC, Lin YC, Yeh YH. Treatment and prognosis of pyogenic liver abscess. Int J Emerg Med. 2010;3(4):381–4.
68. Hope WW, Vrochides DV, Newcomb WL, Mayo-Smith WW, Iannitti DA. Optimal treatment of hepatic abscess. Am Surg. 2008;74(2):178–82.
69. Yu SCH, Ho SSM, Lau WY, Yeung DTK, Yuen EHY, Lee PSF, et al. Treatment of pyogenic liver abscess: prospective randomized comparison of catheter drainage and needle aspiration. Hepatology. 2004;39(4):932–8.
70. Tan YM, Chung AYF, Chow PKH, Cheow PC, Wong WK, Ooi LL, et al. An appraisal of surgical and percutaneous drainage for pyogenic liver abscesses larger than 5 cm. Ann Surg. 2005;241(3):485–90.
71. Johannsen EC, Sifri CD, Madoff LC. Pyogenic liver abscesses. Infect Dis Clin N Am. 2000;14(3):547–63.
72. Olivera MA, Kershenobich D. Pyogenic liver abscess. Curr Treat Options Gastroenterol. 1999;2(2):86–90.

73. Solitary and multiple pyogenic liver abscesses: characteristics of the patients and efficacy of percutaneous drainage. PubMed. [cited 2023 Jun 24]. https://pubmed.ncbi.nlm.nih.gov/9040204/.
74. Sharara AI, Rockey DC. Pyogenic liver abscess. Curr Treat Options Gastroenterol. 2002;5(6):437–42.
75. Sugiyama M, Atomi Y. Pyogenic hepatic abscess with biliary communication. Am J Surg. 2002;183(2):205–8.
76. Prost à la Denise J, Kianmanesh R, Castel B, Flamant Y, Msika S. Syndrome septique après cholécystectomie…. J Chir (Paris). 2008;145(3):278–83.

Descending Necrotizing Mediastinitis: A Life-Threatening Infection

Nissar Shaikh, Umme Nashrah, Umm E Amara, Shoaib Nawaz, and Muhmmad Muslim

Abstract Descending necrotizing mediastinitis (DNM) is a severe, life-threatening necrotizing infection of the fascial covering of the mediastinum. An oropharyngeal cavity is connected to the mediastinum through the retropharyngeal space. It is also called the "dangerous space" as along this fascial connection infection from the mouth and neck rapidly spreads to the mediastinum. The negative intrathoracic pressure and gravity facilitate the spread of infection into the thoracic cavity. Common clinical presentations are dysphonia, dry cough, fever, odynophagia, neck swelling, and pain. The imaging studies mainly computerized tomography (CT) of the neck and chest will be diagnostic of DNM. A crucial treatment component is the use of early broad-spectrum antibiotic medication in conjunction with surgical drainage. A combination of trans-cervical and transthoracic drainage has a survival benefit than the trans-cervical drainage alone.

Keywords Antibiotics · Computerized tomography · Descending necrotizing mediastinitis (DNM) · Dysphonia · Dry cough · Fever · Odynophagia · Surgical drainage

---

N. Shaikh (✉)
Surgical Intensive Care Unit, Hamad Medical Corporation, Doha, Qatar
e-mail: smaheboob@hamad.qa

U. Nashrah
Surgical Intensive Care Unit, Hamad Medical Corporation, Doha, Qatar

Thoracic Surgery/Hamad Medical Corporation, Doha, Qatar

U. E Amara
Deccan College of Medical Sciences, Hyderabad, Telangana, India

S. Nawaz
Apollo Institute of Medical Sciences, Hyderabad, Telangana, India

M. Muslim
New York City Health and Hospitals, New York, USA

N. Shaikh, A. H. Chanda (eds.), *Applied Microbiology in Intensive Care Medicine*, https://doi.org/10.1007/978-981-97-4006-2_8

1 Introduction

Descending necrotizing mediastinitis is a serious rapidly progressive and potentially fatal infection of the mediastinum which originates from the oropharyngeal areas and tracks down the fascial planes of the neck. The most common source of infection is from the odontogenic areas particularly second and third molar teeth. Other common sources include retropharyngeal abscesses and tonsillar and pretracheal areas. Nevertheless, the infection spreads downwards through the pretracheal, visceral, and prevertebral spaces to the mediastinum.

As DNM is a severe and rapidly spreading infection, it carries high mortality which was initially reported to be as high as 60% [1]. Although early surgical procedures and recent improvements in imaging have reduced mortality to about 30%, the infection still raises the risk of septic shock and a protracted course of the disease [2]. The pathologic process of DNM includes tissue edema, effusion, necrosis, and abscess, which may result in pleural and pericardial effusion. The infection can spread in the blood vessels and bloodstream leading to a severe inflammatory response.

2 Clinical Presentation

DNM's mild and nebulous first symptoms can cause delays in infection detection and treatment. Most patients have a primary infection in the oropharyngeal area which may be an odontogenic abscess, retropharyngeal abscess, or tonsillar infection [3]. The symptoms include neck pain, dysphagia, dyspnea, anorexia, chest discomfort, and fever. Inflammatory markers are usually high, including leukocytosis, C-reactive protein, and procalcitonin. Tachycardia, hypotension, and rapid progression to septic shock are not uncommon, so early imaging in suspected cases is the key to diagnosis.

3 Risk Factors

Any immunocompromised state can be a risk factor for developing DNM. Diabetes mellitus is one of the main risk factors for developing descending necrotizing mediastinitis with up to 30% of the patients with the diagnosis having diabetes mellitus [4]. Additional contributing variables include being older, using glucocorticoids, and having an untreated oropharyngeal infection that lasts for a long time [5].

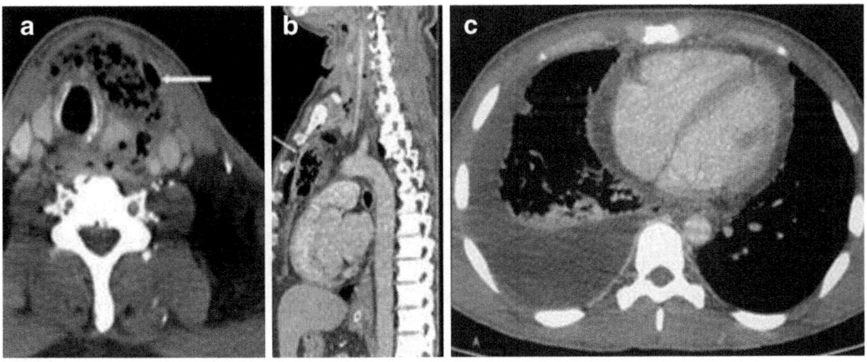

Fig. 1 Showing various stages of extension of DNM

4 Microbiology

The infection is usually polymicrobial including both aerobes and anaerobes. Oral flora including *Streptococcus* sp. (*S. constellatus*, *S. intermedius*, *S. agalactiae*, *S. mitis*) and anaerobes particularly *Peptostreptococcus*, *Bacteroides fragilis*, *Prevotella*, and *Fusobacterium* are the most common causative organisms [4, 6]. Due to the polymicrobial nature of the infection, broad-spectrum antibiotics with anaerobic coverage are indicated as first-line therapy for DNM.

5 Diagnosis (Fig. 1)

Clinical diagnosis of DNM includes a high suspicion in patients presenting with acute hemodynamic compromise with a background of oropharyngeal infection or abscess. High inflammatory markers in a patient presenting with vague septic symptoms with an infection source in the oropharyngeal area is also a suspecting sign. Clinical imaging including radiographs is helpful in diagnosis which may show widened mediastinum and loculated air spaces from gas-forming bacteria. Computed tomography is the gold standard as a diagnostic as well as a therapeutic tool. Multiplanar images can localize the extent of infection, guide surgical management, and also serve for CT-guided drainage of fluid collections [7]. The critical diagnostic fact is the evidence of the continuation of infection from the oropharyngeal region downwards to the mediastinum.

Gram stain and cultures from the infected site are essential to direct antibiotic coverage. The samples may be taken during surgery or CT-guided aspiration of fluid collections. Nevertheless, broad-spectrum antibiotics are initially indicated. Histology gives pinpoint diagnosis of necrosis and inflammation.

Diagnostic criteria for DNM were proposed by Estrera et al. [8] in 1983: (1) clinical evidence of severe oropharyngeal infection, (2) characteristic radiographic features of mediastinitis (Fig. 1), (3) documentation of necrotizing mediastinal infection during operation or autopsy, and (4) establishment of the relationship between DNM and the oropharyngeal process. Modern imaging has reduced this criterion to more of a historical importance.

DNM has been classified based on its severity and extent of infection. Type 1 is only a localized infection above the level of the carina. Type 2A involves only the anterior mediastinum, while type 2B involves both the anterior and posterior mediastinum (Table 1) [9]. This classification helps in deciding the surgical approach for debridement. A drawback of this classification is that it does not include only posterior mediastinal involvement which does occur rarely.

6 Management

The treatment of DNM relies on early diagnosis, broad-spectrum antibiotics, surgical source control, and management of anticipated complications. These patients are typically treated in an intensive care setup. Once the diagnosis is suspected, the patient should be started on broad-spectrum intravenous antibiotics covering gram-positive, negative, aerobic, and anaerobic bacteria. Different regimens can be used according to hospital protocol and antibiogram. Typical antibiotics are clindamycin, piperacillin + tazobactam, vancomycin, and carbapenams [10]. The patient should get a detailed diagnostic CT scan to classify the extent of the disease process.

There is a high chance of airway compromise in severe disease. Airway equipment should be ready, and difficult intubation should be suspected in all cases of DNM. Awake fiberoptic intubation may be considered in such cases to improve safety. Due to the nature of the pathophysiology in these cases, supraglottic airway devices play a little role. In case of failed intubation, cricothyroidotomy is indicated in emergency followed by tracheostomy as the definitive airway.

The patients have a high risk of developing hypotension, hemodynamic instability, and septic shock. Vasopressors should be employed in such cases. Sepsis should be treated according to international Surviving Sepsis Guidelines with aggressive fluid management, antibiotics, and vasopressors as indicated. Early surgical debridement and source control is the mainstay of treatment and improved prognosis.

Different surgical techniques have been employed over time to drain the infected fluids of DNM which include the trans-cervical approach, mediastinoscopy,

Table 1 Showing classification of DNM

Classification of the DNM	
I	The infection is circumscribed to the superior aspect of the mediastinum above the carina Can be treated conservatively
II A	Infection of the lower anterior mediastinum (Prevascular space in the mediastinum classification)
II b	Infection of the lower anterior and posterior mediastinum (Prevascular, previsceral or paratracheal spaces)

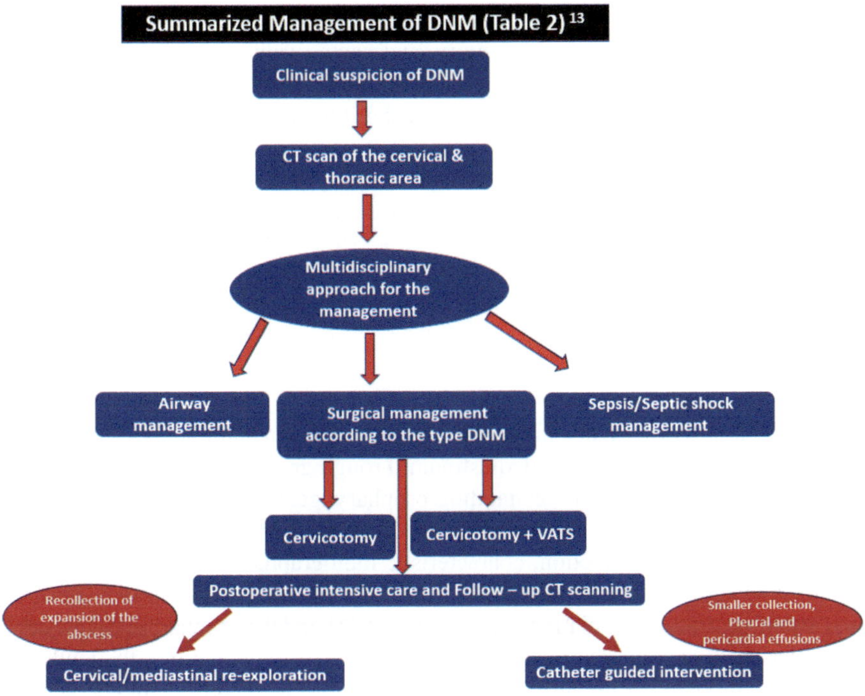

Fig. 2 Summarized management of DNM

video-assisted thoracoscopic surgery (VATS), open thoracotomy, and minimally invasive catheter-guided drainage. Relook surgery is often common in these patients as clearing a continuously spreading infection in a single surgery is limited. The surgical approach is often guided by CT images and the extent of the infection into the anterior or posterior mediastinum [11, 12]. Depending upon the surgical approach used, the patient may end up with open infected wounds which need to be dressed daily, and aggressive wound care is indicated [11–13].

6.1 Management of DNM Summarized in Fig. 2

7 Complications and Prognosis

DNM is a rapidly spreading severe infection and can lead to several regional as well as systemic complications. Local complications include but are not limited to thrombophlebitis of regional veins, vascular erosions, broncho-mediastinal fistula, and vessel rupture and empyema formation. Systemic complications include bacteremia, septic shock, acute kidney injury, and acute respiratory distress syndrome (ARDS) in severe cases.

The prognosis depends on the extent of the infection, early antibiotic therapy, aggressive surgical debridement, and prompt sepsis management and organ support. Historically, these patients had a very high mortality rate of 60%. Recent advances in imaging as well as new minimally invasive surgical approaches have led to a more favorable prognosis for DNM.

8 Conclusion

The DNM usually originates in the oropharyngeal region and travels via the retropharyngeal space and into the mediastinum through gravitation. The immunocompromised status, DM, and long-standing oropharyngeal infections are the risk for complicating the DNM. The diagnostic criteria for DNM are clinical evidence of severe oropharyngeal infection, characteristic radiographic features of mediastinitis, documentation of necrotizing mediastinal infection during operation or autopsy, and establishment of the relationship between DNM and the oropharyngeal process. DNM has been classified according to the extension of infection into the mediastinum. The DNM above the level of carina can be managed conservatively, and the DNM class II requires drainage.

References

1. Panda NK, Simhadri S, Sridhara SR. Cervicofacial necrotizing fasciitis: can we expect a favourable outcome? J Laryngol Otol. 2004;118:771–7.
2. Deu-Martín M, Saez-Barba M, López Sanz I, et al. Mortality risk factors in descending necrotizing mediastinitis. Arch Bronconeumol. 2010;46:182–7.
3. Shaikh N, Maslamani M, Mazhar R, Ahmed AH. Descending necrotizing mediastinitis: report of three cases and review of literature. Arch Int Surg. 2013;3:173–7.
4. Sumi Y. Descending necrotizing mediastinitis: 5 years of published data in Japan. Acute Med Surg. 2014;2(1):1–12. https://doi.org/10.1002/ams2.56. PMID: 29123684; PMCID: PMC5667189.
5. Petitpas F, Blancal JP, Mateo J, et al. Factors associated with the mediastinal spread of cervical necrotizing fasciitis. Ann Thorac Surg. 2012;93:234–8.
6. Brook I, Frazier EH. Microbiology of mediastinitis. Arch Intern Med. 1996;156:333–6.
7. Scaglione M, Pinto A, Giovine S, Di Nuzzo L, Giuliano V, Romano L. CT features of descending necrotizing mediastinitis—a pictorial essay. Emerg Radiol. 2007;14:77–81.
8. Estrera AS, Landay MJ, Grisham JM, Sinn DP, Platt MR. Descending necrotizing mediastinitis. Surg Gynecol Obstet. 1983;157:545–52.
9. Endo S, Murayama F, Hasegawa T, et al. Guideline of surgical management based on diffusion of descending necrotizing mediastinitis. Jpn J Thorac Cardiovasc Surg. 1999;47:14–9.
10. Lanisnik B, Cizmarevic B. Necrotizing fasciitis of the head and neck: 34 cases of a single institution experience. Eur Arch Otorrinolaringol. 2010;267:415–21.

11. Freeman RK, Vallieres E, Verrier ED, Karmy-Jones R, Wood DE. Descending necrotizing mediastinitis: an analysis of the effects of serial surgical debridement on patient mortality. J Thorac Cardiovasc Surg. 2000;119:260–7.
12. Kocher GJ, Hoksch B, Caversaccio M, Wiegand J, Schmid RA. Diffuse descending necrotizing mediastinitis: surgical therapy and outcome in a single-Centre series. Eur J Cardiothorac Surg. 2012;42:e66–72.
13. Ma C, Zhou L, Zhao J-Z, et al. Multidisciplinary treatment of deep neck infection associated with descending necrotizing mediastinitis: a single-Centre experience. J Int Med Res. 2019;47:6027–40.

Acute Adult Supraglottitis: An Update

Majid S. Al-Abdulla, Ahmad Obeidat, Mamoun Ahmed, Bhagyasree Sankar, Arshad Hussain Chanda, and Nissar Shaikh

Abstract Supraglottitis (or epiglottitis) is inflammation of the epiglottis and the supraglottic structures. Without treatment, epiglottitis can progress to a fatal airway obstruction. Prompt recognition and management of epiglottitis are detrimental and essential to achieve good outcomes.

Epiglottitis is a life-threatening condition that needs early and active medical and possible surgical intervention. It is important to be familiar with the anatomy of the upper airway in different age groups. While the physiology of breathing might vary a little, the etiology and infecting organisms, in supraglottitis, are diverse in different age groups.

Clinical identification and diagnosis are sufficient to start treatment measures early. It is advised to encourage a multidisciplinary approach to managing this condition by involving related specialties, to be prepared and to have a strategy in case the scenario turns into an airway emergency, which may lead to serious consequences, such as permanent hypoxic injury, or even death, if airway and oxygenation were not restored promptly.

Alongside managing the airway, physicians and allied health care providers should know how to approach these patients. Physicians should be aware of medical and surgical management options based on their region, infection prevention, and treatment recommendations to start antimicrobial treatment. Maintaining oxygenation is a priority and health care providers should be aware that the airway might

M. S. Al-Abdulla (✉) · M. Ahmed
ENT Section—Department of Surgery—Hamad Medical Corporation, Doha, Qatar
e-mail: MAlabdulla9@hamad.qa

A. Obeidat · A. H. Chanda · N. Shaikh
Surgical Intensive Care Unit, Hamad Medical Corporation, Doha, Qatar
e-mail: achanda@hamad.qa; smaheboob@hamad.qa

B. Sankar
Pharmacy Department, Hamad Medical Corporation, Doha, Qatar
e-mail: BSankar1@hamad.qa

N. Shaikh, A. H. Chanda (eds.), *Applied Microbiology in Intensive Care Medicine*, https://doi.org/10.1007/978-981-97-4006-2_9

quickly be compromised leading to a complete airway obstruction, if appropriate management was delayed or incomplete.

Keywords Epiglottitis · Supraglottitis · Airway obstruction · *Haemophilus* influenza · Antibiotics

1 Introduction

In this chapter, we will discuss larynx anatomy alongside physiological and pathophysiological aspects related to supraglottitis, and we will detail management and treatment options as they may have an essential role in understanding and managing these patients for the best outcomes. Epiglottitis (or supraglottitis) is diagnosed based on clinical findings. Some patients, mainly children and adolescents, may present with severe respiratory compromise and possibly shock. With these age groups, certain entities of diagnosis, anatomy, and treatment have to be considered.

2 Anatomy of the Larynx

The larynx lies in front of the hypopharynx opposite the third to sixth cervical vertebrae. It moves vertically and in an anteroposterior direction during swallowing and phonation. It can be divided into three regions: supraglottic, glottis, and subglottic by the false and true vocal cords. The supraglottis consists of the epiglottis and aryepiglottic folds superiorly as they sweep down to the arytenoids. Its lower border is the ventricular bands (false vocal cords), which form the upper border of the glottis. The glottis includes the vocal cords, anterior commissure, and posterior commissure. The definition of the junction between the glottis and the subglottic has been debated in the literature at some length and is either defined as at the level of the vocal folds or 5–10 mm below. The subglottis becomes the trachea at the lower border of the cricoid [1].

The larynx consists of three unpaired and three paired cartilages. The unpaired cartilages include the thyroid, cricoid, and epiglottis, while the paired are arytenoids, corniculates, and cuneiforms. The thyroid cartilage is the largest of all; and it has two alae that meet anteriorly forming angles of 90° in males and 120° in females. The True vocal cords are attached to the middle of the thyroid angle.

The cricoid cartilage is the only cartilage forming a complete ring, the posterior part expanded to form an arm of the lamina while the anterior part narrowed forming an arch.

The epiglottis is a leaf-like structure, which is made up of elastic cartilage, forming the anterior wall of the laryngeal inlet. It attaches to the body of the hyoid bone by hyoepiglottic ligament, which divides it into suprahyoid and infrahyoid epiglottis. A stalk-like process of epiglottis, the Petiole, attaches the epiglottis to the thyroid angle. The anterior surface of the epiglottis is separated from the thyrohyoid membrane and the upper part of thyroid cartilage by a potential space filled with fat in the pre-epiglottic space, which may be invaded in carcinoma of supraglottic larynx or the base of the tongue.

Furthermore, the paired arytenoid cartilages are pyramidal in shape. Each of them has a base that articulates with the cricoid cartilage; a muscular process directed laterally to give attachment to intrinsic laryngeal muscles; a vocal process directed anteriorly, giving attachment to the vocal cord; and an apex that supports the corniculate cartilage. The corniculate cartilages (of Santorini) are paired and articulate with the apex of the arytenoid cartilages, while the cuneiform cartilages (of Weisberg) are rod shroud–shaped arches in the aryepiglottic fold in front of corniculate cartilage and provide passive support to the fold [1].

The thyroid, cricoid, and most of the arytenoid cartilages are hyaline cartilages, whereas the epiglottis, corniculate, cuneiform, and a tip of arytenoid near the corniculate cartilage are fibroelastic in nature. Hyaline cartilages undergo ossification, which begins at the age of 25 years in the thyroid and a little later in cricoid and arytenoids by 65 years of age.

In addition, the larynx has a thyrohyoid membrane that connects the thyroid cartilage to the hyoid bone; and is pierced by superior laryngeal vessels and internal branch of the superior laryngeal nerve, while the cricothyroid membrane connects the thyroid cartilage to the cricoid cartilage; and cricoid-tracheal membrane which connects the cricoid cartilage to the first tracheal ring.

Intrinsic membranes of the larynx include the cricovocal membrane, which is a triangular fibroelastic membrane. Its upper border is free and stretches between the middle of the thyroid angle to the vocal process of the arytenoid and forms the vocal ligament. The quadrangular membrane is the second intrinsic membrane that lies deep into the mucosa of aryepiglottic folds and is not well defined. It stretches between the epiglottic and arytenoid cartilages. Its lower border forms the vestibular ligament which lies in the false cord.

There are two types of laryngeal muscles: intrinsic, which attach laryngeal cartilages to each other, and extrinsic, which attach the larynx to the surrounding structures. Intrinsic muscles act on vocal cords including posterior arytenoid, lateral cricoarytenoid, inter-arytenoid, thyroarytenoid, vocalis, and cricothyroid. On the other hand, extrinsic muscles connect the larynx to the adjacent structures they are divided into elevators or depressors of the larynx.

The inlet of the larynx is an oblique opening bounded anteriorly by the free margin of the epiglottis, on the sides by aryepiglottic folds, and posteriorly by the inter-arytenoid fold. The vestibule extends from the laryngeal inlet to the vestibular folds. Its anterior wall is formed by the posterior surface of the epiglottis, sides by the aryepiglottic folds, and the posterior wall by mucous membrane over the anterior surface of arytenoids.

The ventricle (sinus of the larynx) is a deep elliptical space between vestibular and vocal folds, also extending a short distance above and lateral to the vestibular fold. The saccule is a diverticulum of mucous membrane which starts from the anterior part of the ventricular cavity and extends upwards between vestibular folds and lamina of the thyroid cartilage, and there are many mucous glands in the saccule [2].

Subglottic space extends from the lower surface of true vocal cords to the lower border of the cricoid cartilage. Vestibular folds (false vocal cords) are two in number, each one consists of a mucous membrane fold that extends anteroposteriorly across the laryngeal cavity, and it contains vestibular ligament, a few fibers of thyroarytenoid muscle, and mucous glands.

Vocal folds (true vocal cords) are two pearly white sharp bands extending from the middle of the thyroid angle to the vocal processes of arytenoids. Each vocal cord consists of a vocal ligament which is the true upper edge of the cricovocal membrane covered by a closely bound mucous membrane with scanty subepithelial connective tissue. Glottis (rima glottidis) is the elongated space between vocal cords anteriorly and vocal processes and the base of arytenoids posteriorly, the anteroposterior glottis is about 24 mm in adult males and16 mm in adult females, and it is the narrowest part of laryngeal cavity.

Moreover, the larynx is composed of many spaces that have clinical importance; of these spaces is the pre-epiglottic space (space of Boyer) filled with fat areolar tissue and some lymphatic. It is bounded by the upper part of the thyroid cartilage and thyrohyoid membrane in front, the hyoepiglottic ligament above, and the infrahyoid epiglottis and quadrangular membrane behind, and laterally, it continues with para-glottic space. Other spaces include para-glottic space, which is bounded by thyroid cartilage laterally, conus elastics anteromedially, the ventricle and quadrangular membrane medially, and mucosa of pyriform fossa posteriorly and continues with pre-epiglottic space. Reinke's space lies under the epithelium of vocal cords, it is a potential space with scanty subepithelial connective tissues, and edema of this space causes fusiform swelling of the membranous cords (Reinke's edema).

3 Embryology of the Larynx

Laryngeal mucosa develops from the endoderm of the cephalic part of the foregut. Laryngeal cartilages and muscles develop from the mesenchyme. Epiglottis is derived from a hypobranchial eminence and thyroid cartilage from the fourth arch, while arytenoid, corniculate, cuneiform, and intrinsic muscles of the larynx are derived from the sixth arch. The superior laryngeal nerve, a branch of the vagus, is the fourth arch nerve and supplies the cricothyroid and constrictors of the pharynx.

An infant differs considerably from that of an adult and has great clinical significance. The infant's larynx is small, conical, and positioned high in the neck opposite C3 or C4 at rest and reaches C1 or C2 during swallowing. This high position allows the epiglottis to meet the soft palate and make a nasopharyngeal channel for nasal breathing during suckling. The diameter of the cricoid cartilage is smaller than the size of the glottis making the subglottic the narrowest part. Laryngeal cartilages are soft and collapse easily with the omega-shaped epiglottis and arytenoids relatively large.

4 Physiology of the Larynx

Protection of the lower airway is one of the functions of the larynx, and phylogenetically, this is the earliest function to develop; voice production is secondary.

The larynx protects the lower passages in three different ways:

(a) Sphincteric closure of the laryngeal opening.
(b) Cessation of respiration.
(c) Cough reflex.

When food is swallowed, its entry into air passage is prevented by the closure of three successive sphincters consisting of the laryngeal inlet, false cords, and true cords which close the glottis [1].

The phonation function of the larynx occurs when vocal cords are kept adducted. Infra-glottic air pressure is generated by the exhaled air from the lungs due to the contraction of thoracic and abdominal muscles, and then, the air force opens the vocal cords to produce a sound that is amplified by the mouth, pharynx, nose, and chest.

Respiration is also considered a function of the larynx which regulates the flow of air into the lungs. Vocal cords abduct during inspiration and adduct during expiration. When the larynx is closed, the chest wall gets fixed, and various thoracic and abdominal muscles can then act best. This is the fixation of chest function which is important in digging, pulling, and climing. Coughing, vomiting, defecation, micturition, and childbirth also require a fixed thoracic cage against a closed glottis [1].

5 Pathophysiology

Infectious supraglottitis is cellulitis of the epiglottis, aryepiglottic fold, and other adjacent tissue. The posterior nasopharynx is the primary source of the pathogens which cause supraglottitis. Microscopic trauma to the epithelial surface (e.g., mucosal damage during viral infection or food during swallowing) may be a predisposing factor. Less common noninfectious conditions cause a local burn or ecchymosis of the supraglottitis and adjacent structures.

For both infectious and noninfectious etiology, swelling of supraglottitis results from edema. The lingual surface of epiglottis and peri-epiglottic tissues has abundant lymphatics and blood vessels that facilitate the spread of infection and subsequent inflammatory process, and clinical scenarios can rapidly deteriorate and lead to airway obstruction, respiratory distress, and death.

Once the infection begins, swelling rapidly progresses to involve the supraglottic larynx (aryepiglottic fold and arytenoids), and the subglottic regions are generally not affected.

In adults, whose cartilages are stiffer, an isolated epiglottic infection and the resultant increase in epiglottic mass may be resisted by the more rigid laryngeal/epiglottic cartilage, but an infection that encompasses more of the tissues of the supraglottitis, leading to edema, can lead to symptoms and an unstable airway.

The cause of epiglottitis is most commonly infectious, whether bacterial, viral, or fungal in origin. In children, *Haemophilus* influenza type B (HIB) is still the most common cause. However, this has decreased dramatically since the widespread availability of immunizations. Other agents such as *Streptococcus pyogenes*, *S. pneumonia*, and *S. aureus* have been implicated. In immunocompromised patients, *Pseudomonas aeruginosa*, *Mycobacterium tuberculosis*, and *Candida* have been named. Noninfectious causes can be traumatic such as thermal, caustic, or foreign body ingestion.

While viruses can cause epiglottitis, a prior viral infection may allow bacterial superinfection to develop. Viruses that may allow a superinfection include varicella-zoster, herpes simplex, and Epstein-Barr viruses.

6 Clinical Presentation

Unlike children, adult patients with supraglottitis often present with vague upper aerodigestive tract symptoms, which could represent several potential diagnoses Table 1. This fact commonly poses a diagnostic dilemma. Therefore, frontline physicians should have a high index of suspicion when dealing with a patient with symptoms that cannot be explained by the physical findings. Establishing the early correct diagnosis and initiating early interventions are essential to avoid significant morbidity and potential mortality, associated with adult supraglottitis. Symptoms such as severe throat pain, odynophagia, and dysphonia, with or without fever, worsening over hours or days, in the absence of obvious abnormal pharyngeal clinical findings, should alert the clinician to investigate to rule out supraglottitis. Clinical manifestation can vary from adults compared to other age groups as described in Table 2.

Table 1 Differential diagnosis of adult supraglottitisremove

Diagnosis	Examples
Infective	Tonsillitis, pharyngitis, deep neck space abscess, tracheitis, laryngopyocele
Inflammatory	Angioedema, radiation therapy
Caustic	Inhaled or ingested
Neoplastic	Base of the tongue, vallecula, pyriform fossa, epiglottis, aryepiglottic fold
Other causes	Trauma, upper aerodigestive F.B.

Table 2 Comparison in clinical presentation between adults and children

Clinical presentation	Pediatrics	Adults
Presenting symptom	Shortness of breath, drooling	Severe sore throat, odynophagia
Stridor	Often present	Often absent
Fever	High grade	Low grade or absent
Onset	Often rapid progression	Often slow progression
Posture	Tripod position	Unlikely

7 History

Detailed history and thorough physical examination should lead the physician to anticipate supraglottitis and its severity, as well as rule out other pathology. Acute adult supraglottitis is characterized by rapidly progressive symptoms over less than 24 h. Patients commonly present with severe sore throat with painful swallowing, frequent spitting of saliva, and change in voice, usually "hot potato speech." These symptoms are associated with high-grade fever, irritability, difficulty in breathing, and stridor. Due to differences in airway anatomy between adults and children, as discussed previously, stridor occurs less frequently in adults with supraglottitis compared to the pediatric population, and its absence does not rule out the diagnosis in adults [3].

In the more common subacute presentation of supraglottitis, patients have vague less severe symptoms, in the absence of high-grade fever. There is no significant shortness of breath, cough, or stridor in the mentioned presentation. These symptoms develop slowly over the course of days. Patients would often give a history of using over-the-counter medications or having sought previous medical advice (Table 2).

8 Examination

Clinical examination should aim to establish a diagnosis that explains the patient's symptoms. Complete oral cavity, throat, and neck examination should point to a clear diagnosis. In the absence of clear physical findings, an attempt should be made to visualize the larynx in a clinically stable patient. Direct visualization of the supraglottitis using flexible fiberoptic nasopharyngolaryngoscopy often confirms the diagnosis by revealing edema and hyperemia of the epiglottis, aryepiglottic folds, and false folds with narrowing of the supraglottic airway. Obliteration of the pyriform fossae with the pooling of saliva is also often found. Visualization also helps to rule out base of the tongue or vallecular mass, parapharyngeal or retropharyngeal abscess, and supraglottic or hypopharyngeal lesions. In sick unstable patients with stridor, shortness of breath, and drooling of saliva, caution should be exercised when considering performing direct visualization. In this situation, other diagnostic tools and initiation of treatment may precede visualization to avoid abrupt airway compromise.

Friedman has described different stages of supraglottitis (epiglottitis). Friedman's staging has been used to standardize the approach to managing adult patients with epiglottis [4, 5]. This four-stage system incorporates respiratory complaints and respiratory efforts, respiratory rate, the presence or absence of stridor, and CO_2 retention and its complications (delirium, cyanosis, loss of conciseness, etc.). According to the stage, Friedman and others advocated appropriate interventions ranging from observation and intravenous medications to aggressive airway management [4, 6].

9 Investigations

Serology such as complete blood count, CRP, procalcitonin, ESR, and blood cultures can be obtained. Elevated white cell count with neutrophilia is a nonspecific finding which is often seen, along with elevated CRP and ESR.

Lateral neck soft tissue X-ray (Fig. 1) has reasonable sensitivity and specificity. The presence of a "thumbprint sign" may suggest the diagnosis of supraglottitis. Thickening of the aryepiglottic folds is another radiographic finding to support the diagnosis. CT scan Image showing a more specific description of the edema and narrowing as seen in Fig. 2.

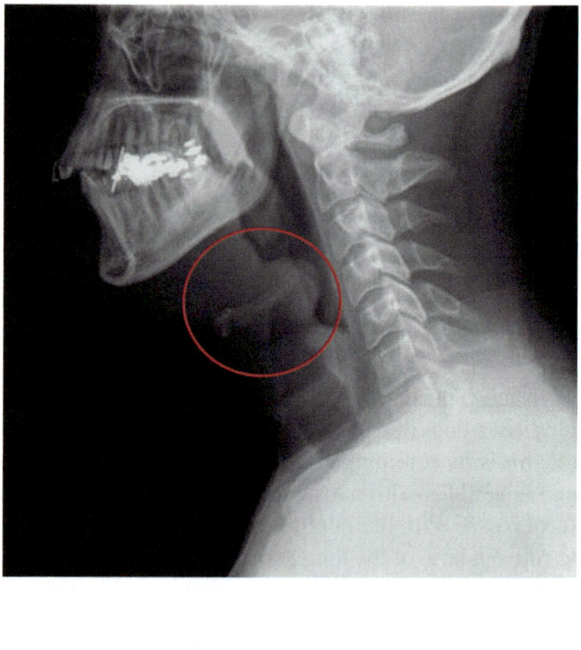

Fig. 1 The thickening of the epiglottis and aryepiglottic folds, with overdistension of the pharynx and subglottic tracheal narrowing

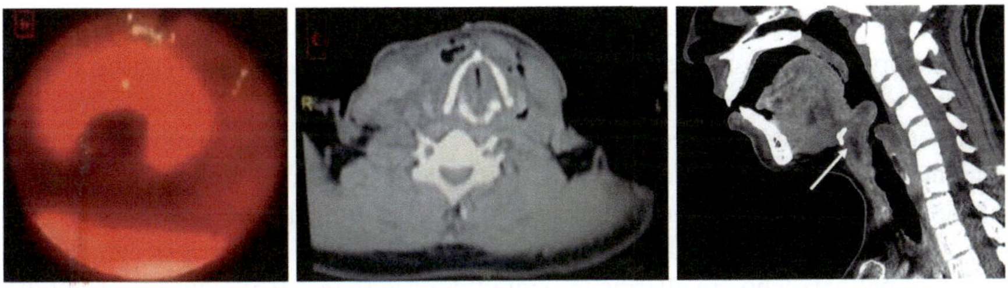

Fig. 2 Thickening and enhancement of the epiglottis, aryepiglottic folds, and obliteration of the pre-epiglottic space

10 Management of Supraglottitis

As acute adult supraglottitis (ASS) is known to cause acute sudden and severe airway obstruction and airway loss, it is mandatory to admit these patients in a well equipped monitored area like a High-Hependency Unit (HDU) or Intensive Care Unit (ICU) with all the advanced airway management and tracheostomy set in near proximity of the patients. The overall management steps of adult supraglottitis should be undertaken by an experienced intensive care physician, and involves using supplementary oxygen, nebulized adrenaline (1:1000 dilution), and intravenous steroids.

10.1 Antimicrobial Therapy

In a stable patient condition or at-least once the airway is secured, a culture swab should be performed, and antibiotics should be started.

The common used antimicrobial agent is ceftriaxone, followed by piperacillin + tazobactam, amoxicillin + clavulanic acid, meropenem, azithromycin, and antifungals (anidulafungin).

Soon after sending the swab culture, patients with infectious supraglottitis should receive empirical antibiotic therapy once the culture and sensitivity toward the most likely organism and the suitable antimicrobial therapy regimen should be commenced.

The major organisms that cause supraglottitis in adults are:

- *Haemophilus* influenza type B.
- *Klebsiella pneumoniae.*
- *Candida* pneumonia.
- *Streptococcus pneumoniae*, including strains that may be penicillin-resistant.
- Group A *Streptococcus.*
- *Staphylococcus aureus*, including community-acquired methicillin-resistant *S. aureus* (MRSA) strains.

Most common antibiotics that are usually administered should cover common respiratory and oral cavity flora including third-generation cephalosporins (cefuroxime, ceftriaxone, and cefotaxime). Patients, who were admitted without risk factors (immunosuppression or diabetes) or severe symptoms, can be treated with amoxicillin with clavulanic acid, and those who present with severe symptoms will better be treated with third-generation cephalosporin as monotherapy. If the patient is immunosuppressed or diabetic or has epiglottic abscesses they can be treated with third-generation cephalosporin with metronidazole.

Other antibiotics like quinolones and piperacillin with tazobactam can be preferred in patients who are allergic to penicillin or with resistant cultures.

Regimens of antibiotics are further described with initial and maintenance doses in Table 3.

10.2 Therapy with Glucocorticosteroids

Supraglottitis is an inflammation of epiglottitis and supraglottis including the pharynx, uvula, base of the tongue, aryepiglottic folds, or the false vocal cord. Controversy still exists with the use of corticosteroids in supraglottitis. The use of corticosteroids is to stabilize endothelial permeability and therefore decrease extracellular and intracellular edema.

If sepsis and/or septic shock is diagnosed in supraglottitis, the use of corticosteroids is recommended while in the intensive care unit to monitor for adrenal suppression from inflammatory responses. The use of corticosteroids dampen the production of inflammatory markers like C-reactive protein, procalcitonin, febrile response, and white cell count.

Corticosteroids are not proven to be beneficial in such patients, and their use may mask clinical signs and physiological variables (hydrocortisone 100 mg bolus followed by 40 mg of methylprednisolone every 12 h [1] and intravenous dexamethasone 4 mg, every 6 h [9]).

Table 3 Intravenous antimicrobial treatment regimens [7, 8]

Antibiotics regimen	Dosing
Ceftriaxone **OR**	2 g daily
Cefotaxime	2 g Q 4–8 h
Vancomycin **OR**	Loading dose: 15–35 mg/kg Consider renal dose adjustments for maintenance only
Clindamycin **OR**	600–900 Q8 h
Oxacillin **OR**	1–2 g Q4–6 h
Nafcillin **OR**	2 g Q4 h
Cefazolin	1–2 g Q6 h
Intravenous piperacillin/tazobactam	3.375 g Q6 h
Micafungin	100 mg once daily

10.3 Antipyretic Agents

Antipyretics (acetaminophen) is used only in patients with an uncontrollable high fever. The excess use of antipyretics suppresses useful proteins such as heat shock protein D, which is a significant inhibitor of proinflammatory cytokines D and increases antibiotics' minimum inhibitory concentration (MIC).

Artificial suppression of a natural response to infection may harm the progress of the illness and mask the febrile response and other clinical variables.

In the case of measuring the severity and prognosis of acute hepatic failure by monitoring the prothrombin time (PT), hepatologists would not correct the PT unless the clinical situation warranted reversal.

10.4 Adrenaline Nebulization

Adrenaline nebulization helps by decreasing local tissue edema.

Dosing: 1 mg of 1 adrenaline (1 mL of 1:1000 diluted to 5 mL with 0.9% saline) over 10 min to avoid cardiovascular adverse effects and administered via a nebulizer is of benefit in some cases of upper airway obstruction [10].

11 Airway Management in Supraglottitis

AAS is characterized by an upper airway infection with significant upper airway edema causing not only airway narrowing but potential complete airway loss, which necessitates prompt recognition and intervention [11]. When patients present with severe symptoms and decompensation or fail medical therapy as described above, urgent airway management must be initiated by an expert anesthesiologist or intensivist with possible emergent tracheostomy by a surgeon. Next will address the indications and considerations for airway management in supraglottitis, as well as potential complications arising from airway edema, infection, inflammation, and loss of airway.

11.1 Indications for Securing Airway in AAS

In the literature, there are no clear indications defined regarding when ASS patients should be intubated or tracheostomized to secure the airway. However, it is suggested to keep a very low threshold for early airway intubation in all patients with ASS, and particularly in AAS patients with stridor [11]. Clinical signs of severe airway compromise that should prompt urgent intervention are stridor and respiratory distress, drooling, patient insisting on maintaining positioning, agitation, and inability to speak or swallow. Delays in intubation in these scenarios may lead to respiratory arrest followed by cardiac arrest requiring emergency tracheostomy and impossible intubation. Soft-tissue neck x-ray in a lateral view may depict "thumb sign" which signifies edema and narrowing the airway. Neck CT scan is more sensitive and specific in showing significant narrowing or total occlusion of the airway [12].

11.2 *Airway Management Strategies*

The airway management ASS needs a rapid but a stepwise approach and organized teamwork. Next we will describe the stepwise approach to airway management. Figure 3 shows an algorithm describing how to approach an airway with a patient with epiglottitis [13].

11.2.1 Airway Management in Mild AAS

Mild AAS patients can be managed with non-invasive measures in high acuity areas like HDU and ICU under close observation. In an intensive care situation, a patient with a supraglottic airway diameter of more than 50% can initially be closely monitored without the use of an artificial airway [14]. The strategies include

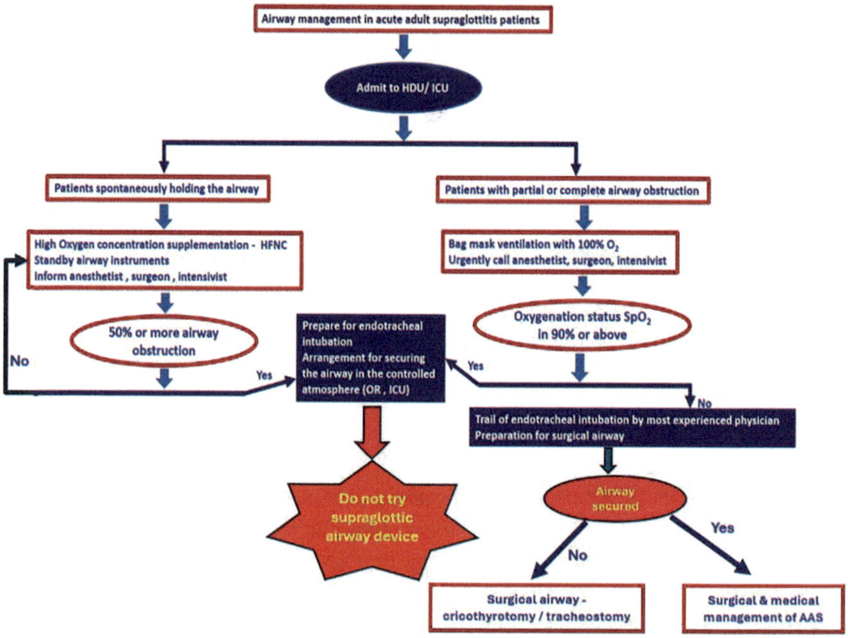

Fig. 3 Airway management in AAS

humidified oxygen by high-flow nasal cannula, continuous positive airway pressure (CPAP) along with antibiotics, nebulized epinephrine, and corticosteroids.

Airway Evaluation: ASS is a known difficult airway. It is crucial to assess the airway before any attempt to intubation to predict difficulty. Factors such as prominent uvula, severe upper airway edema, limited mouth opening, and signs of impending airway obstruction are the signs of a potentially challenging airway.

11.2.2 Initial Stabilization

The goal is to maintain oxygenation and ventilation while preparing for definitive airway management. This can be achieved by providing high-flow oxygen and helix if available. Fluid resuscitation is also important as these patients are often fluid deficient due to odynophagia.

11.2.3 Expertise and Skill

Anesthesiologist's, intensivist's, and surgeon's early involvement is essential and beneficial. Their expertise in advanced airway techniques and equipment can significantly improve patient outcomes.

11.2.4 Definitive Airway Management

In patients with supraglottitis, endotracheal intubation is the method of choice for maintaining the airway. The occluded supraglottic structures are bypassed, ensuring a patent airway for oxygenation and ventilation. Direct laryngoscopy or video laryngoscopy by experienced operator is highly recommended.

Surgical airway intervention such as cricothyroidotomy or tracheostomy are lifesaving procedures if intubation fails (cannot ventilate, cannot intubate) or is impossible. Supraglottic airway devices like laryngeal mask airway are not effective in supraglottitis. "Do **not** attempt these devices in supraglottitis" (Fig. 3).

11.3 Complications of Airway Management

Impossible Intubation: Difficult intubation may become impossible due to edematous airway structures and multiple collections in the supraglottic area limiting the visualization. Awake fiberoptic intubation or video laryngoscopy is now becoming the standard of care in these situations but may not be helpful in the visualization of the vocal cord due to multiple collections. In this scenario, surgical airway is the only option. Failure to secure the airway in these situations can lead to hypoxia and hypoxia-related organ damage [14].

11.3.1 Complication During Airway Manipulation

Bleeding and flare-up of infection are common complications. This is due to injury to inflamed airway mucosa or due to surgical airway interventions [15].

11.3.2 Tube Dislodgment

As the endotracheal tube needs to traverse a relatively longer path due to inflammatory edema of the airway, this can risk accidental extubation with neck movement during nursing care and patient transfer. Secure fixation and close monitoring are necessary to prevent tube dislodgment.

12 Morbidity and Mortality

With proper and prompt detection and management, the majority of epiglottitis cases will have good outcome [16]. Even patients who need intubation are usually extubated within a few days complication-free. However, when the diagnosis and management are delayed, death can occur.

The cause of death is usually due to sudden upper airway obstruction and difficulty in intubating the patient, with extensive swelling of the laryngeal structures. Added to that, every patient with acute epiglottitis is to be reviewed by a team of ENT surgeons and anesthesiologists, to be prepared for all possible scenarios of airway compromise. Internationally, a mortality rate of 3–7% has been reported in patients with compromised airways [16, 17].

13 Conclusion

Epiglottitis is a life-threatening condition that necessitates prompt and effective medical and possible surgical intervention. Patients who present with epiglottitis must seek urgent care. Oropharyngeal examination or oral manipulation should be minimized, or even any stress-inducing interventions, such as IV canulation in children, as they may exacerbate airway obstruction. The diagnosis of epiglottitis can be clinical or assisted by radiograph, fiberoptic, or direct laryngoscopy.

The management of epiglottitis involves securing the airway, typically by endotracheal intubation via the oral or nasal route. Early and specific antibiotic administration is crucial. The prognosis of epiglottitis depends on the severity of the condition and the response to treatment, and delayed identification and management lead to a near-fatal or fatal outcome.

Epiglottitis can be prevented by immunization against Hib, which is recommended for all children under 5 years of age. Hand hygiene, smoking cessation, and avoiding contact with people who have respiratory infections can also reduce the risk of epiglottitis.

Multi-disciplinary team management and adequate preparation before approaching the airway usually end with the best outcomes. The surgeon, anesthetists, intensivists, and pediatricians all have a role in establishing a successful plan of management and treatment.

References

1. Watkinson JC, Clarke RW, Patel LN, Lester S. Anatomy of the larynx and tracheobronchial tree (Chapter 58). In: Scott-Browns otolaryngology head and neck surgery, vol. 3. 8th ed. p. 884–9.
2. Gugatschka M, Hortobagyi D, Ker L. Anatomy and microanatomy of the larynx. In: Remacle M, Eckel HE, editors. Textbook of surgery of larynx and trachea. Cham: Springer; 2022. https://doi.org/10.1007/978-3-031-09621-1_3.
3. Wadie M, Adam SI, Sasaki CT. Development, anatomy, and physiology of the larynx. In: Shaker R, Belafsky P, Postma G, Easterling C, editors. Principles of deglutition. New York, NY: Springer; 2013. https://doi.org/10.1007/978-1-4614-3794-9_13.
4. Frantz TD, Rasgon BM, Quesenberry CP Jr. Acute epiglottitis in adults. Analysis of 129 cases. JAMA. 1994;272(17):1358–60. PMID: 7933397
5. Friedman M, Toriumi DM, Grybauskas V, Applebaum EL. A plea for uniformity in the staging and management of adult epiglottitis. Ear Nose Throat J. 1988;67:873–80.
6. Lagier A, Giovanni A. Physiology of the larynx. In: Remacle M, Eckel HE, editors. Textbook of surgery of larynx and trachea. Cham: Springer; 2022. https://doi.org/10.1007/978-3-031-09621-1_1.
7. Fujimura S, Kojima T, Okanoue Y, Shoji K, Inoue M, Hori R. Discrimination of "hot potato voice" caused by upper airway obstruction utilizing a support vector machine. Laryngoscope. 2019;129:1301–7. https://doi.org/10.1002/lary.27584.
8. Madhotra D, Fenton J, Makura Z, et al. Airway intervention in adult supraglottitis. Ir J Med Sci. 2004;173:197–9. https://doi.org/10.1007/BF02914550.
9. Ward MA. Emergency department management of acute respiratory infections. Semin Respir Infect. 2002;17(1):65–71. https://doi.org/10.1053/srin.2002.31692. PMID: 11891520.
10. Shah RK, Roberson DW, Jones DT. Epiglottitis in the Hemophilus influenza type B vaccine era: changing trends. Laryngoscope. 2004;114(3):557–60. https://doi.org/10.1097/00005537-200403000-00031. PMID: 15091234.
11. Baxter FJ, Dunn GL. Acute epiglottitis in adults. Can J Anaesth. 1988;35(4):428–35. https://doi.org/10.1007/BF03010869. PMID: 3042183.
12. Shaikh N, Nawaz S, Ahmad K, Al MM. Acute adult Supraglottitis: an impending threat to patency of airway and life. Cureus. 2020;12(8):e9976. https://doi.org/10.7759/cureus.9976. PMID: 32983678; PMCID: PMC7510521.
13. Sajeermohammed A, Mathias R, Shaikh N, Idayathullah S, Balal A, Tharayil A, Shallik N. Acute adult supraglottitis: difficult intubation & potential airway loss. Trends Anaesth Crit sCare. 2023;50:101256. https://doi.org/10.1016/j.tacc.2023.101256.
14. Mayo-Smith MF, Spinale JW, Donskey CJ, et al. Acute epiglottitis. An 18-year experience in Rhode Island. Chest. 1995;108:1640.
15. Visvanathan T, Lee J. Epiglottitis. In: StatPearls. Treasure Island, FL: StatPearls Publishing; 2022.
16. Ng HL, Sin LM, Li MF, Que TL, Anandaciva S. Acute epiglottitis in adults: a retrospective review of 106 patients in Hong Kong. Emerg Med J. 2008;25(5):253–5. https://doi.org/10.1136/emj.2007.050153. PMID: 18434453
17. McEwan J, Giridharan W, Clarke RW, Shears P. Paediatric acute epiglottitis: not a disappearing entity. Int J Pediatr Otorhinolaryngol. 2003;67(4):317–21. https://doi.org/10.1016/s0165-5876(02)00393-2. PMID: 12663101

Meningitis and Neurosurgical Treatment an Update

Muhammad Mohsin Khan, Muhammad Mohsin Arshad, Kazim Mohammed, Javeed Iqbal, and Abdulnasser Thabet

Abstract Meningitis and ventriculitis, serious infections causing inflammation of the protective membranes of the brain and spinal cord, pose significant challenges in neurosurgical practice. The management of these conditions often requires interventions beyond antibiotic administration and supportive care. We discuss the role of external ventricular drainage (EVD), a neurosurgical technique used for cerebrospinal fluid drainage, in managing these infections and other neurosurgical techniques for treating these infections. The EVD is particularly beneficial in controlling elevated intracranial pressure, a complication of these infections, and facilitating the administration of intrathecal antibiotics.

Moreover, EVD presents a method for the procurement of cerebrospinal fluid samples for microbiological examination. Notwithstanding the potential perils and difficulties linked with EVD, including infection, hemorrhaging, catheter misplacement, and cerebrospinal fluid seepage, its employment can avert grave neurological sequelae in patients grappling with meningitis and ventriculitis. The intrathecal delivery of antibacterial agents, for instance, ampicillin, ceftazidime, ceftriaxone, gentamicin, meropenem, and vancomycin, has demonstrated efficacy in managing these disorders. The current manuscript emphasizes the imperative for judicious choice and observation of antibiotic utilization, predicated on the root cause of infection and patient-centric clinical determinants.

Keywords Antimicrobial treatment · Coated ventricular catheters · Intrathecal antibiotics · Intraventricular antibiotics · Multidrug-resistant gram-negative · Post-neurosurgical meningitis

M. M. Khan (✉)
Neurosurgery Department, Hamad Medical Corporation, Doha, Qatar

University of Dresden, Dresden, Germany
e-mail: MKhan51@hamad.qa

M. M. Arshad · K. Mohammed · J. Iqbal · A. Thabet
Neurosurgery Department, Hamad Medical Corporation, Doha, Qatar

N. Shaikh, A. H. Chanda (eds.), *Applied Microbiology in Intensive Care Medicine*, https://doi.org/10.1007/978-981-97-4006-2_10

1 Introduction

Meningitis transpires when inflammation occurs in the meninges—the pia, arachnoid, and dura, which are jointly termed the leptomeninges, the tissues encapsulating the nervous system. It is characterized by an unusual accumulation of WBCs in the CSF [1]. On the other hand, ventriculitis relates to the inflammation of the ependymal lining of the central nervous system's ventricular system. This can be a consequence of infection or post-neurosurgical procedures, culminating in meningitis. It is a severe infection that complicates central nervous system operations or is associated with neurosurgical apparatuses or drainage catheters. The meninges are composed of three segments: the pia, arachnoid, and dura mater. Bacterial meningitis signifies infection of the arachnoid mater and the CSF in both the subarachnoid space and the cerebral ventricles.

The clinical manifestations of meningeal inflammation have been chronicled in various historical texts, with surgeon John Abercrombie first elucidating it in 1828. Prior to the antibiotic era, the disease was typically fatal. However, even with significant advancements in healthcare, the mortality rate for the disease hovers around 25%. In 2015, the prevalence of meningitis was 8.7 million, with fatalities reaching 379,000. The primary culprits of community-acquired bacterial meningitis in young adults in developed nations are *Streptococcus pneumonia*, *Neisseria meningitides*, and *Listeria monocytogenes*, predominantly in patients over 50 years of age or those with defects in cell-mediated immunity [2].

Systemic infection or meningitis causes ventriculitis. It is also known as pyogenic ventriculitis, epididymitis, ventricular empyema, or pyocephalus. It is an idle but deadly infection that persists after meningitis treatment. In order to be effective, treatment must be initiated as soon as possible. Particularly at risk are patients with external ventricular drains or intraventricular catheters.

2 Epidemiology

During the era before antibiotics, bacterial meningitis was practically always fatal, with a mortality rate nearing 100%. However, the introduction of antibiotics substantially reduced the mortality rates associated with *H. influenzae* and *Neisseria meningitides* infections to less than 10% and the mortality rate for *S. pneumoniae* to 30% (Table 1). Mortality from *S. pneumoniae* meningitis has further decreased in high-income countries with the utilization of adjunctive steroids. Despite the effectiveness of current antibiotics and the availability of conjugate vaccines for the three most common meningeal pathogens, meningitis continues to be a significant global health disease.

In the USA, the incidence of meningitis is 1.38/100,000 with a case fatality rate of 14.3%. The incidence of meningitis is alarmingly high in Africa, particularly in

Table 1 Characteristic features of common causes of bacterial meningitis (modified and adapted from Rodrigo Hasbun et al. Epidemiology of bacterial meningitis in adults. http://www.uptodate.com)

Organism	Site of entry	Age	Predisposing conditions
Neisseria meningitidis	Nasopharynx	All ages	Usually none, rarely complement deficiency
Streptococcus pneumoniae	Nasopharynx, direct extension across skull fracture, or from contiguous or distant foci of infection	All ages	All conditions that predispose to pneumococcal bacteremia, fracture of the cribriform plate, cochlear implants, cerebrospinal fluid otorrhea from basilar skull fracture, defects of the ear ossicle (Mondini defect)
Listeria monocytogenes	Gastrointestinal tract, placenta	Older adults and neonates	Defects in cell-mediated immunity (e.g., glucocorticoids, transplantation [especially renal transplantation]), pregnancy, liver disease, alcoholism, malignancy
Coagulase-negative staphylococci	Foreign body	All ages	Surgery and foreign body, especially ventricular drains
Staphylococcus aureus	Bacteremia, foreign body, skin	All ages	Endocarditis, surgery, and foreign body, especially ventricular drains; cellulitis, decubitus ulcer
Gram-negative bacilli	Various	Older adults and neonates	Advanced medical illness, neurosurgery, ventricular drains, disseminated strongyloidiasis
Haemophilus influenzae	Nasopharynx, contiguous spread from local infection	Adults, infants, and children if not vaccinated	Diminished humoral immunity

the region known as the meningitis belt, which includes Ethiopia, Senegal, and Nigeria.

The incidence of some of the organisms in 2010 per 100,000 [3] is as follows:

1. *Streptococcus pneumoniae*: 0.3.
2. Group B *Streptococcus*.
3. *Neisseria meningitidis*: 0.123.
4. *Haemophilus influenzae*: 0.058.
5. *Listeria monocytogenes*.

Staphylococcus aureus is the bacterium that is most frequently identified following surgical operations in relation to central lines and trauma; specifically, these bacteria are relatively rare, and their presence could suggest an elevated risk of infection. In immunocompromised patients who have decreased immunity, tuberculosis should be taken into account. Lyme disease prevalent in endemic areas is caused by *Borrelia burgdorferi*. There is evidence suggesting that *Treponema pallidum* (TP) can also contribute to HIV/AIDS that is in individuals with multiple sexual partners; TP is a very prevalent cause of sexually transmitted infections

(STIs) and can present challenges in treatment. If TP is left untreated, HIV/AIDS can result in fatality.

Escherichia coli is a very common pathogen in newborns. They are non-polio enteroviruses (group B coxsackievirus and echovirus) and are the most common viral causes of meningitis. Other viruses, such as mumps, parechovirus, herpesviruses (including Epstein-Barr virus, herpes simplex virus, and varicella-zoster virus), measles, influenza, and arboviruses (West Nile, La Crosse, Powassan, Jamestown Canyon) can also cause meningitis. Fungal meningitis typically occurs in patients who are immunocompromised, such as those with HIV/AIDS, those undergoing chronic corticosteroid therapy, and cancer patients.

Fungal meningitis is caused by the following pathogens:

1. *Cryptococcus neoformans.*
2. *Coccidioides immitis.*
3. *Aspergillus.*
4. *Candida.*
5. Mucormycosis.

On the flip side, there is no clear definition for ventriculitis and no acknowledged diagnostic criteria. It needs to be clarified if ventriculitis or catheter-related infections and positive CSF cultures all describe the same condition.

In ventriculitis, catheter-related infection, shunt-related infections, or hematogenous spread, colonization or contamination is hard to distinguish.

The ambiguity of the term means it is hard to measure the incidence. Ventriculitis secondary to meningitis is more common in infants. Risk factors are associated with a lowered immunity to various diseases. The causative organism is more virulent. Meningitis can often fail to respond to antibiotics, or it may relapse. If ventriculitis is suspected, it should be considered. There are several suggested mechanisms for the hematogenous spread of the virus to the choroid plexus, including direct contact with the blood, the release of virus particles from the eye, or the presence of an open wound on the head. Septations within the ventricles can build up, which can lead to multiloculated hydrocephalus. This condition is more widespread with bacterial infections, and the prognosis is generally worse. Gram-negative bacteria include species such as *Escherichia coli* and *Staphylococcus aureus*; these are typically followed by *S. aureus*.

There has been an increasing incidence of gram-negative meningitis, which is likely due to an increase in meningitis, which is a challenge because it is an indolent disease and tends to recur. The incidence of catheter-related ventricular infection (or healthcare-associated ventricular infection) ranges from 0 to 45% depending on the insertion technique and management.

Because most of the studies are done in a single center with a small amount of patience and are retrospective, there is difficulty in assessing the CSF pleocytosis. And there is a lack of clear definition of the severity of the underlying illness.

A multi-center study done recently found that rates of EVD infection fluctuate between 3 and 18% and from 4.8 to 12.7/1000 EVD [4]. Catheter-related ventriculitis is a serious complication associated with catheter use and can be particularly harmful to people who are infected with certain types of bacteria. Gram-positive cocci found on skin flora in 50–60% of infections are most likely coagulase-negative *Staphylococcus*. Other common bacteria include *Corynebacterium*, *Bacillus*, *Micrococcus*, and *Propionibacterium*.

Due to prolonged hospitalization and increased use of antibiotics targeting gram-negative bacteria. There is a rise in drug-resistant gram-negative bacteria causing meningitis. *Streptococcus pneumoniae* and gram-negative bacteria are the most common cause of meningitis and ventriculitis in patients following head trauma and in patients who have fractures on the skull base and with persistent unresolvable CSF leak. Meningitis can be caused by oral flora [5].

3 Etiology

Meningitis is a disease of the meninges, which is a layer of protective covering around the brain and spinal cord. It can be caused by a variety of infective organisms causing infection in the other parts of the body, including pneumonia and meningococcal disease. Meningitis can be serious and lead to brain damage or even death. The meninges are membranes that line the central nervous system, protecting the brain and spinal cord. Encephalitis is a disease in which the brain is inflamed. This can be caused by a variety of factors but is most commonly caused by infection. Meningitis can also be caused by infection or by other medical conditions. Infectious meningitis is often caused by viruses, such as the common cold, while non-infectious meningitis can be caused by various diseases, such as cancer or paraneoplastic syndromes.

4 CNS Infections and Their Neurosurgical Treatments

Central nervous system (CNS) infections represent a crucial focus area within neurology and neurosurgery, significantly impacting patient health outcomes. Among these disorders, meningitis denotes an inflammation of the membranes encasing the brain and spinal cord, generally incited by bacterial or viral infections. In severe cases, when accompanied by elevated intracranial pressure, external ventricular drainage (EVD) becomes a viable option. Rarely, due to edema and midline shift, a decompressive craniectomy may be called for.

Encephalitis, another such condition, encompasses inflammation of the brain tissue itself, often instigated by viral infections. In these instances, a brain biopsy may be required for diagnosis confirmation or infected tissue excision. A brain abscess, conversely, is a pus accumulation in the brain, typically stemming from bacterial infections. Here, treatment typically demands surgical drainage or abscess resection and the abscess wall, paired with antibiotics to eradicate the infection.

Subdural empyema manifests as a pus accumulation between the brain and the surrounding membrane, often engendered by bacterial infections. The foremost treatment in this case is the surgical drainage of the empyema, succeeded by antibiotic treatment. In certain scenarios, the removal of a portion of the affected brain tissue may be necessitated. A parallel condition, cerebral or spinal epidural abscess, involves pus collection in the space between the skull and the brain or between the vertebrae and the spinal cord. Here, surgical drainage and evacuation of the abscess or the infectious collection are typically succeeded by antibiotic treatment.

Hydrocephalus is a condition characterized by an excess of cerebrospinal fluid (CSF) in the brain, often triggered by obstruction or infection. The usual treatment involves the placement of a shunt, a slender tube that enables surplus CSF to drain from the brain and be reabsorbed by the body. In certain situations, an endoscopic third ventriculostomy, which creates a new conduit for the CSF to exit the brain, may be performed by a surgeon.

Neurocysticercosis is a brain parasitic infection caused by the larvae of the pork tapeworm. Here, the treatment may involve surgical removal of the cysts, along with medications to exterminate the remaining larvae. In some cases, a shunt procedure to alleviate brain pressure may be needed. Lastly, tuberculosis meningitis is a meningitis form incited by the bacteria causing tuberculosis. Treatment typically comprises a combination of antibiotics and corticosteroids to alleviate inflammation. In certain scenarios, a procedure to drain surplus fluid and alleviate brain pressure may be necessary. These diverse CNS infections and their respective treatments underscore the intricacy and diversity of neurosurgical intervention in infectious diseases. Table 2 provides a summary of various infections and their treatment.

5 Role of EVD in Treating Meningitis and Ventriculitis

External ventricular drainage (EVD) is a commonly used neurosurgical technique that involves placing a catheter into the ventricular system of the brain to drain cerebrospinal fluid (CSF).

This technique is used in the management of a variety of neurological conditions such as hydrocephalus, traumatic brain injury, and subarachnoid hemorrhage. In addition, EVD can also be used in the management of meningitis and ventriculitis, a serious infection that can lead to inflammation of the meninges, the protective membranes that surround the brain and spinal cord.

The most common cause of bacterial meningitis and ventriculitis is *Streptococcus pneumoniae*, followed by *Neisseria meningitidis* and *Haemophilus influenzae*. The clinical presentation of meningitis can vary depending on the age of the patient, the severity of the infection, and the causative organism. However, common signs and symptoms include fever, headache, neck stiffness, photophobia, and altered mental status.

The management of meningitis and ventriculitis typically involves prompt administration of antibiotics and supportive care. In severe cases, such as those

Table 2 CNS infections and their respective treatments

CNS infection	Description	Neurosurgical treatment
Meningitis [6]	Inflammation of the membranes surrounding the brain and spinal cord, usually caused by bacteria or viruses	If the infection is severe and causes high intracranial pressure, EVD could be a viable option with an increase in edema, and midline shift decompressive craniectomy might be needed in rare cases
Encephalitis [7]	Inflammation of the brain, often caused by a viral infection	Brain biopsy to confirm the diagnosis or remove any infected tissue
Brain abscess [8]	A collection of pus in the brain, usually caused by a bacterial infection	Treatment typically involves surgical drainage or resection of the abscess and abscess wall, as well as antibiotics to clear the infection
Subdural empyema [9]	A buildup of pus between the brain and the membrane surrounding it, often caused by a bacterial infection	Surgical drainage of the empyema is the primary treatment, followed by antibiotics to clear the infection. In some cases, a surgeon may need to remove a portion of the affected brain tissue
Cerebral or spinal epidural abscess [10]	A collection of pus in the space between the skull and the brain or between the vertebrae and the spinal cord, usually caused by a bacterial infection	Treatment typically involves surgical drainage and evacuation of the abscess or infective collection, followed by antibiotics to clear the infection
Hydrocephalus [11]	A buildup of cerebrospinal fluid (CSF) in the brain, often caused by an obstruction or infection	Treatment may involve placing a shunt, a thin tube that allows excess CSF to drain from the brain and be reabsorbed by the body. In some cases, a surgeon may need to perform an endoscopic third ventriculostomy, which creates a new pathway for CSF to flow out of the brain
Neurocysticercosis [12]	A parasitic infection of the brain caused by the larvae of the pork tapeworm	Treatment may involve surgical removal of the cysts, as well as medications to kill the remaining larvae. In some cases, a surgeon may need to perform a shunt procedure to relieve pressure in the brain
Tuberculosis meningitis	A form of meningitis caused by the bacteria that causes tuberculosis	Treatment typically involves a combination of antibiotics and corticosteroids to reduce inflammation. In some cases, a surgeon may need to perform a procedure to drain excess fluid and relieve pressure in the brain

complicated by hydrocephalus, intracranial pressure (ICP) monitoring and management may also be required. EVD is a key tool in the management of hydrocephalus and can also be used to monitor and manage ICP in patients with meningitis.

The use of EVD in meningitis and ventriculitis is primarily aimed at controlling elevated ICP, which can lead to serious neurological complications such as brain herniation and administration of intrathecal antibiotics and clearing the debris that can lead to obstructive hydrocephalus. ICP monitoring is typically performed using an external ventricular drain (EVD), which involves placing a catheter into the lateral ventricle of the brain and connecting it to an external drainage system. The EVD allows for the measurement and management of ICP by draining excess CSF and maintaining a normal pressure gradient within the cranial vault.

In addition to its role in managing ICP, EVD can also be used to obtain CSF samples for microbiological analysis and to administer intrathecal medications such as antibiotics. Intrathecal antibiotic administration via EVD has been shown to be an effective method of delivering high concentrations of antibiotics directly to the site of infection in the CSF. This method of antibiotic administration has been shown to be effective in the treatment of meningitis and ventriculitis caused by a variety of pathogens, including drug-resistant strains.

The use of EVD in meningitis is not without risks, and complications can occur. These can include infection, bleeding, catheter malposition, and CSF leaks. Careful monitoring of the patient is required to detect and manage these complications.

6 Intrathecal Drug

Intrathecal drug administration involves delivering drugs directly into the cerebrospinal fluid (CSF) through the spinal cord. This method of drug administration is used to treat a variety of conditions such as meningitis, cancer-related pain, and spasticity.

Some commonly used antibiotics that can be given intrathecally for the treatment of meningitis or ventriculitis are as follows:

1. Ampicillin
2. Ceftazidime
3. Ceftriaxone
4. Gentamicin
5. Meropenem
6. Penicillin G
7. Vancomycin

It is important to note that the choice of antibiotic may vary depending on the underlying cause of the infection, the sensitivity of the causative organism, and the patient's individual clinical factors. The use of intrathecal antibiotics should be carefully monitored. The doses of these antibiotics are given in Table 3.

Table 3 For dosage and antibiotics for meningitis and ventriculitis

Antibiotic	Dosage for meningitis	Dosage for ventriculitis
Ampicillin	50–200 mg/day	100–500 mg/day
Cefotaxime	2–3 g/day	2–3 g/day
Ceftazidime	2–3 g/day	2–3 g/day
Gentamicin	4–8 mg/day	4–8 mg/day
Meropenem	1–2 g/day	1–2 g/day
Vancomycin	15–30 mg/day	15–30 mg/day

7 Conclusion

In managing serious infections such as meningitis and ventriculitis, it is critical to identify the causative organisms, understand their common routes of entry, and be aware of the predisposing conditions that increase susceptibility to these infections. The utilization of neurosurgical treatments, including external ventricular drainage (EVD), is critical in mitigating the impact of these infections. EVD plays an integral role in managing elevated intracranial pressure, providing an avenue for intrathecal antibiotic administration, and facilitating cerebrospinal fluid sampling for microbiological analyses. Intrathecal drug administration offers an effective means of delivering high concentrations of antibiotics directly to the site of infection, with various antibiotics such as ampicillin, ceftazidime, and vancomycin among others being frequently employed. The selection of the specific antibiotic to be used is largely dependent on the type of infecting organism and the patient's unique clinical factors. Despite the potential risks associated with EVD, its strategic use can prevent severe neurological complications in patients. The role of EVD in managing meningitis and ventriculitis highlights the importance of comprehensive, multi-disciplinary care in treating these infections, emphasizing the need for careful consideration and administration by skilled neurosurgeons following established protocols.

References

1. Erdem H, Ozturk-Engin D, Cag Y, et al. Central nervous system infections in the absence of cerebrospinal fluid pleocytosis. Int J Infect Dis. 2017;65:107.
2. Giovane RA, Lavender PD. Central nervous system infections. Prim Care. 2018;45(3):505–18.
3. Castelblanco RL, Lee M, Hasbun R. Epidemiology of bacterial meningitis in the USA from 1997 to 2010: a population-based observational study. Lancet Infect Dis. 2014;14(9):813–9.
4. Martin RM, Zimmermann LL, Huynh M, Polage CR. Diagnostic approach to health care-and device-associated central nervous system infections. J Clin Microbiol. 2018;56(11):10.
5. Humphreys H, Jenks P, Wilson J, Weston V, Bayston R, Waterhouse C, Moore A. Healthcare Infection Society working party on neurosurgical infections. Surveillance of infection associated with external ventricular drains: proposed methodology and results from a pilot study. J Hosp Infect. 2017;95(2):154–60.
6. Tunkel AR, et al. IDSA's clinical practice guidelines for acute bacterial meningitis in children and adults. Clin Infect Dis. 2011;53(12):e143–61.

7. Venkatesan A, et al. Case definitions, diagnostic algorithms, and priorities in encephalitis: consensus statement of the international encephalitis consortium. Clin Infect Dis. 2013;57(8):1114–28.
8. Brouwer MC, et al. Clinical characteristics and outcome of brain abscess: systematic review and meta-analysis. Neurology. 2016;86(9):806–13.
9. King NKK, et al. Subdural empyema: a 10-year review. J Neurosurg. 2014;120(5):1368–75.
10. Darouiche RO. Spinal epidural abscess. N Engl J Med. 2006;355(19):2012–20.
11. Drake JM, et al. Shunt complications in the first year: mechanical failure versus infection. Neurosurgery. 2005;56(4):741–6.
12. Carpio A, et al. Guidelines for the diagnosis and treatment of neurocysticercosis in humans. PLoS Neg Trop Dis. 14(8):1.

MRSA to VRE: Management and Prevention

Sarah Salameh and Muna Al Maslamani

Abstract Antimicrobial resistance is an urgent healthcare threat. It kills millions of patients worldwide. Antimicrobial resistance is when the bacteria and fungi are not killed by the medications meant for them. Antimicrobial resistance is not only a threat in healthcare, it is also seen in the field of poultry and agriculture as well.

In humans, the bacterial resistance involves MRSA (methicillin-resistant *Staphylococcus aureus*), VRE (vancomycin-resistant enterococci), CRE (carbapenem-resistant enterobacteria), ESBL (extended-spectrum B-lactamase-producing bacteria), drug-resistant pneumococci and mycobacterium, and multi-drug-resistant *Pseudomonas* and *Neisseria gonorrhea.*

Antimicrobial resistance is rising due to improper prescriptions with inappropriate antibiotics, dose and duration of administration, the agricultural and poultry use of tonnes of antimicrobial agents to get better livestock. Delays or no newer antibiotics in the pipeline are due to a lack of investment, incentives and various administrative approval barriers.

Antibiotic and antimicrobial resistance can be managed and prevented by increasing public awareness of the proper use of antibiotics, promoting antibiotic stewardship, rapid identification of microbes to avoid empirical antibiotic administration, using vaccines, detecting and preventing the spread of infections, and utilizing national and international initiatives in the countries.

Keywords Agricultural abuse · Antibiotics · Antimicrobial bacterial resistance · Antibiotic stewardship · Pre-antibiotic era · Improper prescription · New antibiotics · MRSA (methicillin-resistant *Staphylococcus aureus*) · VRE (vancomycin-resistant enterococci) · CRE (carbapenem-resistant enterobacteria) · ESBL (extended-spectrum B-lactamase-producing bacteria) · Multi-drug-resistant bacterial · Education · Infection prevention

---

S. Salameh · M. Al Maslamani (✉)
Center for Disease Control (CDC)/Infectious Disease Department/Hamad Medical Corporation, Doha, Qatar
e-mail: malmaslamani@hamad.qa

N. Shaikh, A. H. Chanda (eds.), *Applied Microbiology in Intensive Care Medicine*, https://doi.org/10.1007/978-981-97-4006-2_11

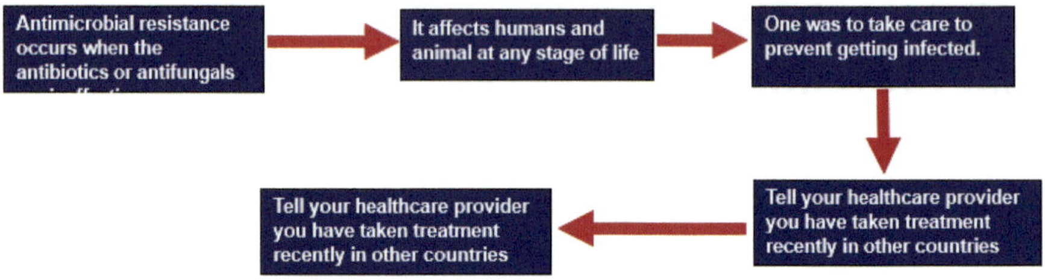

Fig. 1 Five things to know about antimicrobial resistance

1 Introduction

Antimicrobial resistance occurs when microbes develop the ability to defend or defeat the medications that are supposed to kill them. Antimicrobial resistance is one of the world's most urgent public health problems and threats. Antimicrobial resistance kills millions of patients worldwide regularly. Antimicrobial resistance affects not only humans but animals as well, and it is a major concern for healthcare, veterinary, and agricultural industries.

Five things to know about antimicrobial resistance are shown in Fig. 1 [1].

2 Advantages of Antibiotics

The modern days of antimicrobial therapy have started with the discovery of penicillin in the early forties. Before this, the ancient Egyptians, Greeks, and Chinese were known to practice antimicrobial agents to control infections. The discovery of penicillin changed medicine, and millions of patients, particularly during World War 2, were saved and treated [2]. The antibiotics, apart from saving lives, played a crucial role in major advances in surgery and medicine. Antibiotics are essential in preventing the occurrence and treatment of infections in patients on chemotherapy or patients with chronic diseases such as end-stage renal disease and diabetes mellitus. They play an important role in complex surgeries such as cardiothoracic, joint replacement, and transplant surgeries. Antibiotics are useful to extend the expected life span. Antibiotics were useful in treating water-born acute illnesses in countries with endemic water-borne diseases [3].

3 Use and Abuse Leading to Antibiotic and Antimicrobial Resistance

3.1 Abuse of Antibiotics

There is a direct relation between the overuse and consumption of antibiotics and antimicrobial agents and the development of resistant strains. Antibiotics are easily sold over the counter in many countries, leading to abuse and antimicrobial resistance. In countries where antibiotics are regulated, the online purchase of antibiotics is concerned as it again leads to abuse of antimicrobial agents.

3.2 Inappropriate Prescription

Inappropriate and inadequate antimicrobial and antibiotic therapy is an important factor in developing antimicrobial resistance. According to the literature, the indication, duration, and antimicrobial choice are incorrect in up to 50% of antibiotic prescriptions. In critically ill ICU (intensive care unit) patients, antimicrobials are prescribed unnecessarily, suboptimally, and inappropriately [4]. Incorrectly prescribed antibiotics are very harmful, without any therapeutic benefits, leading to more complications. The sub-inhibitor and subtherapeutic antimicrobial dosage support genetic alteration and mutations, thus promoting and developing antimicrobial resistance.

3.3 Agricultural and Poultry Abuse

Approximately 80% of antibiotics used in developed countries are in animals primarily to promote animal growth and prevention of infection, as treating the livestock with the antimicrobials is proven to improve the overall animal health with a longer yield and improved qualities of the product. Hence, the highest resistance rate is found in both forms: animal and insect guts.

These antimicrobials used in livestock will be consumed by humans, thus transmitting resistant microbes. The transmission of resistant microbes from agricultural and meat products is described by four points:

1. Antimicrobial use in livestock will kill or suppress the sensitive microbes, allowing at the same time resistant bacteria to grow.
2. These resistant bacteria are transmitted to humans with food.
3. These bacteria cause infections in humans.

The widespread use of antimicrobials in animal poultry and agriculture is causing dominance of resistance in environmental microbiome. Using antimicrobial agents as hygiene or cleaning agents is also a concern and contributes to limiting the development of immunity in humans [5].

3.4 The Newer Antibiotics

Few new antimicrobials are in the pipeline for the resistant gram-negative bacteria [6], although gram-negative bacteria rapidly increase in resistance compared to gram-positive resistance. The pharmaceutical merge, a lack of funds and investments in antimicrobial growth, no economic incentives to invest in antibiotics research, and as antibiotics are curative and used for shorter duration compared to medication for chronic diseases are the important factors for not having much newer antibiotics in the pipeline or practice [6].

3.5 Red File Barrier

There is no encouragement and appeal from the authorities for the development of newer antibiotics. This is mainly due to ineffective communication channels, changes in the rules, different requirements in various countries, and the absence of clarity in the bureaucracy [7].

4 Antibiotic-Resistant Bacterial Infections

Antibiotic resistance is a global phenomenon. MRSA (methicillin-resistant *Staphylococcus aureus*), the respiratory pathogen, particularly mycobacteria and *Streptococcus*, kills more people than HIV, emphysema, homicide, and parkinsonism in developed countries.

In all systems, the gram-negative multi-drug-resistant bacterial infections are increasing. Sometimes, these gram-negative bacteria are resistant to all antibiotics, causing a crisis like a pre-antibiotic era.

According to the CDC, the antibacterial resistance threats can be classified into urgent threats, including *Clostridium difficile*, carbapenem-resistant enterobacteria, and drug-resistant *Neisseria gonorrhea*. The serious threat resistance includes Actinobacteria, fluconazole-resistant *Candida*, extended-spectrum B-lactamase-producing enterobacteria (ESBL), vancomycin-resistant enterococci (VRE), methicillin-resistant *Staphylococcus aureus* (MRSA), and drug-resistant *Salmonella*, *Shigella*, *Mycobacterium*, and *Streptococcus pneumoniae*, whereas the concerning threat of resistance includes clindamycin-resistant group B-*Streptococcus*, clindamycin-resistant group A-*Streptococcus*, and VRSA (vancomycin-resistant *Staphylococcus aureus*). The following table describes various resistant bacteria and their infections (Table 1).

Table 1 Various resistant bacteria and their infections

Drug-resistant bacteria	Resistance to antibiotics	Effective antibiotics
Methicillin-resistant *Staphylococcus aureus* (MRSA), discovered decades ago, remains a worldwide problem and is most frequent in the antibiotic resistance group There is a decline in hospital-acquired MRSA infections, but community-acquired MRSA infections recently increased	B-lactam	Teicoplanin, vancomycin, tigecycline, linezolid, daptomycin, ceftaroline, ceftobiprole
Vancomycin-resistant enterococci (VRE), detected in late 1980, still causes serious hospital-acquired bloodstream, surgical site, and urinary tract infections	Vancomycin	Oritavancin, tigecycline, daptomycin
Extended-spectrum B-lactamase-producing bacteria (ESBL) significantly cause hospital-acquired infections	Penicillin, cephalosporins	Carbapenems
Carbapenem-resistant Enterobacteria (CRE). Few of the generations of NDM-1 (New Delhi metals beta-lactamase) CRE bacteria became extremely difficult to treat and resistant to all available antibiotics	B-lactam antibiotics, carbapenem	Colistin, tigecycline, fosfomycin
Drug-resistant *Mycobacterium tuberculosis*: Causing hundreds of deaths. It can affect any body part, but it is common in the lungs Extensive drug-resistant TB (XDR-TB) is almost resistant to almost all anti-tuberculous medications and injectable second-line therapy	Isoniazid, rifampicin, amikacin, kanamycin capreomycin	Fluoroquinolone, bedaquiline, linezolid
Drug-resistant *Streptococcus pneumoniae* is a major etiology of pneumonia, meningitis, meningitis, and bloodstream infections	Macrolides, amoxicillin	Pneumococcal conjugate vaccine
MDR *pseudomonas aeruginosa* and Actinobacteria are resistant to all antibiotics, mainly causing pneumonia and bloodstream infections in healthcare settings, causing hundreds of deaths	Aminoglycoside, cephalosporines, fluoroquinolones carbapenems	Ceftazidime-avibactam Ceftolozane-tazobactam
Drug-resistant *Neisseria gonorrhea*, causing recently widespread gonorrheal epididymitis and pelvic inflammatory disease	Penicillin, Tetracycline, fluoroquinolones	Ceftriaxone +/− doxycycline or azithromycin

5 Management of Antibiotic Resistance and Prevention

The following steps will reduce the development and transmission of antimicrobial antibiotic resistance throughout hospital, national, and international levels.

5.1 Optimization of Antimicrobial Prevention

In both inpatient and outpatients, improper prescriptions, incorrect choices, and inappropriate doses of antimicrobial administration were found to be prevalent. Data also confirms that as frequently as 50% of the time, antimicrobial prescriptions are unnecessary or incorrect [8].

In up to 75% of upper respiratory tract viral infections, healthcare practitioners prescribe unnecessary antibiotics, which can harm the patients and favor antimicrobial resistance. Educational campaigns to raise awareness about the need for correct dosing and choice with optimal duration of antibiotics for the healthcare practitioner have improved the prescriptions and awareness about antimicrobial resistance [5].

5.2 Optimizing the Therapeutic Regimens

Antibiotics are prescribed for a fixed duration involving specific doses and frequency. A typical antibiotic regimen varies from 5 to 14 days or longer, depending on the pathology. Recently, the literature shows that the short duration of antibiotic therapy, seven days for pyelonephritis, ten days for septic arthritis, and three days of antibiotic therapy in community-acquired pneumonia, was uniformly effective than the longer antibiotic therapy [9].

The chance of resistant bacteria colonization increases with longer antibiotic use. When symptoms subside, patients may stop taking the antibiotics before finishing the lengthier course and resume them if symptoms return. Procalcitonin-guided antibiotic treatment has the advantage of being shown to reduce antibiotic use by 51% [9]. Keeping the duration and use of antibiotics for other purposes, such as prokinetics, to a minimum duration will be helpful in the prevention of resistance development [10].

5.3 Quicker and Improved Diagnosis

Empirical use of antibiotics and antimicrobial agents with diagnostic uncertainty is the cause of administering multiple antimicrobial agents, which will cause intense pressure on the patient's micro bacteria, which can increase the tendencies for antibiotic resistance. Empirical use of antibiotics can be minimized by rapid, accurate diagnostic methods such as polymerase chain reaction (PCR), matrix-assisted laser desorption/ionization, and time-of-flight mass spectrometry, instead of traditional tests, which may take days to weeks for diagnosis and detection of bacteria and bacterial growth [10].

Early detection of bacteria in the abdominal or infected collection (other organs) is helpful. Instead of repeated image-guided drainage, a proper laparotomy and wash may reduce the longer duration of antimicrobial therapy by properly eradicating the source of infection.

5.4 Prevention of Transmission of Bacterial Infections

Prevention of infections will eliminate the use of antimicrobials, thus significantly decreasing or eliminating resistant bacterial growth. Following the infection control and prevention guidelines is essential in preventing most hospital-acquired infections.

Using automated infection control approaches through robotic and automatic disinfection is quite helpful. Automated hand-washing technology will improve hand washing rates, and self-cleaning hospital rooms can reduce the risk of hospital-acquired infection. Control tracing also is essential in tracing the individual who may have been infected by patients, and treatment is important in preventing the spread of infection. The *Streptococcus pneumoniae* vaccine has shown that vaccination can decrease the transmission of resistant bacterial infections.

5.5 National Initiative

There should be a clear definition and comprehensive national action plan for managing and preventing bacterial resistance. This included data collection and reducing antimicrobial use not only in healthcare but also in agriculture, livestock, and poultry as well. Antibiotic stewardship optimizes the use of newer diagnostic tests and supports the search and research for newer antimicrobial agents.

5.6 *Fostering and Adopting the Antibiotic Stewardship Program*

Antibiotic stewardship in hospitals is essential for the correct use and prevention of the abuse of antimicrobial agents. Antibiotic stewardship can be the handshake method by the presence of infectious disease specialists in the hospital and critical care multidisciplinary rounds. Antibiotic stewardship involves the appropriate use of antibiotics when needed and its administration in proper dose and duration. An antibiotic stewardship program has been shown to improve patient care, shorten hospital stays, and reduce costs [11].

5.7 *International Efforts*

The World Health Organization (WHO) is actively involved in raising awareness about antimicrobial resistance. WHO has organized several conferences under the banner of antibiotic resistance initiatives (ReAct). The British Society of Antimicrobials Chemotherapy launched a global initiative with antibiotic action, which coordinates worldwide to raise awareness about antibiotic resistance and funding for antimicrobial resistance [11]. International workers and the WHO have initiated a few decisions to control antibiotic use in healthcare as well as agricultural industries.

6 Conclusion

The rapidly emerging antimicrobial resistance has a risk of taking humans to the pre-antibiotic era, where millions of people used to die from infections. Antibiotics have saved the human life for decades by controlling infections. But the inappropriate use of antibiotics has a fear of an era where the microbes will be resistant to all available medications. The important factors for antimicrobial resistance are improper prescriptions, longer duration, agricultural use, few newer antimicrobial agents, and lack of investment and research funds for newer antimicrobial agent discoveries. It is essential that we control and prevent further progress in antimicrobial resistance through antibiotic stewardship, awareness, and education with national and international strategies.

References

1. Ventola CL. The Antibiotic resistance crisis. Pharm Ther. 2015;40(5):344–52.
2. Golkar Z, Bagasra O, Pace DG. Bacteriophage therapy: a potential solution for the antibiotic resistance crisis. J Infect Dev Ctries. 2014;8(2):129–36. https://jidc.org/index.php/journal/article/view/24518621.
3. Rossolini GM, Arena F, Pecile P, Pollini S. Update on the antibiotic resistance crisis. Curr Opin Pharmacol. 2014;18:56–60.
4. Luyt CE, Bréchot N, Trouillet JL, Chastre J. Antibiotic stewardship in the intensive care unit. Crit Care. 2014;18(5):480.
5. Michael CA, Dominey-Howes D, Labbate M. The antimicrobial resistance crisis: causes, consequences, and management. Front Public Health. 2014;2:145. https://www.ncbi.nlm.nih.gov/pmc/articles/PMC4165128/.
6. Lushniak BD. Antibiotic resistance: a public health crisis. Public Health Rep. 2014;129(4):314–6. https://doi.org/10.1177/003335491412900402.
7. Piddock LJV. The crisis of no new antibiotics—what is the way forward? Lancet Infect Dis. 2012;12(3):249–53.
8. Gould IM, Bal AM. New antibiotic agents in the pipeline and how they can help overcome microbial resistance. Virulence. 2013;4(2):185–91. https://www.ncbi.nlm.nih.gov/pmc/articles/PMC3654619/.
9. Bartlett JG, Gilbert DN, Spellberg B. Seven ways to preserve the miracle of antibiotics. Clin Infect Dis. 2013;56(10):1445–50.
10. Shaikh N, Nainthramveetil M, Nawaz S, Hassan J, Shible AA, Karic E, et al. Optimal dose and duration of enteral erythromycin as a prokinetic: A surgical intensive care experience. Qatar Med J. 2020;2020(3):36. https://www.ncbi.nlm.nih.gov/pmc/articles/PMC7802089/.
11. Ventola CL. The antibiotic resistance crisis: part 2: management strategies and new agents. P T Peer-Rev J Formul Manag. 2015;40(5):344–52.

Future of Applied Microbiology in Critical Care

Ahmad A. Al Bishawi, Ans Alamami, Ali S. Omrani, and Nadir Kharma

Abstract Sepsis is one of the most commonly encountered clinical entities in the intensive care unit. Despite that, the diagnosis and management of sepsis are challenging due to the lack of a clear and practical definition. In this chapter, we go over the hurdles of obtaining a positive diagnostic test in a suspected septic patient. Timely initiation and correct choice of antimicrobials are vital to reducing sepsis-related mortality. As such, the problem of culture-negative sepsis is highlighted with its potential consequences. This includes delays in the identification of organisms and their antimicrobial resistance patterns, prolonged antimicrobial therapy, longer hospital stays, and missing alternative diagnoses. The advances in the early detection of organisms and the detection of resistance/virulence patterns are summarized. These include matrix-assisted laser desorption/ionization-time-of-flight mass spectrometry (MALDI-TOF MS) and Nucleic Acid Amplification Test (NAAT) technologies with several subtypes and applications in the fields of bacteriology, mycology, virology, and mycobacterial. While these new technologies are promising with earlier detection time and high diagnostic accuracy, more research is needed to further define their exact role and utility in the diagnosis and management of sepsis in critically ill patients.

Keywords Culture-negative sepsis · Sepsis · Early identification of organism · MALDI-TOF · NAAT · PCR · Microarray multiplex · T2 magnetic resonance

A. A. Al Bishawi · A. S. Omrani
Communicable Diseases Center, Infectious Diseases Department, Hamad Medical Corporation, Doha, Qatar

A. Alamami · N. Kharma (✉)
Department of Critical Care, Hamad Medical Corporation, Doha, Qatar
e-mail: nkharma@hamad.qa

N. Shaikh, A. H. Chanda (eds.), *Applied Microbiology in Intensive Care Medicine*, https://doi.org/10.1007/978-981-97-4006-2_12

1 Introduction and Current Microbiologic Challenges in ICU

Sepsis is one of the main causes of morbidity and mortality in the intensive care unit, making it a significant healthcare burden with mortality ranging from 20–40% [1]. Despite the declining sepsis-associated mortality, the incidence of sepsis and especially severe sepsis has been increasing over the last several years [2].

The International Sepsis Forum, the Society of Critical Care Medicine (SCCM), and the European Society of Intensive Care Medicine (ESICM) developed the Surviving Sepsis Campaign (SSC), which was introduced at the ESICM annual meeting in Barcelona in 2002 [3]. The timely identification and management of the septic patient have been the cornerstone of this evolving paradigm shift over the last two decades.

In their last 2021 update, the SSC continues to recommend that appropriate routine microbiologic cultures, including blood cultures, be obtained before beginning antimicrobial therapy in patients with suspected sepsis and septic shock if the delay is not significant (i.e., more than 45 min) [4].

The gold standard for diagnosing bloodstream infection continues to be the detection of pathogens from blood culture samples, in the lack of a viable alternative test. Regrettably, the process of obtaining results from normal blood cultures can exceed a duration of 72 h before a detectable organism is identified. Furthermore, more time is necessary to conduct identification and antibiotic susceptibility testing. The duration of time under consideration is deemed unsatisfactory within the context of sepsis, as any delays in initiating treatment have been observed to exacerbate both morbidity and death rates. Such delays also add to the necessity of employing empirical antibiotic therapy that is non-specific in nature.

Even if cultures are negative, careful de-escalation of antimicrobials based on acceptable clinical improvement is appropriate given the negative implications of continuing unneeded antimicrobial therapy. If infection has been ruled out, it is advised to stop all antimicrobial therapy as soon as possible. The SSC statements refer to the potential role of rapid diagnostic procedures as they may make it easier to de-escalate the use of antibiotics [4].

2 Culture-Negative Sepsis

Indeed, one of the major challenges of sepsis diagnosis and management highlighted by SSC is the culture-negative sepsis (CNS), slowly growing and antimicrobial-resistant organisms.

Although an infectious etiology is essential to the definition of sepsis, it is frequently difficult to isolate certain organisms by culture. Culture-negative sepsis (CNS), often known as severe sepsis without a microbiologically confirmed infection, is the condition. No particular organism can be located in a sizable majority of severe sepsis episodes [5–10]. A quarter of patients with severe sepsis are culture-negative even in well-designed randomized controlled trials [11]. Multiple multi-center studies have documented that the percentage of CNS admissions ranges from 28% to 49% among those diagnosed with severe sepsis [5–10]. Despite the prevalence of CNS, there is a notable scarcity of studies that have explored its epidemiology and consequences.

In recent years, there has been a growing focus on the management of CNS. The results of studies comparing culture-negative sepsis with culture-positive sepsis are inconclusive, and there is a lack of clarity about whether culture-negative sepsis may be considered as a unique entity. Determining the underlying causes of cultural negativity is a challenge; however, several potential etiologies can be identified. These include the administration of antibiotics prior to the collection of cultures, infections caused by fastidious organisms, and limitations of current bacterial culture growth techniques.

Several studies have attempted to address the issue of CNS by estimating the proportion of septic patients with CNS. They also looked at trends over time and associated clinical outcomes like mortality, length of hospital stay, and antibiotic usage.

In a big study [12] that utilized a Nationwide Inpatient Sample database spanning the years 2000–2010, Gupta et al. focused on analyzing the proportion of hospitalizations related to severe sepsis and examined the impact of CNS on mortality.

Out of a total of 6,843,279 admissions with severe sepsis in US hospitals, a significant proportion of 3,226,406 cases (47.1%) were found to be culture-negative. The age-adjusted proportion of individuals with CNS has exhibited a significant increase, rising from 33.9% in the year 2000 to 43.5% in 2010. Individuals diagnosed with CNS exhibited a higher prevalence of comorbidities, as well as acute dysfunctions in many organs such as the respiratory, cardiac, hepatic, and renal systems. Additionally, the in-hospital mortality rate was significantly higher in this group compared to those without CNS (34.6% vs. 22.7%).

However, the incidence of acute kidney injury necessitating dialysis was lower in individuals with CNS compared to those without (5.3% vs. 6.1%). CNS was found to be an independent predictor of mortality in individuals diagnosed with severe sepsis (odds ratio of 1.75). While this big database study showed negative outcomes associated with CNS, not all studies reached the same conclusion. Phua and colleagues [13] conducted a study to examine the prevalence and fatality rates of CNS. Their findings revealed that 41.5% of patients admitted with severe sepsis had negative culture results. Moreover, in their research, those who tested negative for the presence of a particular culture had lower severity of illness scores and a mortality rate of 35.9%, which was lower compared to the group with positive cultures at 44.0%. The research conducted by Phua et al. was carried out at a university hospital that served as a single center for the study. Patients with proven viral, fungal, and parasitic infections were not part of this study.

An attempt to resolve the controversy with a systematic review and meta-analysis by Li et al. included seven studies with a sample size of 22,655 patients [14]. The main finding of this meta-analysis indicated that there was no statistically significant disparity in all-cause mortality between the culture-positive and culture-negative sepsis groups. Furthermore, there was no statistically significant difference observed in the length of stay in the intensive care unit, mechanical ventilation requirements, and renal replacement requirements between the two groups. The duration of hospitalization for patients who tested positive for the culture was found to be significantly longer compared to those who tested negative for the culture. The duration of mechanical ventilation in the culture-positive group was found to be longer compared to the culture-negative group.

Another more recent meta-analysis by Afzal and colleagues [15] included 10 studies and found no statistically significant disparities between the culture-positive and culture-negative sepsis groups in terms of overall mortality, a requirement for mechanical ventilation, a need for renal replacement treatment, and duration of stay in the intensive care unit. A notable disparity was observed in the duration of hospitalization, as individuals who tested positive for the culture had a lengthier stay. The implementation of instructions from the Surviving Sepsis Campaign has been associated with improvements in the proportion of blood cultures being obtained prior to the administration of antibiotics, as demonstrated by many studies [16, 17]. It was anticipated that there would be an improvement in culture-positive rates for these admissions. However, Gupta et al.'s findings indicated a decline in culture positivity rates throughout the course of time. This could be attributed to the use of empiric outpatient therapy for infections, the early initiation of antibiotic treatment in patients with severe sepsis prior to obtaining cultures, and the readmission of individuals with severe sepsis. Additional possible explanations include viral infections masquerading as bacterial sepsis.

The potential correlation between incorrect or insufficient antibiotic treatment and increased mortality in patients with sepsis has been observed in previous studies [18–20]. Consequently, this association may also contribute to the higher fatality rates observed in the CNS cohort. The implementation of SSC recommendations advocating for antibiotic de-escalation in culture-negative cases may pose several challenges. Potential strategies include limiting the duration of antibiotic treatment, implementing antibiotic stewardship programs, promptly considering the use of narrow-spectrum antibiotics, and utilizing rapid diagnostic technology.

The available data from another study indicate that individuals diagnosed with pathogen-negative sepsis could potentially serve as a significant target for antimicrobial stewardship efforts, under the assumption that a considerable portion of these patients have received needless treatment [21]. The excessive utilization of antibiotics facilitates the emergence of bacteria that are resistant to many drugs. For example, prolonged exposure to cefepime, meropenem, and piperacillin-tazobactam is linked to a higher likelihood of developing new resistance [22]. These studies demonstrate the potential negative consequences of administering antibiotics when they are not medically essential. This problem is likely more prevalent in culture-negative sepsis.

3 Delay in Identification of Organism

Sepsis is defined as a life-threatening condition caused by a dysregulated severe systemic inflammatory response triggered by an infection that spreads throughout the body. The timely initiation of appropriate antimicrobial therapy is a key factor in reducing mortality and morbidity in patients with sepsis. With mortality ranging between 10 and 22% [23, 24], sepsis is a huge and devastating health problem, for which the need for rapid and effective management is key. One of the important constituents of sepsis management is timely mannered and efficacious antimicrobial therapy, and this depends primarily on the detection of both the microorganism and its resistance toward the treatment.

The rapid identification of the causative microorganism in patients with sepsis is crucial for effective treatment and improved patient outcomes. However, the identification of the causative microorganism can often be delayed due to the intricate and time-consuming nature of traditional microbiological techniques. This lag can result in a delay in the initiation of appropriate antimicrobial therapy, which is associated with increased mortality and morbidity in patients with sepsis [25, 26]. Several studies have investigated the impact of delayed microbial identification on patient outcomes in sepsis and it correlates with increased mortality and morbidity [27, 28]. A systematic review and meta-analysis of a sixteen study that included rapid molecular technique in the process of microbial identification conducted by Buehler et al showed a significant and consistent reduction in mortality when rapid molecular testing is utilized in conjunction with direct communication [29]. Another systematic review by D'Onofrio et al. found that rapid identification and susceptibility testing of bloodstream pathogens including different rapid molecular microbiological detection techniques had a positive impact on patient outcomes. The study found that rapid identification and susceptibility testing were associated with a reduction in the length of the hospital, and intensive care (ICU) stays, along with decreased antimicrobial cost [30].

4 Delay in Identification of Antimicrobial Resistance

Antimicrobial resistance (AMR) pertains to the capacity of microorganisms, including bacteria, viruses, fungi, and parasites, to withstand the impacts of antimicrobial medications that were before efficacious in managing infections induced by these microorganisms. The global impact of AMR on patient outcomes is substantial and growing. Antimicrobial resistance is a growing global health problem, with the emergence of drug-resistant bacteria posing a significant threat to public health. In the management of bacterial infections, the early detection of antimicrobial resistance is essential for effective treatment and improved patient outcomes. In some cases, infections caused by resistant microorganisms can be untreatable, leading to death. As per the World Health Organization (WHO), AMR is accountable for approximately 700,000 fatalities worldwide each year, and this figure is projected to escalate to ten million by 2050 if prompt and effective measures are not implemented.

The magnitude of the AMR problem is substantial and is likely due to the absence of a comprehensive and standardized surveillance system for monitoring antimicrobial resistance and consumption. Additionally, the deficiency in quality data generated from the current surveillance efforts can be attributed to the lack of standardization and harmonization. According to WHO (2012), only 30% of the monitoring program they supervise has a nationally coordinated body. A significant financial investment is necessary to cover the expenses associated with training interdisciplinary staff, hiring expert consultants, collecting data, conducting joint analysis and sharing results, and expanding the surveillance program to include human, animal, and environmental sectors [31]. Infrastructure development is crucial in low- and middle-income countries to improve transport facilities for the efficient delivery of biological samples to laboratories. Additionally, establishing local laboratories to generate primary data, training and recruiting staff, and developing a center for excellence are essential. It is also important to foster collaboration between medical and veterinary colleges, as well as health institutes, to ensure the effectiveness of the program. However, obtaining ethical approval for interdisciplinary health research work poses a challenge, as it involves multiple institutes from different countries with varying ethical parameters [32].

Several studies have investigated the impact of delayed detection of antimicrobial resistance on patient outcomes, and it was found that AMR is a significant outcome-independent mortality predictor [33]. Van Heck et al. conducted a systematic review and meta-analysis, including 26 studies with a total of 5659 patients. The study findings demonstrated that individuals with laboratory-confirmed antibiotic-resistant urinary and respiratory tract infections are prone to encountering delays in clinical improvement subsequent to antibiotic treatment [34].

AMR also has significant economic impacts, as it increases healthcare costs due to the need for longer hospital stays and the use of more costly drugs. It also reduces productivity as people are unable to work while they are sick which can result in long-term health problems. Several studies have focused on the economic outcome as a repercussion of AMR; in a study by Carmeli et al., it was estimated that hospital admission costs $7340 for a patient with emergent resistance of *Pseudomonas aeruginosa* in the year 1999 [35].

A key factor in the battle against AMR is the rapid susceptibility and resistance detection. Previously, the main tool to discover antimicrobial susceptibility used to be the determination of the antimicrobial effect on the positive growth of the culture media; this method is time-consuming and requires microbial growth to be tested, and unfortunately, in the occasion of a negative culture, it cannot be utilized. Currently, there is a promising and significant advancement in the field of early microbial identification and AMR resistance through molecular genetics, which provides a way of directed therapeutic intervention (objective) and safeguards toward the reduction of unnecessary antimicrobial administration that by itself can lead to a reduction in the rate of AMR and eventually the patient's morbidity and mortality.

Rapid antimicrobial susceptibility testing (rAST) has the potential to advance the care of bloodborne infections, with the rapid detection, adjustments, de-escalation of therapy, and abbreviated time to start effective antibiotics when compared with conventional antimicrobial susceptibility AST results [36]. Time-wise, the rAST can be divided into a rapid test which yields the result within 8 h or ultra-rapidly in which the test has the potential to extract the result in less than 4 h [29], and with regard to the functional purposes and results, the rAST has been subdivided into a rapid single test and the accelerated full AST such as the utilization of the matrix-assisted laser desorption/ionization-time-of-flight mass spectrometry (MALDI-TOF MS) to detect the carbapenem-resistant gram-negative bacteria using a novel direct-on-target microdroplet growth assay [37]. On the contrary, the accelerated full AST, such as the VITEK 2 and Phoenix systems which are automated identification and susceptibility testing systems, has been evaluated in one study through a total of 307 fresh isolates tested including 51 *Enterococcus* spp., 141 Enterobacteriaceae, 93 *Staphylococcus* spp., and 22 nonfermenters. The average time to obtain results for all bacterial groups was around 700 min for the Phoenix system and nearly 500 min for the VITEK 2 system. Regarding the process of identification, it was shown that both Phoenix and VITEK 2 systems produced almost identical outcomes for staphylococci, nonfermenters, and enterococci. Out of the 140 Enterobacteriaceae strains that were assessed, Phoenix successfully identified 135 strains (96%), while VITEK 2 correctly identified 137 strains (98%) with no significant statistical difference. Both methods had 97% agreement for all isolates [38].

In conclusion, the rapid identification of the causative microorganism in patients with sepsis in addition to the early detection of antimicrobial resistance is crucial for effective treatment and improved patient outcomes. Delayed microbial identification is associated with increased mortality and morbidity in septic patients. The emergence of quick and effective tools for microbial identification is therefore essential for the management of septic patients.

5 Advances in Early Detection of Organisms and Detection of Resistance/Virulence Patterns [39–49]

For decades, microbiological growth detection and antimicrobial susceptibility testing (AST) have relied heavily on using conventional methods such as microscopic examination, culture on microbiological media, and identification by serial biochemical phenotypic tests and agglutination kits. While these techniques are sensitive, accurate, and inexpensive, at least for the most commonly encountered pathogens, it is still laborious and time-consuming and requires experienced laboratory staff skills.

In the field of microbiology, many advances have emerged, and the development and implementation of innovative diagnostic techniques which are increasingly used have had a paramount effect on microbiology laboratory services enhancing existing diagnostic algorithms and accuracy guiding clinical judgment.

6 Bacteriology: Early Growth Identification

6.1 MALDI-TOF MS

Matrix-assisted laser desorption/ionization-time-of-flight mass spectrometry (MALDI-TOF MS) has become a popular instrument for the quick and accurate identification not only of bacteria but also of mycobacteria and several fungal organisms in the clinical microbiology laboratory.

In clinical practice, testing can be conducted on colony growth on the plates as well as directly on a direct growth when it is signaled from the incubator, where a specific amount is added to a target plate which is then pre-treated with a chemical reagent (i.e., matrix) and subjected to a mass spectrometer for additional processing inside the MALDI-TOF device that uses laser to ionize and disperse the sample into different molecules. These molecules will move in a vacuum creating a specific fingerprint for each species as determined by the movement time of the different molecules depending on their mass as defined according to the ribosomal protein and their electric charge.

These created fingerprint data are put together, creating a specific spectrum that is compared to an available database. This in turn will result in rapid identification of the infectious organism.

Many studies showed that the diagnostic accuracy for MALDI-TOF has comparable and even superior results to the most commonly used automated machines for microbiological identification like Phoenix and VITEK 2 systems.

The technology's limitation lies in its ability to identify new isolates. This is only viable if the spectrum database has peptide mass fingerprints of the specific strains or subspecies.

6.2 Nucleic Acid Amplification Test (NAAT)

PCR, or polymerase chain reaction, is a recent breakthrough in molecular biology that facilitates the identification of pathogen genetic material by amplifying specific regions of interest. This technique has revolutionized the detection of viruses, as well as certain bacteria and parasites, ushering in a new era of quick and precise diagnosis.

When it comes to microbiological diagnostics, it is important to take into account multiple parameters while detecting and measuring viral, bacterial, and parasite agents. These considerations are determined by the type of genetic material being targeted (DNA or RNA), the ability to grow the pathogen in a lab, the interpretation of the results, and the clinical importance of polymerase chain reaction (PCR) results. PCR is the most commonly used nucleic acid amplification technique (NAAT), where the principle is based on enzymatically replicating the DNA of the specific pathogen being targeted. PCR is the amplification of a small section of DNA using enzymes that are mediated by primers. DNA polymerase synthesizes new DNA strands that are complementary to the template DNA. DNA polymerase can only attach a nucleotide to the existing 3′-OH group. Thus, a primer is necessary. Therefore, more nucleotides are appended to the 3′ end of the DNA polymerase. DNA polymerase facilitates the incorporation of nucleotides to create fresh DNA strands.

The notion of each cycle consists of numerous heat cycles that involve denaturation, primer annealing (hybridization), and extension. During these cycles, DNA polymerase catalyzes the addition of nucleotides to create new DNA strands. After the first cycle, each DNA molecule is composed of one new and one old DNA strand. In subsequent cycles, the newly synthesized DNA segments act as templates, leading to exponential amplification of the DNA target by millions of folds.

6.3 *Types of PCR*

6.3.1 Real-Time PCR

This method involves the detection of DNA amplification in real time using a fluorescent reporter. The intensity of the fluorescent reporter is directly correlated with the quantity of amplified DNA molecules.

6.3.2 Nested PCR

The purpose of this technique is to enhance the accuracy and precision of the results by increasing sensitivity and specificity. They decrease the occurrence of non-specific binding of products caused by the amplification of unanticipated primer binding sites.

6.3.3 Multiplex PCR

This is a technique used to amplify many targets simultaneously in a single PCR experiment. It simultaneously amplifies multiple DNA sequences (as seen below).

6.3.4 Quantitative PCR

This technique utilizes the linear amplification of DNA to identify, describe, and measure a specific sequence in a given sample.

6.3.5 Arbitrary Primed PCR

This is a DNA fingerprinting technology that relies on the polymerase chain reaction (PCR). The DNA sequence of the primers is selected arbitrarily. Currently, PCR detection methods are commonly utilized in practice for screening MRSA (methicillin-resistant *Staphylococcus aureus*) by targeting the Meq A gene, as well as for detecting *C. difficile* by targeting the tcdA and tcdB genes.

6.4 Microarray Multiplex

Using multiplexing technologies with the same principle as illustrated above (multiplex PCR), this technology is dependent on the ability to detect and identify more than one pathogen simultaneously from the same specimen with easy, rapid, and accurate identification. This technology has recently emerged as a substitute for blood culture in diagnosing bacterial bloodstream infections. Most molecular diagnostics rely on PCR to detect bacterial DNA in the blood, but they require separate PCR tests for each bacterial species. Various multiplexed PCR formats have been developed to simultaneously detect various microorganisms in blood samples. PCR testing is more rapid than blood culture, while its sensitivity often falls within the same range as blood culture and has limited repeatability. Additionally, it is important to note that there exists a constraint on the quantity of bacterial species that can be concurrently examined. This limitation arises from the technological complexities associated with multiplexing several polymerase chain reaction (PCR) tests. These complexities stem from variations in the amplification efficiency of distinct primer combinations and the restricted availability of fluorescent labels. In essence, a microarray refers to an assemblage of minuscule elements, typically composed of DNA, that may be examined using target molecules in order to generate either quantitative data pertaining to gene expression or qualitative data for diagnostic purposes.

6.5 Gene Sequencing

16S ribosomal RNA (rRNA) sequencing is a method of sequencing that amplifies and analyzes the genetic material of bacteria in a sample to determine and compare the different species present.

16S rRNA gene sequencing is employed to investigate the phylogeny and taxonomy of samples derived from intricate microbiomes or habitats that are challenging or unfeasible to examine.

Unlike PCR, 16S metagenomics is an assumption-free detection approach. This means that instead of targeting a specific bacterial species, it can identify all the various bacteria present in a clinical sample in a single assay. This is possible because 16S metagenomics utilizes parallel deep sequencing of the prokaryote-specific 16S ribosomal RNA gene. This gene exhibits both conserved and varied sequence regions. The conserved sequence enables the pre-amplification of the variable region using a single pair of primers. Subsequently, the species can be recognized by analyzing the variable region sequence. 16S metagenomics is commonly employed to analyze the composition of bacterial communities in many environments, including the human gut, as well as in clinical microbiology for non-sterile body samples. Nevertheless, recent research has utilized 16S metagenomics to analyze blood cultures and has observed a wide range of bacterial species with a satisfactory level of accuracy. However, the reliability of this method for analyzing sterile samples is still being investigated.

6.6 Bacterial AST

Classical techniques for antimicrobial susceptibility testing (AST), such as broth microdilution, gradient tests, disk diffusion, breakpoint tests, and agar dilution, involve exposing a bacterial isolate to a range of antimicrobials for a prolonged period of time and then visually observing the growth of the bacteria to determine their susceptibility. Traditional AST methods typically utilize a fixed amount of a single strain of bacteria and follow a standard incubation period lasting between 16 and 20 h. However, certain bacterial species may necessitate extended incubation times of 72 h or more. Consequently, the cutoff points established by AST committees (EUCAST and CLSI) were adjusted based on these specific incubation times.

6.7 MALDI-TOF

Using the same principle illustrated above, recently, MALDI-TOF MS has been utilized to determine drug resistance in bacteria using several principles, mainly by the analysis of bacterially induced hydrolysis of β-lactam antibiotics. Hydrolysis is identified by observing particular changes in mass after incubating the pathogen with the tested β-lactam antibiotic for 30–180 min.

6.8 rAST

EUCAST (European Committee on Antimicrobial Susceptibility Testing) has developed a rapid phenotypic antimicrobial susceptibility testing (rAST) method based on the disk diffusion technique which can be done directly from blood culture bottles. After inoculation on the disk diffusion, the diameter of the zone of inhibition can be read at 4, 6, and 8 h), the interpretation of the breakpoints has been adapted to the time of incubation, and A buffer area, known as the area of technical uncertainty (ATU), was implemented to minimize the likelihood of inaccurate results due to technical variance caused by the shortened incubation period. This buffer area lies between the susceptible and resistant categories. The approach is now validated for the following species: *Acinetobacter baumannii*, *Enterococcus faecium*, *Enterococcus faecalis*, *Klebsiella pneumoniae*, *Escherichia coli*, *Pseudomonas aeruginosa*, *Staphylococcus aureus*, and *Streptococcus pneumoniae*.

7 Multiplex Microarray

Using the same principle of NAAT in multiplying certain DNA targets for pathogen identification, the multiplex microarray technique was invested also in microbial resistance detection where the DNA-based assay requires the detection of specific conserved regions of the responsible and determinant resistant genes in the tested isolates (e.g., meqA for MRSA) reporting the genotype where many studies showed that in combination with its high diagnostic performance, the assay offers a rapid and dependable method to anticipate the resistance (phenotype) of clinical isolates.

8 Mycology

The diagnosis of invasive fungal disease infection is difficult due to the limitations of current diagnostic methods such as culture, direct microscopy, and histopathology. These methods either lack accuracy or require a long time to produce clinically useful results.

However, traditional culture obtained from a clinical sample remains the most reliable method for diagnosing fungal infections. This method has the advantage of identifying the exact cause of the disease if the result is positive. Additionally, it allows for evaluating the susceptibility of the pathogen to treatment. However, it should be noted that this method has a poor success rate in diagnosing the disease, and it takes a long time to obtain results.

8.1 I-MALDI-TOF

Using the same principle with bacterial isolates, MALDI-TOF is used as well in the rapid and accurate detection of fungal infection (as aforementioned).

8.2 II-T2Candida

T2Candida is an FDA-approved non-culture-based platform for diagnosing candidemia. This panel is a rapid diagnostic method that can specifically and sensitively detect *Candida* pathogens in whole blood without the need for culture or nucleic acid extraction steps. The T2MR technology is utilized in this process. After collecting whole blood specimens, they are introduced into the fully automated T2Dx device. The *Candida* DNA is then amplified and recognized as supermagnetic particles. The identification is achieved using T2 magnetic resonance (T2MR) measurement.

The final outcome is documented as either positive or negative to identify the five prevalent *Candida* species (*C. albicans*, *C. tropicalis*, *C. parapsilosis*, *C. glabrata*, and *C. krusei*), which collectively make up more than 95% of candidemia cases in most medical facilities.

The results are categorized based on their vulnerability to antifungal drugs, especially fluconazole. They are reported as *C. albicans/C. tropicalis*, *C. parapsilosis*, and *C. krusei/C. glabrata.* The T2Candida panel has a minimum detectable limit of 1 colony-forming unit (CFU) per milliliter of whole blood and an average turnaround time of less than 5 h. The positive predictive value (PPV) of the test relied significantly on the initial likelihood of the condition being tested (pretest probability) and the prevalence of the condition in the patient population. On the other hand, the negative predictive value (NPV) remained consistently high (98–100%) regardless of the prevalence values, which ranged from less than 1–10%. However, additional research is required to further confirm these findings.

8.3 III-Serological Testing

(1,3)-Beta-D-Glucan: Beta-D-glucan is a polysaccharide present in the cell wall of most fungi but not bacteria or viruses. The test is routinely used for *Candida* species but has significant cross-reactivity. The interpretation depends on the detected beta-D-glucan concentrations. It is worth noting that the beta-D-glucan assay is always negative in subjects with mucormycosis or *Cryptococcus* because of their inability to produce B-D-glucan. The test is useful for detecting candidiasis and invasive aspergillosis in immunocompromised hosts with appropriate clinical context. In addition, *Histoplasma* spp. and *Pneumocystis jirovecii* can yield positive test results.

The diagnostic performance is variable among different studies with sensitivity ranging from 55% to 95% (higher sensitivity on BAL specimens) and specificity ranging from 77% to 96% (higher specificity for invasive fungal infections in hematologic malignancies) with an overall excellent NPV that can reach up to 100%.

8.4 Galactomannan Antigen Detection

Galactomannan is a polysaccharide that is present in the *Aspergillus* cell wall and is secreted throughout its growth. The diagnostic procedures utilized a double-sandwich ELISA that includes the B 1–5 galactofuranose-specific EBA2 monoclonal antibody. Galactomannan is a polymer found in the cell wall of *Aspergillus* fungi that is secreted during their growth. The diagnostic procedures employed a double-sandwich ELISA that integrates the B 1–5 galactofuranose-specific EBA2 monoclonal antibody. The test can be conducted and validated for blood and BAL samples so far. The main limitation of the test is the clinical interpretation of the results as the diagnostic performance depends heavily on the pretest probability with sensitivity and specificity for diagnosis of invasive aspergillosis of 82% and 81%, respectively. PPV is <50%, while NPV is >90%. However, in most clinical situations, the low pretest probability limits the use of this test; hence, clinical context should be considered to determine the probability of infection.

Other limitations include its cross-reactivity with other fungi like *Penicillium*, *Fusarium*, and *Histoplasma* due to the presence of galactomannan in their cell walls and that its detection in BAL cannot differentiate colonization from invasive disease as well as negative results in recent antifungal use. However, the test is still most useful in diagnosing invasive aspergillosis, especially in patients with hematologic malignancies to categorize them into possible and probable invasive aspergillosis infection.

8.5 NAAT

Molecular identification methods are a type of diagnostic tool that can be successful without the necessity for live fungal cells. Typically, molecular techniques for identifying fungi are most effective when there are pure cultures present. PCR is a crucial component in most molecular identification techniques. This means that even if active cells are not present, molecular identification can still be carried out using patient materials that contain template nucleic acid, including fixed tissue. Currently, only a limited number of assays have received FDA approval for fungal identification. These assays utilize a PCR-based reaction and are specifically designed to be run on positive blood cultures. They mostly detect *Candida* spp. PCR-based techniques, which involve amplifying a specific nucleic acid target, might produce false-positive results when used on specimens that may be contaminated with human tissue or fluids.

8.6 Fungal AST

Microdilution methods and agar dilution–based tests are considered standard approaches for fungal antimicrobial susceptibility testing (AST). Both the European Committee on Antibiotic Susceptibility Testing (EUCAST) and the Clinical Laboratory Standards Institute (CLSI) have standardized two methodologies for conducting antifungal susceptibility testing. Microdilution procedures are primarily utilized in reference laboratories because of their labor-intensive nature. Furthermore, the utilization of 96-well cell culture trays in the microbroth type is infrequently employed in clinical laboratories. Various commercially available procedures, such as agar diffusion or the use of colorimetric markers in Etest, Fungitest, Sensititre YeastOne, or VITEK, have been developed for routine use.

One of the assays used is the Sensititre YeastOne assay, developed by Thermo Scientific and TREK Diagnostics. This assay utilizes broth microdilution and involves the use of a colorimetric dye, namely, Blue, that is transformed into a dark pink to red color by fungal cells that are metabolically active. The wells exhibit a blue color when metabolic activity is suppressed, and the transition from pink/red to blue is the designated endpoint for determining the minimum inhibitory concentration (MIC) in this test. The disk and strip diffusion approaches are characterized by their simplicity, speed, cost-effectiveness, and ability to yield comparable findings to the reference methods for yeasts. Automated methods greatly decrease the amount of time that biologists need to spend directly on a task, as well as the time it takes to complete the task and the amount of variation that can occur due to the use of a standardized format. To evaluate these procedures, it is necessary to determine the level of agreement between breakpoint categories and reference methods. It is important to note that interpretive breakpoints are only defined for a limited number of species such as *Candida*, *Aspergillus*, and *Cryptococcus*. However, for many other fungal species that might cause invasive infections, clinical breakpoints have not been identified. Therefore, clinical laboratories are unable to determine whether a patient is susceptible or resistant, particularly for numerous molds that have the potential to cause serious mycoses.

9 Virology

Viruses are only able to replicate within live cells, known as host cells, that are capable of supporting a certain type of virus. Traditional tube culture techniques have historically been the primary method in clinical virology for cultivating and identifying viruses from clinical samples. Nevertheless, molecular techniques, specifically the polymerase chain reaction (PCR), have taken over the function of viral culture in numerous laboratories, either restricting its application for virus identification or completely replacing it.

Viral infections can be detected using three main methods: direct virus detection, which is done shortly after infection (e.g., COVID-19 PCR); detection of antibodies against viral proteins, which are typically observed during or after the incubation period; and detection of viral cytopathic effects. The diagnostic tests, such as fast testing, utilize several methods to detect the presence of antibodies and antigens. These methods include the production of antibody-antigen complexes, detection of IgG/IgM antibodies, multiplex PCR, viral antigen detection, and viral culture. Nevertheless, the most delicate and commonly employed techniques are real-time or quantitative PCR procedures. The selection of a viral or illness detection method is contingent upon various aspects, including the characteristics of the virus, the location of the infection, the incubation period of the virus, and the duration of activity and viability on different types of surfaces.

9.1 Viral AST

Most antiviral medications focus on proteins that are particular to the virus and play a crucial role in its replication. Antiviral resistance arises when mutations occur in the viral genes that are specifically targeted by therapy. The measurement of antiviral resistance can be done using two primary methods: genotypic antiviral resistance testing and phenotypic drug susceptibility testing.

9.2 Genotypic Antiviral Resistance Testing

This utilizes sequencing techniques to identify mutations that are linked to medication resistance. Genotypic interpretation systems (GISs) are computational tools that use a database of sequence, clinical, and laboratory information to convert the pattern of mutations in an amplified sequence into drug susceptibility levels. This is done by assigning arbitrary scores.

9.3 Phenotypic Drug Susceptibility Testing

This is a method that directly assesses the capacity of a viral isolate to reproduce in the presence of a particular drug utilizing a cell culture system in a laboratory setting.

9.4 Plaque Reduction Assay

This is used to determine the IC_{50} of antiviral drugs. A known amount of virus is challenged with dilutions of a drug, and the number of plaques is counted until the number of plaques in the presence of the drug is half the number of plaques in wells not containing the drug.

9.5 Relevant Viral Resistance

The emergence of treatment resistance in CMV infections is a significant challenge in individuals with weakened immune systems. In the context of clinical practice, drug resistance is characterized by a consistently high level of viral particles in the body despite undergoing therapy with antiviral medications for the recommended length, often two weeks. The plaque-reduction assay is considered the benchmark for evaluating the effectiveness of antiviral drugs against viruses. However, this approach is not suitable for use in clinical laboratories due to its long processing time and lack of consistent testing standards. The genotype approach is an alternative laboratory technique used in clinical practice to accurately identify antiviral drug resistance. It effectively analyzes CMV mutations and their relationship with treatment resistance.

9.6 Mechanisms of CMV Resistance

- UL97 gene mutation: It is the most common resistance mechanism and occurs prior to UL54 gene mutation. It causes a deficiency in viral kinase responsible for ganciclovir phosphorylation/activation. With this mutation, cidofovir and foscarnet usually retain efficacy.
- UL54 gene mutation: It is often clinically observed as co-existing with the UL97 gene mutation. It results in changes to the viral DNA polymerase; however, cross-resistance to cidofovir is common, and cross-resistance to foscarnet is rare but possible.

10 Mycobacterium

Identification for all mycobacterial species is as follows:

10.1 Smear Microscopy

Mycobacteria have significant quantities of mycolic acids in their cellular walls, which results in poor staining when subjected to Gram stain. They stain well with carbol fuchsin, however. In the process of identification, firstly, an auramine-rhodamine fluorescent stain is used to stain direct specimens as an initial screening on the submitted sample. Afterward, the Kinyoun acid-fast stain or the Ziehl-Neelsen stain is used to confirm positive auramine-rhodamine stains and is used to stain positive culture growth. In the context of active pulmonary TB, the sensitivity of sputum was 38.8%, and the specificity was 96.7%.

10.2 MTB/Rif PCR

This method uses PCR detection of MTB and rifampin resistance (targeting rpoB) from clinical specimens. However, it must be used in conjunction with culture due to false positivity post treatment (can remain positive for 2 years post achieving cure). The diagnostic performance varies among the submitted specimens, e.g., smear-positive specimens: 97–99% sensitivity, 98–99% specificity, while smear-negative specimens: 61–74% sensitivity, 98–99% specificity.

10.3 Culture

This is still considered the gold standard for diagnosis. The specimen can be inoculated on liquid and solid media where some species grow better in liquid media, while others grow better in solid media. Liquid media is incubated for 6 weeks, while solid media is incubated for 8 weeks.

10.4 Classification

Rapidly growing mycobacteria (RGM) versus slowly growing mycobacteria (SGM) refers to the growth rate from a dilute specimen or upon subculture, where growth within 7 days means that more likely it is a RGM. However, MTBC and other SGM from a high-burden specimen may grow from the original specimen in as little as 3–4 days. On the other hand, growth after 7 days means it is likely an SGM. However, RGM from a low-burden specimen may require a week or longer to grow from the original specimen.

10.5 Lipoarabinomannan (LAM) Point-of-Care (POC) Testing

LAM, or lipoarabinomannan, is a type of glycolipid found in the cell wall of the bacteria *Mycobacterium* TB. It is released as a soluble form in urine. WHO has proposed LAM as a biomarker for the development of diagnostic tests that detect its presence in urine. LAM is a leading contender for the creation of point-of-care diagnostic tests due to the ease of collecting urine samples. In addition, LAM has the ability to control the immunological response in the host and may be detected in the blood of individuals with tuberculosis (TB). It likely impacts several types of host cells, hence impacting the effectiveness of both the innate and adaptive immune responses during TB infection. The urine LAM test demonstrated a sensitivity and a specificity of 65% and 89%, respectively, for diagnosing tuberculosis (TB). The urine LAM for TB had a negative predictive value of 97.85%, which is worth mentioning.

10.6 Culture Identification Methods

10.6.1 DNA Probes

This is for use with positive cultures (broth or solid media), where commercially available DNA probes were collected for *Mycobacterium tuberculosis* complex (MTBC), *Mycobacterium kansasii*, *Mycobacterium gordonae*, and *Mycobacterium avium-intracellulare* complex (MAIC).

10.6.2 MALDI-TOF MS

This is for use with positive cultures (broth or solid media), where it requires inactivation and extraction steps to break down mycolic acids and inactivate potential MTBC before performing MALDI-TOF, while the rest of the process is the same as non-acid-fast bacteria. The preparatory/inactivation steps must be applied in the biosafety level 2 or 3 laboratories, and after inactivation, work-up can resume in the routine lab (BSL 2). The main limitation is that it is not able to differentiate all species or subspecies within the MTBC, MAIC, *Mycobacterium abscessus* group, and *M. fortuitum* group, which require sequencing for full identification.

10.6.3 High-Performance Liquid Chromatography (HPLC)

This is for use with positive cultures (broth or solid media). The makeup of mycolic acids in the cell wall varies by *Mycobacterium* species. Mycolic acids are extracted and subjected to HPLC, which creates a profile. HPLC profiles are compared to a library and identified based on similarity to the library. However, as in MALDI-TOF, it cannot differentiate species or subspecies within the MTBC, MAIC, *M. abscessus/chelonae* group, and *M. fortuitum* group, which still require sequencing for full identification.

10.6.4 Sequencing

This may be performed on positive cultures or direct specimens (usually a combination of PCR and sequencing). It can be regarded as the gold standard for the diagnosis of mycobacteria along with culture but is more technically challenging and has a longer turnaround time compared to other methods. 16S rRNA gene is the most commonly used method.

10.7 Mycobacterial AST

10.7.1 Agar Proportion

It is the reference method for MTBC susceptibility testing, but it is being replaced by broth methods, which are easier to perform. The turnaround time is up to 28–35 days (MTBC), where testing is performed using established antimicrobial concentrations and growth on antimicrobial-containing plates is compared to growth on antimicrobial-free plates. If there is $\geq 1\%$ growth on the antimicrobial-containing plate compared to the antimicrobial-free plate, the isolate is considered resistant to that drug. However, each antimicrobial is reported as susceptible or resistant. No MIC is determined or reported.

10.7.2 Broth Method

This is where the microbroth or macrobroth method may be used, where microbroth testing is performed in a 96-well plate. Doubling dilutions are used, and a MIC is determined. In macrobroth testing, the test is performed with designated concentrations, similar to agar proportion. No MIC is determined or reported, and many commercial instruments can be used like BD BACTEC MGIT, bioMerieux BacT/ALERT, ThermoFisher, and VersaTREK. Turnaround time is up to 12–35 days for SGM and 6–12 days for RGM.

11 Tropical Diseases

Disease	Most used test	Other tests
Malaria	Rapid diagnostic test (RDS) and thick blood film	NAAT (not widely available)
Dengue fever	NAAT	Paired antibody sera (IgM/IgG)
Ebola and other hemorrhagic fever viruses	NAAT	Paired antibody sera (IgM/IgG)
Leishmaniasis	Culture or histopathology is still the gold standard	Serology (not in cutaneous) PCR on samples
Tropical rickettsia diseases	NAAT	Paired antibody sera (IgM/IgG)
Schistosomiasis	Detection of *Schistosoma* eggs on the submitted samples(low sensitivity)	ELISA serology
Leptospirosis	Paired antibody sera (IgM/IgG) especially in the early phase	Microscopic agglutination tests, NAAT, culture (very laborious and difficult)

12 Conclusions

Sepsis remains a major diagnostic and management dilemma in critically ill patients. Currently, there is a lack of a diagnostic tool that possesses the capability to promptly ascertain or rule out the presence of infection on presentation. The absence of a dependable diagnostic tool is a contributing factor to the widespread use of broad-spectrum antimicrobial therapy on a global scale. This situation is likely to result in the under-recognition and subsequent over-treatment of noninfectious inflammatory conditions. With the escalating occurrence of antibiotic resistance, there exists a continuous and urgent demand for innovative interventions.

Novel methodologies for pathogen identification present a promising alternative to traditional approaches. Matrix-assisted laser desorption/ionization-time-of-flight mass spectrometry (MALDI-TOF MS) is a high-speed diagnostic technique that has the ability to precisely identify various microorganisms, including bacteria, fungi, and mycobacteria species, immediately upon detection of growth on culture. They may result in the reduction of hospitalization by several days as well as an enhancement in overall survival rates. These findings underscore the need to promptly identify pathogens.

At present, MALDI-TOF MS lacks the capability to identify antimicrobial resistance pathways. Consequently, the assessment of antibiotic susceptibility continues to depend on conventional laboratory methods. Ongoing research is being conducted to discover spectral peaks that exhibit correlation with enzymatic pathways of antimicrobial resistance. The findings of this research are expected to enhance the value of this technology in customizing antibiotic therapy.

Integrating the polymerase chain reaction (PCR) for microbe amplification with mass spectrometry (MS) detection is another promising approach to enhance the expeditious identification of medically significant bacteria and yeast species. These advanced systems boast a turnaround time of 6–8 h and offer a greater diagnostic efficacy compared to conventional blood culture methods.

It is crucial to remember that the recognition and management of sepsis are still dependent on comprehensive patient assessment and informed clinical decision-making. It is imperative that novel procedures serve as a supplement rather than a substitute for these essential components of patient care.

References

1. Melamed A, Sorvillo FJ. The burden of sepsis-associated mortality in the United States from 1999 to 2005: an analysis of multiple-cause-of-death data. Crit Care. 2009;13(1):R28.
2. Kumar G, Kumar N, Taneja A, et al. Nationwide trends of severe sepsis in the 21st century (2000-2007). Chest. 2011;140(5):1223–31.
3. https://www.sccm.org/SurvivingSepsisCampaign/About-SSC/History#:~:text=SSC%20was%20formed%20by%20SCCM,each%20of%20the%20three%20societies.
4. Evans L, Rhodes A, Alhazzani W, Antonelli M, Coopersmith CM, French C, Machado FR, Mcintyre L, Ostermann M, Prescott HC, Schorr C, Simpson S, Wiersinga WJ, Alshamsi F, Angus DC, Arabi Y, Azevedo L, Beale R, Beilman G, Belley-Cote E, Burry L, Cecconi M, Centofanti J, Coz Yataco A, De Waele J, Dellinger RP, Doi K, Du B, Estenssoro E, Ferrer R, Gomersall C, Hodgson C, Møller MH, Iwashyna T, Jacob S, Kleinpell R, Klompas M, Koh Y, Kumar A, Kwizera A, Lobo S, Masur H, McGloughlin S, Mehta S, Mehta Y, Mer M, Nunnally M, Oczkowski S, Osborn T, Papathanassoglou E, Perner A, Puskarich M, Roberts J, Schweickert W, Seckel M, Sevransky J, Sprung CL, Welte T, Zimmerman J, Levy M. Surviving sepsis campaign: international guidelines for management of sepsis and septic shock 2021. Intensive Care Med. 2021;47(11):1181–247. https://doi.org/10.1007/s00134-021-06506-y. Epub 2021 Oct 2. PMID: 34599691; PMCID: PMC8486643.
5. Martin GS, Mannino DM, Eaton S, Moss M. The epidemiology of sepsis in the United States from 1979 through 2000. N Engl J Med. 2003;348(16):1546–54.
6. Kumar A, Roberts D, Wood KE, et al. Duration of hypotension before initiation of effective antimicrobial therapy is the critical determinant of survival in human septic shock. Crit Care Med. 2006;34(6):1589–96.
7. Blanco J, Muriel-Bombin A, Sagredo V, et al. Incidence, organ dysfunction and mortality in severe sepsis: a Spanish multicentre study. Crit Care. 2008;12(6):R158.
8. Vincent JL, Sakr Y, Sprung CL, et al. Sepsis in European intensive care units: results of the SOAP study. Crit Care Med. 2006;34(2):344–53.
9. Brun-Buisson C, Meshaka P, Pinton P, Vallet B, Group ES. EPISEPSIS: a reappraisal of the epidemiology and outcome of severe sepsis in French intensive care units. Intensive Care Med. 2004;30(4):580–8.
10. Martin CM, Priestap F, Fisher H, et al. A prospective, observational registry of patients with severe sepsis: the Canadian sepsis treatment and response registry. Crit Care Med. 2009;37(1):81–8.
11. Rivers E, Nguyen B, Havstad S, et al. Early goal-directed therapy in the treatment of severe sepsis and septic shock. N Engl J Med. 2001;345(19):1368–77.
12. Gupta S, Sakhuja A, Kumar G, McGrath E, Nanchal RS, Kashani KB. Culture-negative severe sepsis: Nationwide trends and outcomes. Chest. 2016;150(6):1251–9. https://doi.org/10.1016/j.chest.2016.08.1460. Epub 2016 Sep 9.

13. Phua J, Ngerng W, See K, et al. Characteristics and outcomes of culture-negative versus culture-positive severe sepsis. Crit Care. 2013;17(5):R202.
14. Li Y, Guo J, Yang H, Li H, Shen Y, Zhang D. Comparison of culture-negative and culture-positive sepsis or septic shock: a systematic review and meta-analysis. Crit Care. 2021;25(1):167. https://doi.org/10.1186/s13054-021-03592-8. PMID: 33964934; PMCID: PMC8106121.
15. Afzal MS, Nandan Chennuri R, Naveed H, Raveena Bai B, Hanif R, Shahzad Z, Umer M, Saleem F. Comparison of clinical outcomes between culture-positive and culture-negative sepsis and septic shock patients: a meta-analysis. Cureus. 2023;15(2):e35416. https://doi.org/10.7759/cureus.35416. PMID: 36987463; PMCID: PMC10040220.
16. Ferrer R, Artigas A, Levy MM, et al. Improvement in process of care and outcome after a multicenter severe sepsis educational program in Spain. JAMA. 2008;299(19):2294–303.
17. Yokota PK, Marra AR, Martino MD, et al. Impact of appropriate antimicrobial therapy for patients with severe sepsis and septic shock—a quality improvement study. PLoS One. 2014;9(11):e104475.
18. Kollef MH, Sherman G, Ward S, Fraser VJ. Inadequate antimicrobial treatment of infections: a risk factor for hospital mortality among critically ill patients. Chest. 1999;115(2):462–74.
19. Harbarth S, Garbino J, Pugin J, Romand JA, Lew D, Pittet D. Inappropriate initial antimicrobial therapy and its effect on survival in a clinical trial of immunomodulating therapy for severe sepsis. Am J Med. 2003;115(7):529–35.
20. Garnacho-Montero J, Garcia-Garmendia JL, Barrero-Almodovar A, Jimenez-Jimenez FJ, Perez-Paredes C, Ortiz-Leyba C. Impact of adequate empirical antibiotic therapy on the outcome of patients admitted to the intensive care unit with sepsis. Crit Care Med. 2003;31(12):2742–51.
21. Lockhart GC, Hanin J, Micek ST, Kollef MH. Pathogen-negative sepsis-an opportunity for antimicrobial stewardship. Open Forum Infect Dis. 2019;6(10):ofz397. https://doi.org/10.1093/ofid/ofz397. PMID: 31660359; PMCID: PMC6785674.
22. Teshome BF, Vouri SM, Hampton N, et al. Duration of exposure to antipseudomonal β-lactam antibiotics in the critically ill and development of new resistance. Pharmacotherapy. 2019;39:261–70.
23. Dombrovskiy VY, Martin AA, Sunderram J, Paz HL. Rapid increase in hospitalization and mortality rates for severe sepsis in the United States: a trend analysis from 1993 to 2003. Crit Care Med. 2007;35:1244–50. https://doi.org/10.1097/01.CCM.0000261890.41311.E9.
24. Dombrovskiy VY, Martin AA, Sunderram J, Paz HL. Facing the challenge: decreasing case fatality rates in severe sepsis despite increasing hospitalizations. Crit Care Med. 2005;33:2555–62. https://doi.org/10.1097/01.ccm.0000186748.64438.7b.
25. Lin MY, Weinstein RA, Hota B. Delay of active antimicrobial therapy and mortality among patients with bacteremia: impact of severe neutropenia. Antimicrob Agents Chemother. 2008;52:3188–94. https://doi.org/10.1128/AAC.01553-07.
26. Lodise TP, Patel N, Kwa A, Graves J, Furuno JP, Graffunder E, Lomaestro B, McGregor JC. Predictors of 30-day mortality among patients with Pseudomonas aeruginosa bloodstream infections: impact of delayed appropriate antibiotic selection. Antimicrob Agents Chemother. 2007;51:3510–5. https://doi.org/10.1128/AAC.00338-07.
27. Bonine NG, Berger A, Altincatal A, Wang R, Bhagnani T, Gillard P, Lodise T. Impact of delayed appropriate antibiotic therapy on patient outcomes by antibiotic resistance status from serious gram-negative bacterial infections. Am J Med Sci. 2019;357:103–10. https://doi.org/10.1016/j.amjms.2018.11.009.
28. Raman G, Avendano E, Berger S, Menon V. Appropriate initial antibiotic therapy in hospitalized patients with gram-negative infections: systematic review and meta-analysis. BMC Infect Dis. 2015;15:395. https://doi.org/10.1186/s12879-015-1123-5.
29. Buehler SS, Madison B, Snyder SR, Derzon JH, Cornish NE, Saubolle MA, Weissfeld AS, Weinstein MP, Liebow EB, Wolk DM. Effectiveness of practices to increase timeliness of providing targeted therapy for inpatients with bloodstream infections: a laboratory medicine best

practices systematic review and meta-analysis. Clin Microbiol Rev. 2016;29:59–103. https://doi.org/10.1128/CMR.00053-14.

30. D'Onofrio V, Salimans L, Bedenić B, Cartuyvels R, Barišić I, Gyssens IC. The clinical impact of rapid molecular microbiological diagnostics for pathogen and resistance gene identification in patients with sepsis: a systematic review. Open Forum Infect Dis. 2020;7:ofaa352. https://doi.org/10.1093/ofid/ofaa352.
31. Queenan K, Häsler B, Rushton J. A one health approach to antimicrobial resistance surveillance: is there a business case for it? Int J Antimicrob Agents. 2016;48:422–7. https://doi.org/10.1016/j.ijantimicag.2016.06.014.
32. Mfutso-Bengu JM, Taylor TE. Ethical jurisdictions in biomedical research. Trends Parasitol. 2002;18:231–4. https://doi.org/10.1016/S1471-4922(01)02218-8.
33. Kang C-I, Kim S-H, Park WB, Lee K-D, Kim H-B, Kim E-C, Oh M, Choe K-W. Bloodstream infections caused by antibiotic-resistant gram-negative bacilli: risk factors for mortality and impact of inappropriate initial antimicrobial therapy on outcome. Antimicrob Agents Chemother. 2005;49:760–6. https://doi.org/10.1128/AAC.49.2.760-766.2005.
34. van Hecke O, Wang K, Lee JJ, Roberts NW, Butler CC. Implications of antibiotic resistance for patients' recovery from common infections in the community: a systematic review and meta-analysis. Clin Infect Dis. 2017;65:371–82. https://doi.org/10.1093/cid/cix233.
35. Carmeli Y, Troillet N, Karchmer AW, Samore MH. Health and economic outcomes of antibiotic resistance in Pseudomonas aeruginosa. Arch Intern Med. 1999;159:1127–32. https://doi.org/10.1001/archinte.159.10.1127.
36. Anton-Vazquez V, Suarez C, Planche T. Impact of rapid susceptibility testing on antimicrobial therapy and clinical outcomes in gram-negative bloodstream infections. J Antimicrob Chemother. 2022;77:771–81. https://doi.org/10.1093/jac/dkab449.
37. Idelevich EA, Sparbier K, Kostrzewa M, Becker K. Rapid detection of antibiotic resistance by MALDI-TOF mass spectrometry using a novel direct-on-target microdroplet growth assay. Clin Microbiol Infect. 2018;24:738–43. https://doi.org/10.1016/j.cmi.2017.10.016.
38. Eigner U, Schmid A, Wild U, Bertsch D, Fahr A-M. Analysis of the comparative workflow and performance characteristics of the VITEK 2 and Phoenix systems. J Clin Microbiol. 2005;43:3829–34. https://doi.org/10.1128/JCM.43.8.3829-3834.2005.
39. Alizadeh M, Yousefi L, Pakdel F, et al. MALDI-TOF Mass spectroscopy applications in clinical microbiology. Adv Pharmacol Pharm Sci. 2021;2021:9928238. Published 2021 May 7. https://doi.org/10.1155/2021/9928238.
40. Elnifro EM, Ashshi AM, Cooper RJ, Klapper PE. Multiplex PCR: optimization and application in diagnostic virology. Clin Microbiol Rev. 2000;13(4):559–70. https://doi.org/10.1128/CMR.13.4.559.
41. Blaschke AJ, Heyrend C, Byington CL, et al. Rapid identification of pathogens from positive blood cultures by multiplex polymerase chain reaction using the FilmArray system. Diagn Microbiol Infect Dis. 2012;74(4):349–55. https://doi.org/10.1016/j.diagmicrobio.2012.08.013.
42. Åkerlund A, Jonasson E, Matuschek E, et al. EUCAST rapid antimicrobial susceptibility testing (RAST) in blood cultures: validation in 55 European laboratories. J Antimicrob Chemother. 2020;75(11):3230–8. https://doi.org/10.1093/jac/dkaa333.
43. Rutanga JP, Van Puyvelde S, Heroes AS, Muvunyi CM, Jacobs J, Deborggraeve S. 16S metagenomics for diagnosis of bloodstream infections: opportunities and pitfalls. Expert Rev Mol Diagn. 2018;18(8):749–59.
44. Pfaller MA, Wolk DM, Lowery TJ. MR and T2Candida: novel technology for the rapid diagnosis of candidemia and invasive candidiasis. Future Microbiol. 2016;11(1):103–17. https://doi.org/10.2217/fmb.15.111.
45. Verweij PE. Advances in diagnostic testing. Med Mycol. 2005;43(Suppl 1):S121–4. https://doi.org/10.1080/13693780400025245.
46. Consortium OPATHY, Gabaldón T. Recent trends in molecular diagnostics of yeast infections: from PCR to NGS. FEMS Microbiol Rev. 2019;43(5):517–47. https://doi.org/10.1093/femsre/fuz015.

47. Afshar B, Bibby DF, Piorkowska R, et al. A European multi-Centre external quality assessment (EQA) study on phenotypic and genotypic methods used for herpes simplex virus (HSV) drug resistance testing. J Clin Virol. 2017;96:89–93. https://doi.org/10.1016/j.jcv.2017.10.002.
48. Hakki M, Chou S. The biology of cytomegalovirus drug resistance. Curr Opin Infect Dis. 2011;24(6):605–11. https://doi.org/10.1097/QCO.0b013e32834cfb58.
49. Buchan BW, Riebe KM, Timke M, Kostrzewa M, Ledeboer NA. Comparison of MALDI-TOF MS with HPLC and nucleic acid sequencing for the identification of mycobacterium species in cultures using solid medium and broth. Am J Clin Pathol. 2014;141(1):25–34. https://doi.org/10.1309/AJCPBPUBUDEW2OAG.

Importance of Infectious Disease Specialist and/or Clinical Microbiologist During Multidisciplinary Rounds in Intensive Care Unit

Umm E Amara, Umme Nashrah, Siddhartha Banerjee, Wael Khalaf, Arshad Hussain Chanda, and Nissar Shaikh

Abstract Before the year 2010, the clinical rounds for critically ill patients in the intensive care unit (ICU) were typically done by physicians and nurses. There are issues regarding the communication of information about patient's clinical data with other specialties whose input will improve the morbidity and mortality of critically ill patients.

A typical multidisciplinary team (MDT) in the ICU is composed of an intensivist or physician, bedside nurse, clinical pharmacist, respiratory therapist, physical and occupational therapist, dietician, and sometimes family members.

It is recommended to have an infectious disease (ID) expert and or clinical microbiologist in MDT rounds in the ICU as a majority of critically ill patients are admitted in the ICU due to sepsis and septic shock. These groups of patients need antimicrobial drugs, microbiological investigations, proper infection control, and prevention policies along with prompt resuscitation and sustained organ support management to have a better and favorable outcome. The inclusion of these specialists during MDT rounds in the ICU will help in appropriate antimicrobial drug administration with the right dose and duration of therapy, thereby avoiding the development of antimicrobial resistance and following a strict antibiotic stewardship program. At the same time, the presence of ID specialists and/or clinical microbiologists and or infection control practitioners will help in following and implementing hospital infection control practice, thus improving patient outcomes with lesser duration of appropriate antimicrobial agents and reducing the overall cost of the therapy.

U. E Amara
Apollo Institutes of Medical Sciences, Hyderabad, Telangana, India

U. Nashrah
Deccan College of Medical Sciences, Hyderabad, Telangana, India

S. Banerjee · W. Khalaf · A. H. Chanda · N. Shaikh (✉)
Surgical Intensive Care Unit, Hamad Medical Corporation, Doha, Qatar
e-mail: achanda@hamad.qa; smaheboob@hamad.qa

N. Shaikh, A. H. Chanda (eds.), *Applied Microbiology in Intensive Care Medicine*, https://doi.org/10.1007/978-981-97-4006-2_13

Keywords Antibiotics · Antimicrobial agents · Intensive care unit · Multidisciplinary rounds · Infectious disease expert · Clinical microbiologist · Infection control and prevention policies · Improved outcome

1 Introduction

In the late 1990s and early 2000, the intensive care unit (ICU) patients were managed mainly by the ICU physician and nurses. There were fewer options for consultation from team members of different specialties on a daily basis, and there was no concept of multidisciplinary rounds for critically ill ICU patients. In the year 2010, the concept of multidisciplinary team (MDT) rounds in the ICU was proposed and started subsequently [1]. Patients in intensive care settings are unique and complicated and require multi-professional inputs to improve their outcomes. The primary objective of MDT rounds is efficient, direct, and effective communication between the team members taking care of critically ill patients.

2 MDT Rounds in ICU

Multidisciplinary rounds (MDT rounds) in ICU are a form of clinical rounds for critically ill patients in the ICU, where all experts from various specialties provide their valuable inputs to review and discuss the clinical information of patients and thereby formulate a concise care plan on a daily basis. MDT rounds allow experts to put forward their treatment options about the complex critical disease process, and at the same time, experts from different fields can discuss among themselves the best possible therapeutic intervention for these patients by addressing the issues, concerns, and goals of management.

The multidisciplinary ICU rounds should occur at a preset time on a regular schedule. Standardization of the timing for MDT rounds will improve the rounding effectiveness by encouraging all the healthcare staff, and it also facilitates a greater number of other specialties experts. MDT rounds should emphasize a systematic and coordinated approach to build patient data, issues in the preceding 24 hours, and current issues of a critically ill patient that should be presented by ICU team members. The input from different experts in MDT should be documented, and the overall plan and goals of the day should be summarized. These MDT-coordinated care and team efforts will improve patient outcomes [1].

3 MDT Composition and Functions

A typical MDT in ICU is composed of an intensivist, advanced practice provider, bedside nurse, clinical pharmacist, case manager, dietician, physical and occupational therapist, respiratory and physical therapist, and occasionally also a family member. Each MDT team member offers unique information and expression of technical skills, which may improve critically ill patient care as well as the outcome [1].

The primary physician or the intensivist leads the MDT team and functions to determine the direction for therapy and patient management plans depending upon the team member's input. The advance providers direct support to the intensivist's decision.

The bedside nurse along with the intensivist is a frontline healthcare provider for the critically ill patients in ICU. Depending upon hospital staffing, the nurse is assigned as one-to-one sick patients per shift. Bedside nurse typically spends more time with patients and patient relatives and hence get accustomed to patient and family concerns very quickly and should convey them to the MDT team and intensivist during MDT rounds. The bedside nurse also has a responsibility to carry out a daily MDT care plan, which includes medication administration to feeding and protocol adherence.

The clinical pharmacist is an important component of MDT, and they will assist MDT with pharmacotherapeutic decisions about the adverse effects and dosing of different drugs used for critically ill patients. The clinical pharmacist helps to reduce medication errors by advising about correct doses, duration, therapeutic drug level monitoring, and possible medication interactions [2].

The majority of critically ill patients need pulmonary care ranging from invasive mechanical ventilations to chest physiotherapy and non-invasive ventilation support. The respiratory therapist should constantly monitor and adjust ventilator settings depending on arterial blood gas analysis and recommendations from the intensivist.

The role of a dietitian is equally important in MDT, as all ICU patients need feeding, either enteral or parental. The dietitian advises the MDT about the quantity and composition of the feed based on past history of nutrition, present weight, adjusted caloric requirements, disease-specific interventions, NUTRIC (Nutrition Risk in Critically Ill) score, and clinical assessment. As there are more preferences for enteral feeding than parenteral ones, the dietician should monitor the feeding tolerance, advise about prokinetics, and indicate when to start parenteral nutrition.

Physiotherapists can help in MDT by doing passive and active physiotherapy and early mobilization of ICU patients.

Case managers and social workers help with the transition of the patient to the next level of care, thereby coordinating with various teams for intra- or inter-hospital patient transfers. Family members and/or home caregivers are key sources of personal information about patients, giving insight into patient's preferences, likes and dislikes, and home medications. Collaborative decision-making allows the following of medical care as well as a decrease in anxiety and confusion about therapy for the illness.

4 Significance of MDT Round and Patient Outcome

MDT rounds in the ICU improve communication about critically ill patients' care plans which decreases major adverse events. It also avoids delaying appropriate therapeutic interventions. All these factors will lead to improved patient outcomes. MDT rounds decrease the length of stay in the ICU; morbidities such as weaning and ventilation days, stress ulcers, and deep venous thrombosis incidences; and healthcare-associated infections and readmissions in ICU within 48 h of discharge.

MDT rounds improve interpersonal communication as well as a greater understanding of patient care, staff satisfaction, and a better sense of teamwork than the provider's traditional rounds [3].

5 Role of Infectious Disease Specialist in MDT

Almost all patients in the ICU at one stage receive antibiotic therapy. Critically ill ICU patients frequently have complex infections, and most of them suffer from multidrug-resistant microbes. Approximately 50% of antimicrobial usage is unnecessary or inappropriate during hospital stays. Keeping all these things in mind, the proper selection of antimicrobial agents and antibiotic stewardship practice is very important.

Many hospitals and ICUs follow "handshake" antibiotics stewardship (HAS), where infectious disease physician does their clinical rounds in the ICU every 24 h or 48 h in person. During these visits, discussions happen with the intensivist or ICU physician and nurses for the proper selection of antimicrobials based on the hospital antibiogram and the nature of the infection. The concept of HAS lacks restriction and preauthorization; at the same time, HAS needs strong interpersonal communication skills; and it has no fixed timing which may miss the "handshake" with the intensive care team. Still, it works for the antibiotic stewardship program. HAS also needs the coordination and cooperation of non-infectious disease pharmacists. Although many critical care physicians like HAS, but still, it has the aforementioned disadvantages [4].

The infectious disease specialist as a team member in MDT can see the unseen. The antibiotic stewardship program provides valuable benefits to the patients as well as physicians. Antibiotic stewardship is a coordinated intervention designed to improve and measure the appropriate use of antibiotic agents by promoting the selection of the optimal antibiotic drug regimen use, proper dose, appropriate duration of therapy, and route of administration. There are barriers that impede the success of antibiotic stewardship. In ICU settings, such barriers arise as there remains a constant fear regarding inadequate antimicrobial coverage and thereby suboptimal treatment of potential pathogens, and at the same time, there remain certain diagnostic uncertainties depending upon the advancement of laboratory facilities [4].

The presence of infectious disease specialist inputs in the ICU was shown to significantly decrease the antibiotic days and cost of therapy due to antimicrobials. Valuable discussions with ID specialists during MDT help on a daily basis with infection control practice, implementation of antibiotic stewardship program, and clarity of microbiological and radiological diagnostic decisions, and it also helps as a support to the continuous learning platform for various critical care educational programs.

The presence of infectious disease specialists during MDT rounds has also been shown to reduce the total cost of therapy, length of stay in ICU, the likelihood of readmission of elderly patients, and mortality [5].

In critically ill patients, it is expected that the daily assessment of the overall clinical picture and microbiological data will allow the MDT to deescalate or escalate the antibiotic therapy, it takes confidence to deescalate, escalate, or discontinue antimicrobial therapy, and the presence of infectious disease specialist in MDT is a major asset in these situations in interpreting microbiological and imaging data. Moreover, the presence of infectious disease specialists in MDT is quite helpful in the early diagnosis and management of new infections in ICU patients. The field of infectious disease specialty in developed countries as well as infectious disease fellowship programs remains underfilled due to various reasons. The supplementation of ID training with the critical care fellowship program may be the way to regenerate the interest as well as proper management of infections and antimicrobial stewardship. There is an ongoing process of developing a year fellowship program in infectious disease in critical care [6].

6 Clinical Microbiologist Present During MDT Round; Time for Intensivist to Resurrect Lost Art

Sepsis is a life-threatening organ dysfunction caused by dysregulated host response to infections. There are more than 7,000,000 patients who suffer from sepsis, in the USA, every year, and sepsis is on the rise. Sepsis and septic shock remain a potentially life-threatening disease, and despite recent advances, mortality remains around 30% [7].

The direct microbiological input for critically ill septic patients is essential. Early broad-spectrum antibiotic therapy initiation and appropriate diagnostic studies to know the causative organism are well established, and it requires reassessment and follow-up on a daily basis to prevent the development of bacterial resistance and toxicity of unnecessary drugs ultimately reducing the cost. The most frequent and common information missing from the patient's file is the microbiological data. Hence, the presence of microbiologists in MDT in ICUs is of great importance [8].

Clinical or medical microbiologists are part of the ICU, and MDT is common in the Indian subcontinent. Medical microbiologist expertise helps in MDT by assisting intensivists in diagnosis and initiation of empirical treatment by rapid microbiological result analysis and communication, aiding interpretation of microbiological results (colonization or contamination), avoiding unnecessary testing, assisting in correct microbiological sampling at appropriate time, and deciding targeted antibiotics for optimal duration of therapy which ultimately facilitate infection control and prevention practices [9].

7 Conclusion

The multidisciplinary team rounds in ICUs are better than the traditional physician-led clinical rounds for critically ill patients, where the specialists or experts from other specialties from clinical pharmacy to respiratory therapy all are present at a single point of time to exchange their opinions for the best outcome of patient. As the majority of patients in the ICU require antimicrobial agents due to sepsis and are also prone to develop new infections because of the complex nature of critical illness, the presence of an infectious disease expert is desirable. Advanced knowledge of clinical microbiologists regarding interpretations of microbiological results based on in-hospital antibiograms will help patient outcomes.

References

1. Sharma S, Hashmi MF, Friede R. Interprofessional rounds in the ICU. Treasure Island, FL: Stat Pearl; 2023. ncbi.nlm.nih.gov/books.
2. Masood U, Sharma A, Bhatti A, et al. A Successful pharmacist-based quality initiatives to reduce inappropriate stress ulcer prophylaxis use in an academic Medical Intensive Care Unit. Inquiry. 2018;55:46958018759116.
3. Abraham J, Tannampalli TG, Patel VL, et al. Impact of structured rounding tools on time allocation during multidisciplinary rounds: an observational study. JMIR Hum Factors. 2016;3(2):e29. Published online 2016 Dec 9.
4. Evans B, Losar J, Peerrmohammed S. Attitudes and perceptions amongst critical care physicians towards handshake antimicrobial stewardship rounds. Cureus. 2019;11(12):e6419. Published online 2019 Dec 19.
5. Gilbert DN. Influence of an infectious disease specialist on ICU multidisciplinary rounds. Crit Care Res Pract. 2014;2014:307817. Published online 2014 Apr 17
6. Kadri SS, et al. Critical Care Medicine and Infectious diseases: an Emerging Combined Subspecialty in the United States. Clin Infect Dis. 2015;61(4):609.
7. Ogura T, Nakamura Y, Takahashi K. Treatment of patients with sepsis in a closed intensive care unit is associated with improved survival: a nationwide observational study in Japan. J Intensive Care. 2018;6:57.

8. Wilson L, Dempsey G. Medical microbiology ward rounds in critical care. Crit Care. 2007;11(Suppl 2):P75.
9. Schouten J, De Angelis G, De Waele JJ. A microbiologist consultant should attend daily ICU rounds. Intensive Care Med. 2020;46:372–4.

Obstetric Sepsis and Management

Suruchi Mohan, Arabo Ibrahim Bayo, and Gbemisola Okunoye

Abstract Maternal sepsis remains one of the main causes of maternal morbidity and mortality. According to reports, direct obstetric infections are the third most common cause of maternal mortality, accounting for 10.7% of all maternal deaths in low-and middle-income countries and 4.7% in high-income nations. The assessment of maternal sepsis is extremely difficult due to the physiological changes and the variety of sepsis criteria. Early indications of sepsis can be difficult to detect due to typical physiological changes throughout pregnancy and maternal efforts during labor. The symptoms of sepsis such as tachycardia, tachypnea, and leukophilia are, as well, typical physiological changes that occur throughout pregnancy and childbirth. The vaginal, urinary, and respiratory tracts are frequently the sites of infection that can potentially lead to sepsis during pregnancy. Endometritis is the most typical postpartum infection linked to sepsis. In addition to the pregnancy itself, pregnant and puerperal women may be at risk for sepsis for additional reasons. In the obstetric population, sepsis clinical characteristics can differ and may also depend on the infection's origin. A variety of diagnostic scoring systems and methods have been offered to address the challenges in the diagnosis of sepsis in the obstetric population. These scoring systems were created to help clinicians quickly assess the severity of each patient's condition and initiate interventions, thereby lowering morbidity and death. Specific to maternity settings are MEOWS (Modified Early Obstetric Warning System) and SOS (Sepsis in Obstetrics Score). Obstetric sepsis is managed by resuscitating and stabilizing the patient, halting further deterioration, and enhancing tissue perfusion to limit or avoid organ damage.

According to the widely accepted SSC (Surviving Sepsis Campaign) Guideline, the SSC Hour-1 bundle, which consists of several interventions, should be started

S. Mohan · G. Okunoye
Sidra Medicine, Education City North Campus, Qatar Foundation, Doha, Qatar

Weill Cornell Medical College, Doha, Qatar

A. I. Bayo (✉)
Women's Wellness and Research Center, Hamad Medical Corporation, Doha, Qatar

College of Medicine, Qatar University, Doha, Qatar
e-mail: Abayo@hamad.qa

N. Shaikh, A. H. Chanda (eds.), *Applied Microbiology in Intensive Care Medicine*, https://doi.org/10.1007/978-981-97-4006-2_14

within 1 h of the diagnosis of sepsis. Broad-spectrum antibiotics must be chosen to cover a variety of frequently implicated pathogens, including both gram-negative and gram-positive microorganisms and anaerobes, for initial therapy. A change toward a personalized care approach is consequently being developed and utilized more frequently as a result of new technologies and growing evidence.

Keywords Antibiotics · Maternal sepsis · Pathophysiology of sepsis Physiological changes in pregnancy · Scoring systems in sepsis · Sepsis bundle Source control

1 Introduction

Although it is difficult to ascertain the true burden of maternal sepsis globally, due to varying definitions and diagnostic inaccuracies, it remains a leading cause of maternal morbidity and mortality in the world [1, 2]. Direct obstetric infections are reported to be the third most common cause of maternal mortality contributing to 10.7% of all maternal deaths in low-and middle-income countries and at 4.7% in high-income countries [1]. The actual burden of infection-related mortality may be even higher as these reported figures do not include deaths from infections exacerbated by pregnancy. Further, normal gestational physiological changes may not only predispose to infection but may also affect the body's response to infection and hinder identification and management strategies [3]. Recognizing the challenges posed by maternal sepsis, the World Health Organisation (WHO) commenced the Global Maternal Sepsis Study (GLOSS) in 2017, across 52 countries, with the aim of improving the understanding of the epidemiology of maternal infections as well as reviewing current management [4]. The findings of this study revealed that the frequency of health facility admissions with maternal infections remains high, and direct and indirect obstetric infections contribute to maternal mortality to a greater extent than previously understood [2].

Defining sepsis with global uniformity is a key consideration because unless this standardization is adopted, incidence estimation can be challenging. A universal definition can help accurately assess the burden of the condition and inform health strategy as well as research. Therefore, the WHO undertook a consultation with experts and a systematic review to arrive at a consensus definition for application worldwide. Maternal sepsis as defined by the WHO is a "Life-threatening condition defined as organ dysfunction resulting from infection during pregnancy, childbirth, post-abortion or post-partum period." [4].

2 Pathophysiology of Sepsis

Sepsis is a complex pathological process that is not produced by all infections. However, sepsis is said to occur if an infection has triggered a systemic immune response. This systemic response to the physiological is characterized by hyperthermia, tachycardia, tachypnoea, and leucophilia.

In a normal response to infection, the immune system works to contain the infection and prevent systemic spread. Vasodilatation is triggered allowing the advent of immune cells to the site. The complement system is activated containing the infection and triggering of the coagulation pathway resulting in the deposition of fibrin at the site to prevent dissemination of the infectious agent or toxin. Anti-inflammatory factors are also released to keep the magnitude of the immune response in check [2].

However, sepsis represents an exaggerated immune response where there is overproduction and systemic dissemination of immune mediators including cytokines, nitric oxide, and other proinflammatory agents that lead to vasodilatation and hypotension. The ensuing endothelial injury also results in increased capillary permeability resulting in third space extravasation of fluid and decreased intravascular volume. The resultant hemoconcentration and hypotension result in reduced organ perfusion mediating tissue damage.

3 Relevant Physiological Changes of Pregnancy

Maternal physiological changes and immunological adaptations during pregnancy and puerperium can enhance susceptibility to infection and may also impair the response to infection. Furthermore, some of these normal physiological changes may produce effects that are like those of early sepsis, and this may delay diagnosis. Understanding these physiological changes can therefore aid early detection and appropriate management.

Immunological changes: It has long been known that pregnancy induces a state of relative immune tolerance to aid the development of the fetus, and our understanding of the underlying mechanisms is evolving. Both maternal innate and adaptive immune systems are thought to adapt to pregnancy. Immunosuppressive T cells have been shown to have a role in maintaining immunological tolerance [5]. Immune suppression in a normal pregnancy is also mediated by a change in the innate immune phenotype and gene expression [6, 7]. Cytokine levels and cell-mediated immunity also vary throughout the pregnancy [8]. These effects result in a greater susceptibility to infections and an impaired immune response when infections occur.

Cardiovascular changes: The cardiovascular changes in pregnancy are well studied and described, and among the changes pertinent to obstetric sepsis is where pregnancy results in the upregulation of endothelium-derived nitric oxide as well as other estradiol-dependent vasodilators resulting in peripheral vasodilatation. A significant part of the morbidity of sepsis itself is also mediated largely by the cardiovascular effects it mediates including upregulation of nitric oxide resulting in smooth muscle relaxation and vasodilatation. As a result, when sepsis occurs in pregnancy, the potentiated vasodilatory effect may result in a greater risk of abrupt hypotension and hypoperfusion of organs resulting in a greater risk of organ dysfunction.

Further, progesterone in pregnancy mediates hyperventilation as a physiological gestational adaptation. This results in respiratory alkalosis that promotes a compensatory increase in renal bicarbonate secretion, and this may impair the ability of pregnant patients to buffer the metabolic acidosis that occurs in sepsis [9].

Importantly, the normal physiological changes of pregnancy and maternal efforts during labor may impede the recognition of early signs of sepsis. For instance, tachycardia, tachypnea, and leukophilia are signs of sepsis and normal physiological changes in pregnancy and parturition [8]. This also has implications for the use of non-pregnancy-adjusted scoring systems, warning scores, and blood parameter reference ranges.

4 Etiology

During pregnancy, the common sites of infection that can potentially trigger sepsis are the genital, urinary, and respiratory tracts. The most common postpartum infection implicated in sepsis is endometritis. Table 1 lists the potential causes of maternal sepsis.

The main organisms implicated in maternal sepsis in the UK include Group A *Streptococcus*, *Streptococcus pneumoniae*, *Escherichia coli*, and the H1N1 virus [10]. Studies from the USA report the most common infections and pathogens causing obstetric sepsis as chorioamnionitis, endometritis, and pneumonia caused by *E. coli* or group A *Streptococcus* [10]. The WHO states that, overall, the leading causative pathogen for maternal and neonatal sepsis is currently group B *Streptococcus*, and *E. coli* has been identified as an emerging threat [11, 12].

Table 1 Potential causes of sepsis

Site of infection	Infection type	Most frequent causative organism
Genital tract	Septic abortion Chorioamnionitis Endometritis Postpartum wound infections	*Escherichia coli, Proteus, Enterobacter, Staphylococcus,* and anaerobes (e.g., *Clostridium*) *Ureaplasma urealyticum* and *Mycoplasma hominis, Gardnerella vaginalis, Bacteroides,* group B *Streptococcus* Polymicrobial: *Peptostreptococcus, Bacteroides, Clostridium,* group B *Streptococcus, Enterococcus, E. Coli, Streptococcus pyogenes, Staphylococcus aureus* Group A b-hemolytic *Streptococcus, Staphylococcus aureus*
Respiratory tract	Pneumonias	*Streptococcus pneumoniae, Haemophilus influenzae, Chlamydia pneumoniae, Mycoplasma pneumoniae, Legionella pneumophila,* influenza A and B
Urinary tract	Cystitis Pyelonephritis	*E. coli, Klebsiella, Enterobacter, Proteus* species
Systemic	Tuberculosis Malaria	*Mycobacterium* *Plasmodium* species
Other	Appendicitis Cholecystitis Mastitis/breast abscess	*E. coli, Enterobacter, Klebsiella, Enterococcus* *Staphylococcus* species

Group B Streptococcus (GBS) or Streptococcus agalactiae: This is a gram-positive beta-hemolytic coccus that can be found in normal human or animal flora, and in pregnancy, its genitourinary carriage can be frequently detected. The organism can cause chorioamnionitis and endometritis and can lead to neonatal sepsis if contracted by the fetus during parturition. In recent years, it has emerged as a leading cause of maternal sepsis, and its incidence is rising. Proposed explanations for increasing cases include the spread from bovine hosts, the emergence of a virulent strain, and the increasing tetracycline resistance [11]. As a response to this threat, many obstetric professional bodies now recommend screening for GBS carriage during pregnancy with intrapartum antibiotic prophylaxis if detected [13, 14].

Group A Streptococcus or S.pyogenes: This is a beta-hemolytic gram-positive coccus which can form part of normal throat or skin flora and may be spread by droplet transmission or direct contact [10]. The organism produces exotoxin A which mediates the clinical manifestations of the infection. Invasive group A *Streptococcus* infection has the potential to cause life-threatening sepsis with postpartum endometritis, toxic shock syndrome, and necrotizing fasciitis.

Escherichia coli: This gram-negative *Bacillus* is a common cause of antepartum and intrapartum as well as postpartum infections. It is a commensal found in intestinal flora, and certain strains have virulent pathogenic potential. Rectovaginal colonization with *E. coli* in pregnancy may be associated with preterm rupture of membranes, chorioamnionitis, and preterm birth as well as neonatal sepsis and meningitis [15].

It should be noted that postpartum endometritis tends to be a polymicrobial infection with mixed aerobes like *E. coli*, *Klebsiella*, *Enterococcus*, *Staphylococcus*, and anaerobes such as *Peptostreptococcus*, *Fusobacterium*, and *Bacteroides* [16].

5 Risk Factors for Sepsis

Aside from pregnancy itself, pregnant and puerperal women may have other risk factors that may predispose them to sepsis. These are listed in Table 2. The presence of risk factors should alert the obstetric team to maintain surveillance for potential early signs of sepsis and prompt preparedness for appropriate intervention.

Table 2 Risk factors for sepsis

Antepartum factors	Intrapartum factors	Factors unrelated to pregnancy
• Chorionic villus sampling • Amniocentesis • cervical suture placement	• Prolonged labor • Premature or prolonged rupture of membranes • Frequent examinations in labor	• Ethnicity: Black or ethnic minority • Poor socioeconomic status • Obesity • Anemia • Diabetes • Immunosuppression due to any cause

6 Clinical Presentations

Clinical features of sepsis can vary in the obstetric population and may also depend on the source of the infection. Symptoms may include fever, vomiting, diarrhea, rashes, abdominal or pelvic pain, premature contractions, vaginal discharge, sore throat, shortness of breath, and altered mental status. In a pregnant patient, diarrhea along with abdominal pain and altered fetal heart rate may be an indicator of pelvic sepsis [8]. Clinical signs of sepsis are fever (>38 °C) or hypothermia (<36 °C), tachycardia, tachypnea, hypotension, oliguria, reduced capillary refill, and cold extremities. Laboratory tests can reveal leukocytosis or leucopenia, thrombocytopenia, raised CRP, high lactate levels, raised creatinine levels, hyperglycemia without diabetes, hyperbilirubinemia, hypoxia, and coagulopathy. Hypothermia, tachypnea, and neutropenia are features of particular concern [8].

However, pregnant patients may commonly present with symptoms such as abdominal pain, nausea, vomiting, cramps, malaise, and fever which may not always signify sepsis. Further, underlying infections may not always be easily apparent. Signs such as tachycardia and tachypnea may arise from the normal physiological state of pregnancy and delay the recognition of sepsis. Even laboratory tests may be less reliable as parameters due to the physiological changes of pregnancy such as a raised white blood cell count. The real challenge in obstetric sepsis is, therefore, distinguishing the early signs of sepsis from normal physiological presentations in pregnancy. In these situations, a high index of suspicion is required, and the clinician should be wary of disproportionately severe abdominal pain, diarrhea, abnormal fetal heart rate patterns especially tachycardia, and high-grade fever (>38 °C) or hypothermia.

To overcome the impediments to diagnosing sepsis in the obstetric population, several diagnostic scoring systems and tools have been proposed, and these are considered in the following section.

7 Identification and Assessment of Severity of Sepsis

Several scoring systems and tools have been designed to aid early identification of sepsis to facilitate timely intervention. And, while early identification of sepsis in the obstetric patient is challenging as discussed, supporting the use of tools, the very physiological changes that make diagnosis difficult can also affect the applicability of these tools in the maternity patient subset.

The current concept of sepsis as a syndrome, without agreed set diagnostic criteria, necessitates the identification of features that relate to the infection, the host response, and organ dysfunction, and measurement of each of these can help identify the presence and severity of sepsis. Previously used terms of sepsis, severe sepsis, and septic shock have now been replaced by sepsis and septic shock as per the third consensus definition of sepsis given by the European

Society of Intensive Care Medicine and the Society of Critical Care Medicine [17]. Having defined sepsis as organ dysfunction arising from the host's response to infection, septic shock is diagnosed when vasopressors are needed to maintain the mean blood pressure above 65 mmHg or the serum lactate level is >2 mmol/L [18].

To clinically assess the severity of sepsis, scoring systems have been developed:

7.1 *Sequential Organ Failure Assessment (SOFA)*

Scoring systems are based on the quantification of organ dysfunction of sepsis by considering clinical features as well as laboratory parameters. The most commonly used system is the **SOFA or Sequential Organ Failure Assessment** [19]. This system is used in critical care to clinically assess sepsis. Table 3 shows the SOFA parameters and scores.

A SOFA score of ≥2 signifies organ dysfunction and is associated with a 10% mortality risk [17].

Table 3 SOFA parameters and scores

Parameter					
	0	1	2	3	4
Respiration					
PaO_2/FiO_2 (mmHg)	≥400	<400	<300	<200 with respiratory effort	<100 with respiratory effort
Coagulation					
Platelets ($\times10^3$/micL)	≥150	<150	<100	<50	<20
Liver					
Bilirubin (mg/dL)	<1.2	1.2–1.9	2.0–5.9	6.0–11.9	>12.0
Cardiovascular					
MAP (mmHg)	≥70	<70	Dopamine low dose or any dose of dobutamine	Medium dose dopamine or low/medium epinephrine or norepinephrine	High dose dopamine or epinephrine or norepinephrine
Central nervous system					
GCS score	15	13–14	10–12	6–9	<6
Renal					
Creatinine (mg/dL)	<1.2	1.2–1.9	2.0–3.4	3.5–4.9	>5.0
Urine output (mL/day)				<500	<200

7.2 Quick SOFA (qSOFA)

In the absence of available laboratory results, a simple clinical bedside score called qSOFA or quick SOFA (Table 4) was developed as a set of criteria to be used to identify patients who have suspected infections and who are at risk of poor outcomes. This may be used as a trigger for initiating further evaluation and prompt therapeutic measures [17]. The qSOFA considers the three clinical parameters in the SOFA including blood pressure, respiratory rate, and the Glasgow Coma Scale Score with two abnormal results out of the three indicating organ dysfunction.

7.3 Modified SOFA

The above scales have not been validated for use on the pregnant population, and several studies have been conducted to suggest modifications to these scoring systems for use in obstetric sepsis. Notably, the Society of Obstetric Medicine of Australia and New Zealand (SOMANZ) proposed modifications to SOFA and qSOFA criteria in their obstetric sepsis guidelines [20]. Their suggested modification to the SOFA is detailed in Table 5.

Table 4 qSOFA score

Parameter	Score	
	0	1
Systolic blood pressure	≥90	<90
Respiratory rate	<25 per minute	≥25 per minute
Mental status	Alert	Not alert
Scores of ≥2 are associated with higher mortality		

Table 5 Modified SOFA for sepsis in pregnancy

Parameter			
	0	1	2
Respiration			
PaO_2/FiO_2 (mmHg)	≥400	300–<400	<300
Coagulation			
Platelets (×10^3/micL)	≥150	100–150	<100
Liver			
Bilirubin (mg/dL)	<1.2	1.2–1.9	≥2.0
Cardiovascular			
MAP (mmHg)	≥70	<70	Vasopressors needed
Central nervous system			
Assessment	Alert	Responds to voice	Responds to pain
Renal			
Creatinine (mg/dL)	<1.2	1.2–1.9	2.0–3.4

With respect to the qSOFA, SOMANZ recommends that maternal sepsis should be considered from qSOFA assessments that show respiratory rate of ≥25/min, systolic blood pressure ≤ 90 mmHg, or an altered mental state.

7.4 Early warning System

These are scoring systems that have been developed to aid the clinician in rapidly identifying the severity of illness and triggering intervention in all patients admitted to the hospital, thus reducing morbidity and mortality. Examples of early warning scoring systems include the MEWS (Modified Early Warning Score), MEOWS (Modified Early Obstetric Warning System), SOS (Sepsis in Obstetrics Score), and REMS (Rapid Emergency Medicine Score). Among these, the MEOWS and SOS are specific to maternity settings.

The MEOWS Score is commonly used in the maternity setting and incorporates assessment of the heart rate, blood pressure, respiratory rate, oxygen saturation, temperature, and state of consciousness +/− urine output. Each parameter is assigned a score from 0 to 3 where an increasing score indicates further deviation from normal. A total score of 1–4 is considered low, 5–6 medium, and 7 high with higher scores necessitating urgent intervention [21].

The SOS (Table 6) identifies ICU admission risk for patients with obstetric sepsis. This score includes the parameters of heart rate, blood pressure, respiratory rate, temperature, oxygen saturation, white blood cell count, percentage of immature neutrophils, and lactic acid levels. Each parameter is scored, and a score of ≥6 identifies obstetric sepsis patients requiring ICU admission [22].

Table 6 Sepsis in Obstetrics Score

Parameter	score								
	+4	+3	+2	+1	0 (normal)	+1	+2	+3	+4
Temperature (°C)	>40.9	39–40.9		38.5–38.9	36–38.4	34–35.9	32–33.9	30–31.9	<30
Systolic BP (mmHg)					>90		70–90		<70
Heart rate (bpm)	>179	150–179	130–149	120–129	≤119				
Respiratory rate (per minute)	>49	35–49		25–34	12–24	10–11	6–9		<5
SpO_2 (%)					≥92	90–91		85–89	<85
Leucocytes (per mic L)	>39.9		25–39.9	17–24.9	5.7–16.9	3–5.6	1–2.9		<1
Immature neutrophils (%)			≥10		<10				
Lactic acid (mmol/L)			≥4		<4				
Score of ≥6 indicates a higher risk of intensive care admission									

8 Management of Sepsis in Obstetric Patients

Management of obstetric sepsis is directed at resuscitation and stabilization of the patient, prevention of deterioration, and improving tissue perfusion so that organ dysfunction is limited or prevented. The key components of management include resuscitative treatment and infection source control which should be undertaken concurrently.

9 Immediate Management: Care Bundles

Intervention time is critical in sepsis management, and to aid appropriate and timely intervention, bundles of care for sepsis management have been developed. The commonly used SSC (Surviving Sepsis Campaign) Guideline [23] recommends the initiation of the ***SSC Hour-1 bundle*** incorporating a series of interventions to be made within the first hour of the diagnosis of sepsis. These interventions include the following:

1. Measure maternal blood lactate levels and repeat if initial level ≥ 2 mmol/L.
2. Obtain blood culture samples.
3. Commence broad-spectrum antibiotic therapy.
4. Administer 30 mL/Kg crystalloid if there is hypotension or if the lactate levels are ≥4 mmol/L.
5. Give vasopressors if there is hypotension during or after fluid resuscitation to maintain MAP ≥65 mmHg.

The UK Sepsis Trust [24] provides similar guidance on initial sepsis management in the form of the ***sepsis-6 bundle*** (Table 7) which recommends six interventions within 1 h of diagnosis of sepsis, including administering oxygen, taking blood cultures, measuring serum lactate (serially if required), administering broad-spectrum antibiotics, administering 500 mL stats (can repeat 30 mL/Kg) of crystalloid if hypotensive or lactate ≥2 mmol/L, and measuring urine output.

Table 7 Sepsis-6 bundle

		Time	Initials
1	Oxygen		
	Give supplementary oxygen and maintain saturation between 94 and 98%		
2	Blood cultures		
	Take at least one prior to antibiotics and consider genital, urine, throat, and wound cultures		
3	IV antibiotics (according to women sepsis antibiotics algorithm)		
4	Fluid resuscitation		
	Insert two large bore IV lines, give 500 mL of lactated Ringer's solution for hypotension, and give 20 mL/kg bolus of lactated Ringer's solution		
5	Laboratory		
	Obtain serum lactate level initially within one hour NB: If lactate ≥2 mmol/L, repeat again 2 hours after fluid resuscitation Other labs: CBC, CMP, coagulation profile, venous/arterial blood gas		
6	Measure urine output Record urine output every hour Call a physician if no urine output in 2 h		
	Consider bladder catheterization for careful measurement of output		

10 Investigations

10.1 Lactate

Serum lactate levels function as a test for detecting the presence of tissue hypoperfusion and hypoxia as a result of sepsis. Lactate levels in normal pregnancy may be considered like those of the general population; however, labor may cause an elevation of lactate levels due to metabolic load. Sepsis can compromise the pregnant patient's ability to compensate for rising lactate, and elevated levels of maternal sepsis are associated with poor outcomes. Here, lactate concentrations show a correlation with risk of mortality with levels between 2 and 4 mmol/L correlating with 25% mortality and > 4 mmol/L with 40% mortality [25]. A lactate level of 2 mmol/L should trigger critical care input, and levels above 4 mmol/L in spite of fluid resuscitation should prompt central venous pressure and oxygen saturation monitoring [26].

10.2 White Cell Count and C-Reactive Protein

Elevated white cell counts and C-reactive protein levels may help aid the diagnosis of sepsis but cannot distinguish inflammation from infection, and physiological effects on their normal ranges in pregnancy have to be considered.

10.3 Procalcitonin

Procalcitonin has been shown to be a biomarker specific for bacterial infection; however, there appears to be insufficient evidence at present to recommend its use in guiding the treatment of sepsis [23, 27]. Further more, normal ranges in pregnancy have not been established to aid accurate interpretation [28].

10.4 Blood Cultures

Ideally, blood cultures (both aerobic and anaerobic) should be drawn before initiation of antibiotic therapy as sterilization can occur as early as within minutes of antibiotic administration, rendering cultures negative [29]. However, if difficulties in obtaining samples are encountered, antibiotic therapy should not be delayed.

11 Isolating the Source of Infection

Identifying and addressing the source of infection should be part of initial and ongoing care in sepsis. The diagnosis of sepsis should therefore include the identification of possible sources of infection, and the prompt collection of appropriate cultures of urine, vaginal discharge, amniotic fluid, wounds, abscesses, or the placenta is required. The presumed site of infection will also help guide the choice of antibiotics for treatment. Depending on the site of infection, antibiotic therapy alone may not be sufficient, and evacuation of infective material may be indicated such as draining of an abscess, wound debridement, or expediting delivery in suspected chorioamnionitis. A delay in infection source identification and control is associated with a substantial increase in mortality [30].

12 Choice of Empiric Antibiotic Therapy

For initial management, broad-spectrum antibiotics must be selected to cover a wide range of commonly implicated organisms including both gram-negative and gram-positive microbes.

Taking into consideration common pathogens in obstetric sepsis (see Table 1), the following empiric antibiotics are recommended for initial treatment:

1. Co-amoxiclav, cefuroxime, or cefotaxime and metronidazole for patients who are not critically ill. In case of penicillin allergy, clindamycin and gentamicin may be used.

2. In severe cases, piperacillin-tazobactam, meropenem, or ciprofloxacin and gentamicin may be used along with metronidazole for anaerobic cover.
3. In case MRSA cover is required (suspected hospital-acquired infections), teicoplanin and linezolid may be considered.
4. For group A streptococcal infection, clindamycin may show better efficacy.
5. Treatment against fungi (e.g., *Candida*) or viruses (e.g., influenza) is given only if indicated.

Once the culture reports identify the causative pathogen and the antimicrobial sensitivities, antibiotic therapy should be adjusted appropriately.

Doses of antibiotics in pregnancy are the same as for non-pregnant individuals. However, patients with sepsis are more likely to have kidney or liver dysfunction, and fluid resuscitation may rapidly expand extracellular volume, all of which can influence the dose requirements.

Antibiotics are continued for 7–10 days depending on the severity of the infection, but longer treatments may be required in cases of poor clinical responses and immunological deficiencies and where it is not possible to remove infective material.

13 Critical Care Considerations in Maternity Contexts

Critical care considerations relevant to the obstetric patient are considered below.

13.1 General Resuscitation

The pregnant patient should be positioned in the left lateral decubitus position to optimize maternal and fetal circulation. If mechanical ventilation becomes necessary in a pregnant patient, hyperventilation should be avoided as it can decrease uteroplacental blood flow, reducing fetal pH. Further, chest wall compliance may be decreased in pregnancy requiring higher inflation pressures to achieve adequate tidal volume. In pregnant patients, it is recommended that the urine output should be maintained at 0.5 mL/Kg [23].

13.2 Oxygen

In general, it is recommended that oxygen saturation should be maintained at >94% in critically ill sepsis patients [26]. However, it has been proposed that to ensure fetal well-being, saturation should be maintained at >95% in pregnant patients [31].

13.3 Vasopressors

Sepsis can lead to reversible myocardial depression, and inotropes may be used for a good effect if there is a fall in cardiac output. Vasopressors can be used to treat septic shock in pregnant patients receiving critical care, and norepinephrine, which is generally the first-line agent, may be given, having been shown to be safe for the fetus [3]. Norepinephrine may cause a reduction in uteroplacental blood flow, but in septic shock contexts, this risk is outweighed by the benefit of improved perfusion of the vital organs and the uterus.

Dobutamine, which is also used as an inotrope in septic shock has been shown to restrict blood flow to the uterus and is therefore not commonly used in maternal sepsis. Vasopressin has not been adequately studied for use in obstetric sepsis.

Levosimendan is a relatively new inotropic agent that possesses vasodilatory, anti-inflammatory, and anti-oxidative properties and has shown promise in septic shock management. However, there is not enough efficacy and safety data available at present to recommend its use in pregnant patients [32].

13.4 Glycemic Control

Against the background state of relative glucose intolerance in pregnancy and physiological changes to glucose homeostasis, limits of hyperglycemia on sepsis of acute illness in the pregnant patient are not defined. In the absence of robust evidence, there is no consensus on the impact of rigorous glucose control in maternal sepsis on morbidity and mortality [2]. While risks of uncontrolled hyperglycemia have to be considered, a balanced approach considering the potential risk of fetal hypoglycemia is advised. The current approach is to maintain blood sugar levels less than 10 mmol/L (180 mg/dL) [30].

13.5 Steroids

A somewhat controversial area of critical care in the context of sepsis and septic shock is the use of intravenous steroid therapy. While a systematic review in 2018 did report a 2% reduction in mortality among ICU patients who received intravenous corticosteroids, the evidence was not applicable to the obstetric population [33]. Additionally, potential concerns with the use of steroids in pregnancy must be considered.

13.6 IV Immunoglobulins

A dysregulated immune response lies at the heart of the pathogenesis of sepsis where cytokine release mediates endothelial dysfunction and results in edema, hemoconcentration, and hypotension. Regulating the immune response can result in improved outcomes [34]. Administration of IV immunoglobulins here may be useful through neutralization of antigens reducing cytokine release.

IV immunoglobulins have been used in sepsis management in adults and showed a demonstrable reduction in mortality [35]; however, their use in the obstetric population is not supported by robust evidence.

14 Obstetric Considerations and Delivery

Sepsis in pregnancy is associated with an increased risk of complications including miscarriage, preterm birth, and fetal death even when the primary focus of infection is not intraamniotic [35]. Where sepsis affects maternal arterial pressure, it may reduce placental perfusion, and tests of fetal well-being may manifest these effects. However, stabilization of the mother is a priority, and unless chorioamnionitis is the primary infective focus, the presence of sepsis alone may not be an indication for delivery [20]. Therefore, in the absence of chorioamnionitis, indications for delivery remain obstetrical. Further indications for the use of steroids for fetal lung maturity, tocolysis, and magnesium sulfate for prematurity need to be prioritized based on the gestational age and expectation of imminent delivery [18].

15 Isolation and Infection Control Considerations

Depending on the nature of the primary focus of infection, suitable isolation and infection control precautions may have to be applied while caring for the maternity sepsis patient. These may include droplet, airborne, or enteric precautions as appropriate.

16 Future Directions in Maternal Sepsis

16.1 Precision Medicine

Much like many other areas of medicine, attention in sepsis management is shifting toward a personalized "precision" medicine approach to care that considers the interactions between the individual and their environment and how this influences the disease. This is particularly relevant in sepsis where the complex bodily response to infection can be influenced by multiple factors and, even more so, in maternal sepsis where individuals may show varied responses under the influence of gestational physiological changes necessitating a tailored approach.

Precision medicine strategies can be applied in diagnosis and clinical monitoring as well as therapeutics. Biomarkers, bioinformatics, and omics find application in monitoring and diagnosis, while immunoglobulins, immunoparalysis, and endotoxin/cytokine hemoadsorption are examples of novel therapeutic interventions [36].

16.2 "Omics" Technologies in Sepsis

Precision medicine offers several disciplines, i.e., "omics"-based tools, which find application in personalized care of sepsis. Omics-based technologies being applied in sepsis care include genomics, epigenomics, transcriptomics, metabolomics, and proteomics [37].

Within sepsis, these are being employed to understand pathological mechanisms and identify pathogen and host biomarkers that could aid clinical management. There is also a potential to use innovative therapies to modulate these pathological processes.

Genomics and epigenomics: These involve the study of genes and their functions where it is known that presentations and prognosis on sepsis can be influenced by genetic variants. Cytokines and cytokine receptors which play a major role in sepsis are encoded by different genes which show polymorphisms. In addition, epigenomics studies explore factors that may alter gene expression without changes in gene sequence. These represent very promising areas in sepsis care where there is potential to identify high-risk individuals, predict prognosis, and potentially inform the choice of therapies, based on the individual patient's genetic characteristics [38].

Transcriptomics: The transcription of messenger RNA as the mechanism of gene expression is the focus of these studies. Here, the different gene profiles are studied, and these help identify individuals who are more likely to develop sepsis and those in which sepsis is more likely to evolve unfavorably [39].

Metabolomics: Here, the focus of the study is the metabolome or the collective of molecules that cells produce. These molecules may be utilized as markers for diagnosis, assessment of severity and prognostic factors, and changes in response to treatment.

Proteomics: The proteome is the collection of all proteins expressed by a cell at a given time and under given circumstances. Proteomics represents a highly sensitive and specific technology to study cellular processes. In sepsis, this may be applied in diagnosis and prognosis. An example is the use of vimentin, an inflammatory response, and lymphocyte apoptosis modulator. Vimentin levels are higher in patients with sepsis and septic shock and in sepsis non-survivors [40].

17 Conclusions

Sepsis continues to be a leading cause of maternal mortality and morbidity. In response to this as well as initiatives by health organizations like the Global Sepsis Alliance, the World Health Assembly has identified sepsis as a global health priority and has adopted a resolution to improve, prevent, diagnose, and manage sepsis. This has provided a great impetus to the worldwide battle against sepsis. Pregnancy and puerperium represent a unique physiological state where responses to pathological processes are affected. The occurrence and evolution of sepsis in the maternity context present challenges as discussed, and early recognition and intervention remain key. Further, evidence in sepsis diagnosis and management is rapidly evolving, and the world of omics is proving to be a promising area for future research in sepsis care. With emerging evidence and new technologies, a shift toward a tailored individualized care approach is therefore increasingly being developed and adopted.

References

1. Reinhart K, Daniels R, Kissoon N, Machado FR, Schachter RD, Finfer S. Recognizing sepsis as a global health priority—a WHO resolution. N Engl J Med. 2017;377(5):414–7.
2. Greer O, Shah NM, Johnson MR. Maternal sepsis update: current management and controversies. Obstet Gynaecol. 2020;22:45.
3. Plante LA. Management of sepsis and septic shock for the obstetrician–gynecologist. Obstet Gynecol Clin. 2016;43(4):659–78.
4. Maternal G, Neonatal Sepsis Initiative Working Group. The global maternal and neonatal sepsis initiative: a call for collaboration and action by 2030. Lancet Glob Health. 2017;5(4):e390–1.
5. Bonet M, Brizuela V, Abalos E, Cuesta C, Baguiya A, Chamillard M, Dunlop C. Frequency and management of maternal infection in health facilities in 52 countries (GLOSS): a 1-week inception cohort study. Lancet Glob Health. 2020;8(5):e661–71.
6. La Rocca C, Carbone F, Longobardi S, Matarese G. The immunology of pregnancy: regulatory T cells control maternal immune tolerance toward the fetus. Immunol Lett. 2014;162(1):41–8.
7. Blazkova J, Gupta S, Liu Y, Gaudilliere B, Ganio EA, Bolen CR, Furman D. Multicenter systems analysis of human blood reveals immature neutrophils in males and during pregnancy. J Immunol. 2017;198(6):2479–88.
8. Elton RJ, Chaudhari S. Sepsis in obstetrics. BJA Educ. 2015;15(5):259–64.
9. Albright CM, Ali TN, Lopes V, Rouse DJ, Anderson BL. Lactic acid measurement to identify risk of morbidity from sepsis in pregnancy. Am J Perinatol. 2015;32(05):481–6.
10. Bauer ME, Housey M, Bauer ST, Behrmann S, Chau A, Clancy C, Bateman BT. Risk factors, etiologies, and screening tools for sepsis in pregnant women: a multicenter case-control study. Anesth Analg. 2019;129(6):1613.
11. Lawn JE, Bianchi-Jassir F, Russell NJ, Kohli-Lynch M, Tann CJ, Hall J, Seale AC. Group B streptococcal disease worldwide for pregnant women, stillbirths, and children: why, what, and how to undertake estimates? Clin Infect Dis. 2017;65(suppl_2):S89–99.
12. Stoll BJ, Hansen NI, Sánchez PJ, Faix RG, Poindexter BB, Van Meurs KP, Eunice Kennedy Shriver National Institute of Child Health and Human Development Neonatal Research Network. Early onset neonatal sepsis: the burden of group B streptococcal and E. coli disease continues. Pediatrics. 2011;127(5):817–26.

13. ACOG. Prevention of Group B Streptococcal Early-Onset Disease in Newbornsl. https://www.acog.org/clinical/clinical-guidance/committee-opinion/articles/2020/02/prevention-of-group-b-streptococcal-early-onset-disease-in-newborns.
14. CDC. GBS screening guidelines. 2016. https://www.cdc.gov/groupbstrep/guidelines/algorithms-tables.html.
15. Liu TH, Wang HP, Cho FN, Wang JL, Hung CH, Chiou YH, Cheng MF. Rectovaginal colonization with pathogenic Escherichia coli during pregnancy and neonatal outcomes. Infect Drug Resist. 2019;12:3103.
16. Rosene K, Eschenbach DA, Tompkins LS, Kenny GE, Watkins H. Polymicrobial early postpartum endometritis with facultative and anaerobic bacteria, genital mycoplasmas, and chlamydia trachomatis: treatment with piperacillin or cefoxitin. J Infect Dis. 1986;153(6):1028–37.
17. Singer M, Deutschman CS, Seymour CW, Shankar-Hari M, Annane D, Bauer M, Angus DC. The third international consensus definitions for sepsis and septic shock (Sepsis-3). JAMA. 2016;315(8):801–10.
18. Escobar MF, Echavarría MP, Zambrano MA, Ramos I, Kusanovic JP. Maternal sepsis. Am J Obstet Gynecol MFM. 2020;2(3):100149.
19. Bonet M, Nogueira Pileggi V, Rijken MJ, Coomarasamy A, Lissauer D, Souza JP, Gülmezoglu AM. Towards a consensus definition of maternal sepsis: results of a systematic review and expert consultation. Reprod Health. 2017;14(1):1–13.
20. Bowyer L, Robinson HL, Barrett H, Crozier TM, Giles M, Idel I, Makris A. SOMANZ guidelines for the investigation and management sepsis in pregnancy. Aust N Z J Obstet Gynaecol. 2017;57(5):540–51.
21. Cole MF. A modified early obstetric warning system. Br J Midwifery. 2014;22(12):862–8.
22. Albright CM, Ali TN, Lopes V, Rouse DJ, Anderson BL. The sepsis in obstetrics score: a model to identify risk of morbidity from sepsis in pregnancy. Am J Obstet Gynecol. 2014;211(1):39–e1.
23. Levy MM, Evans LE, Rhodes A. The surviving sepsis campaign bundle: 2018 update. Intensive Care Med. 2018;44:925–8.
24. The UK Sepsis Trust. Inpatient maternal sepsis tool. sepsistrust.org; 2018. Inpatient-maternal-NICE-Final-1107-2.pdf.
25. Trzeciak S, Dellinger RP, Chansky ME, Arnold RC, Schorr C, Milcarek B, Parrillo JE. Serum lactate as a predictor of mortality in patients with infection. Intensive Care Med. 2007;33:970–7.
26. Padilla C, Palanisamy A. Managing maternal sepsis: early warning criteria to ECMO. Clin Obstet Gynecol. 2017;60(2):418–24.
27. NICE. Procalcitonin testing for diagnosing and monitoring sepsis, diagnostic guidance. NICE; 2015.
28. Tujula B, Kokki H, Räsänen J, Kokki M. Procalcitonin; a feasible biomarker for severe bacterial infections in obstetrics and gynecology? Acta Obstet Gynecol Scand. 2018;97:505.
29. Zadroga R, Williams DN, Gottschall R, Hanson K, Nordberg V, Deike M, Hansen GT. Comparison of 2 blood culture media shows significant differences in bacterial recovery for patients on antimicrobial therapy. Clin Infect Dis. 2013;56(6):790–7.
30. Cecconi M, Evans L, Levy M, Rhodes A. Sepsis and septic shock. Lancet. 2018;392(10141):75–87.
31. Cole DE, Taylor TL, McCullough DM, Shoff CT, Derdak S. Acute respiratory distress syndrome in pregnancy. Crit Care Med. 2005;33(10):S269–78.
32. van Hagen IM, Cornette J, Johnson MR, Roos-Hesselink JW. Managing cardiac emergencies in pregnancy. Heart. 2017;103(2):159–73.
33. Lamontagne F, Rochwerg B, Lytvyn L, Guyatt GH, Møller MH, Annane D, Siemieniuk RA. Corticosteroid therapy for sepsis: a clinical practice guideline. BMJ. 2018;362:k3284.
34. Schulte W, Bernhagen J, Bucala R. Cytokines in sepsis: potent immunoregulators and potential therapeutic targets—an updated view. Mediators Inflamm. 2013;2013:165974.

35. Alejandria MM, Lansang MAD, Dans LF, Mantaring JB III. Intravenous immunoglobulin for treating sepsis, severe sepsis and septic shock. Cochrane Database Syst Rev. 2013;2013(9):CD001090.
36. Knowles SJ, O'sullivan NP, Meenan AM, Hanniffy R, Robson M. Maternal sepsis incidence, aetiology and outcome for mother and fetus: a prospective study. BJOG Int J Obstet Gynaecol. 2015;122(5):663–71.
37. Ruiz-Rodriguez JC, Plata-Menchaca EP, Chiscano-Camón L, Ruiz-Sanmartin A, Pérez-Carrasco M, Palmada C, Ferrer R. Precision medicine in sepsis and septic shock: from omics to clinical tools. World J Crit Care Med. 2022;11(1):1.
38. Kouskouti C, Evangelatos N, Brand A, Kainer F. Maternal sepsis in the era of genomic medicine. Arch Gynecol Obstet. 2018;297:49–60.
39. Sweeney TE, Azad TD, Donato M, Haynes WA, Perumal TM, Henao R, Bermejo-Martin JF, Almansa R, Tamayo E, Howrylak JA, Choi A, Parnell GP, Tang B, Nichols M, Woods CW, Ginsburg GS, Kingsmore SF, Omberg L, Mangravite LM, Wong HR, Tsalik EL, Langley RJ, Khatri P. Unsupervised analysis of transcriptomics in bacterial sepsis across multiple datasets reveals three robust clusters. Crit Care Med. 2018;46:915–25.
40. Su L, Pan P, Yan P, Long Y, Zhou X, Wang X, Zhou R, Wen B, Xie L, Liu D. Role of vimentin in modulating immune cell apoptosis and inflammatory responses in sepsis. Sci Rep. 2019;9:5747.

Infection and Sepsis Management in Trauma

Gustav Frans Strandvik, Abdulgafoor M. Thayaril, Mohammad Elayan Tawfiq Alabdallat, and Ahmed Obiedat

Abstract Infection may be the cause, consequence, or complication of traumatic injury. Issues concerning the prophylaxis, diagnosis, and management of infection and sepsis in victims of trauma continue to be daily challenges for clinicians, despite the advent of novel diagnostic tests. Up to 50% of fever episodes in trauma patients admitted to an intensive care unit are due to non-infective causes. An understanding of the pathobiology of infection and sepsis in trauma is essential, and a pragmatic approach to evaluate for infection is crucial to avoid poor antimicrobial stewardship. At the same time, appreciating the risk and fatal consequences of undiagnosed sepsis in this patient cohort requires a high index of suspicion and prompt intervention.

Keywords Trauma · Injury · Infection · Fever · Sepsis · Prophylaxis · Treatment · Diagnosis · Antimicrobial stewardship · Procalcitonin · Source control

1 Introduction

Traumatic injuries are a major cause of morbidity and mortality worldwide. Indeed, it has been proposed that trauma should be considered a disease state itself, with attendant approaches in line with chronic disease prevention and management. Fever and leukocytosis are common consequences of trauma, with non-infective fever being particularly common in traumatic brain injury. The pathophysiological consequences of traumatic injury may predispose patients to infection; similarly, the context of the trauma incident as well as the body region involved may make infection more likely. In contrast, antimicrobial prophylaxis and empirical treatment of fever with antibiotics may not be indicated and frequently result in complications

G. F. Strandvik (✉) · M. E. T. Alabdallat
Trauma Intensive Care, Hamad Medical Corporation, Doha, Qatar
e-mail: GStrandvik@hamad.qa

A. M. Thayaril · A. Obiedat
Surgical Intensive Care, Hamad Medical Corporation, Doha, Qatar

N. Shaikh, A. H. Chanda (eds.), *Applied Microbiology in Intensive Care Medicine*, https://doi.org/10.1007/978-981-97-4006-2_15

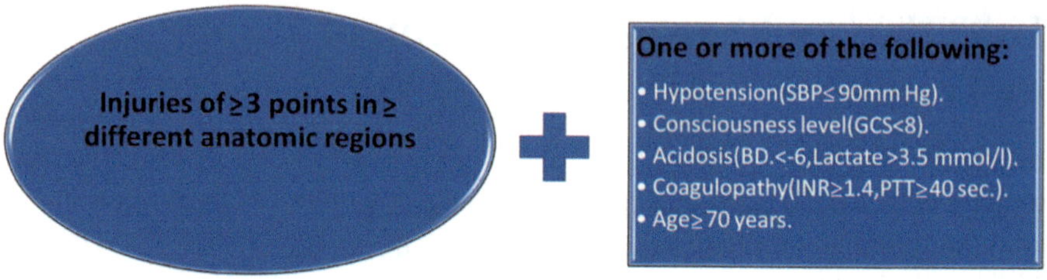

Fig. 1 Definition of polytrauma

related to antimicrobial management. This chapter explores the pathobiology of infection and sepsis in trauma and describes approaches to prophylaxis, diagnosis, and management.

2 Definition of Sepsis and Polytrauma

Polytrauma is defined as serious injuries in two or more body regions, associated with significant physiological impairment [1].

The Abbreviated Injury Scale (AIS) encompasses six body regions (head or neck, face, chest, abdomen or pelvic contents, pelvic girdle, or extremities, external or other) and grades this according to the severity of the injury. The maximum score possible is 75. If two or more regions are seriously injured and paired with either hypotension, alterations in consciousness, acidosis, coagulopathy, or age over 70, polytrauma may be present (Fig. 1).

It is evident from the latest definition of sepsis that a significant overlap exists between the vital sign derangements of polytrauma and sepsis [2].

3 Epidemiology

It is estimated that trauma results in six million deaths per year; this is more than the sum of all contagious diseases including HIV/AIDS, TB, malaria, and COVID-19. Annually, approximately 40 million people are permanently injured, and up to 100 million are temporarily injured [3]. Sepsis is an important complication of trauma and is estimated to contribute to 10% of trauma deaths [4]. The incidence of post-traumatic sepsis has decreased, leading to flattening of the "late death" curve in the "tri-modal pattern of death" as described by the American College of Surgeons in the Advanced Trauma Life Support course. However, the mortality of trauma patients who develop sepsis remains high at 19.5–23% [5]. Approximately 6–10% of trauma victims will develop a septic complication during their illness journey [6].

4 Immunological and Physiological Alterations in Polytrauma and Sepsis

Alterations in metabolism and the immune dysregulation observed in trauma and sepsis are interrelated; immune-metabolic disturbance strongly influences outcomes [7]. Infection risk increases due to the destruction of mechanical barriers (including invasive diagnostic and therapeutic interventions), exogenous bacterial contamination, and local wound factors. The host defense state might be associated with a severe weakening of the patient's immunity, impairing humoral and cell-mediated responses. Opportunistic infections in trauma happen due to any microorganism (bacteria, fungus, or virus) that has the capacity to multiply, despite the activation of the host's initial defense mechanisms [4]. Traumatic injury results in alarmins that prime the immune system to protect against tissue injury and microbe invasion. Excessive inflammatory cascades may be triggered by the primed immune response, resulting in multi-organ failure. This has been termed the "two-hit" response [8]. The "first hit" (organ and soft tissue injuries, fractures, hypoxia, hypotension) primes the immune system for an exaggerated and potentially lethal reaction to a secondary, otherwise nonlethal, stimulus (the "second hit"). The secondary stimulus may be either endogenous or exogenous, with the former including cardiovascular instability, respiratory distress, metabolic derangements, and ischemia/reperfusion. The exogenous second hits include surgical interventions, blood product transfusions, and missed injuries. The two-hit model proposes that this second hit results in destructive inflammation leading to multiple organ failure (MOF) and potentially death.

Systemic inflammatory response syndrome (SIRS) commonly follows traumatic injury. Injury activates the inflammasome pathway in the lymph nodes draining the injury site within 2 h; the activation then spreads systemically. It is characterized by local and systemic release of proinflammatory cytokines, arachidonic acid metabolites, and activation of complement factors, kinins, and coagulation, as well as hormonal mediators. An anti-inflammatory response parallels SIRS and is referred to as the compensatory anti-inflammatory response syndrome (CARS). An imbalance between these responses appears to be responsible for increased susceptibility to infection and organ dysfunction. See Fig. 2 for a visual representation of this process.

CARS and SIRS occur simultaneously. When these two opposing responses are appropriately balanced, the traumatized individual can effectively heal the injury without incurring secondary injury from the autoimmune inflammatory response. Over time, SIRS ceases to exist, and CARS becomes the predominant force. Overwhelming CARS appears responsible for post-traumatic immunosuppression with increased susceptibility to infections and sepsis.

Proinflammatory cytokines, tumor necrosis factor-α (TNF-α), and interleukin-1β (IL-1β) are released within 1–2 h. Secondary proinflammatory cytokines are released in a subacute fashion and include IL-6, IL-8, macrophage migratory factor (MMF), IL-12, and IL-18.

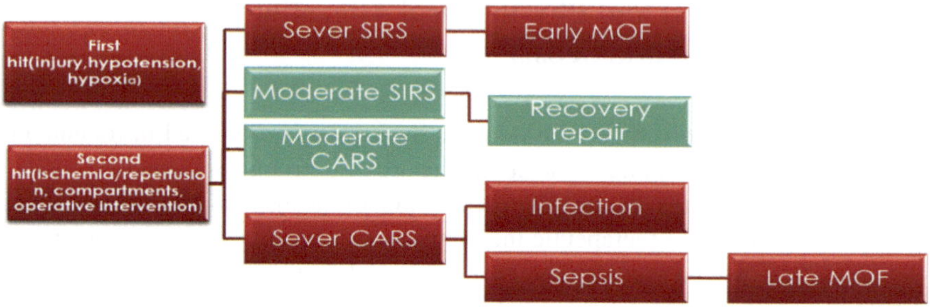

Fig. 2 Post-traumatic MOF due to dysfunctional inflammatory response. *CARS* compensatory anti-inflammatory response syndrome, *MOF* multiple organ failure, *SIRS* systemic inflammatory response syndrome

Clinically, IL-6 levels correlate with Injury Severity Score (ISS), the development of MOF, acute respiratory distress syndrome (ARDS), and sepsis. Interleukin-6 also acts as an immunoregulatory cytokine by stimulating the release of anti-inflammatory mediators such as IL-1 receptor antagonists and TNF receptors which bind circulating proinflammatory cytokines. IL-6 triggers the release of prostaglandin E2 (PGE2) from macrophages. Prostaglandin E2 is potentially the most potent endogenous immunosuppressant; it suppresses T-cell and macrophage responsiveness and induces the release of IL-10, a potent anti-inflammatory cytokine that deactivates monocytes. Serum IL-10 levels correlate with ISS and the development of post-traumatic complications. Trauma results in decreased IL-12 production, stimulating a shift in favor of T_H2 cells and the subsequent production of anti-inflammatory mediators IL-4, IL-10, IL-13, and transforming growth factor beta (TGF-β). This decrease in IL-12 and the resultant increase in T_H2 cells correlates with adverse outcomes.

5 Infection Prevention, Diagnosis, and Management in Trauma

The SIRS response to trauma and its consequences may predispose to infection and mimic sepsis in trauma victims. It is thus not surprising that antimicrobial agents are used liberally in these patients. Antibiotics administered on the *presumption* that injuries are contaminated and have a high probability of infection seem like a sensible approach. Concerns about complications of antibiotic use, however, include the emergence of resistant organisms and overgrowth of usually benign organisms such as *Clostridium*. Furthermore, effective utilization of resources, specific injury patterns, and patient factors mandate a rational, evidence-based decision-making process. Early extensive washout and closure of wounds, particularly contaminated and deep wounds, is essential [9]. The latter should occur in an operating room environment [10]. The use of antibiotic powders and antibiotic-impregnated orthopedic sponges and implants shows promise in reducing trauma-related orthopedic infections [11]. The prevention of tetanus involves confirming up-to-date vaccination status and the administration of tetanus immune globulin in the presence of vaccination uncertainty (Table 1) [12].

Important further considerations are the context of trauma and individual patient characteristics; trauma occurring in saltwater or freshwater is likely to expose the patient to pathogens which do not normally cause infection in humans. Common facultatively anaerobic gram-negative bacilli include *Vibrio* species (seawater) and *Aeromonas hydrophila* (lakes, ponds, streams). Piperacillin-tazobactam is the antibiotic of choice with freshwater, and doxycycline should be added to cover *Vibrio* species with seawater exposure. Carbapenems may be substituted for beta-lactams in the case of allergy. Immunocompromised trauma victims and the elderly should prompt lower thresholds for instituting prophylactic and therapeutic antimicrobial use. Allergy to penicillin or other antimicrobial agents will also influence antibiotic choice.

Severely injured patients may be in a hyperdynamic state, and some authorities recommend increasing the dose of antibiotics to compensate for enhanced tissue clearance [13]. In patients in whom the whole blood volume has been replaced, it may be worth reducing the dosing interval. Finally, it must be remembered that individuals with intercurrent infections or sepsis may present with trauma. Although leucocyte elevation early in major trauma is common, fever is not; effective PPE and an index of suspicion for infection are important weapons in the first physician's arsenal.

Suggestions for specific agents are included below; however, local antibiograms and resistance patterns must inform all antimicrobial decisions.

5.1 *Pre-Hospital Administration of Antibiotics*

There is limited evidence for this strategy in civilian populations. Guidance for open fractures recommends a one-hour window for intravenous antibiotics, and improvements in pre-hospital care provision may show promise for this strategy in the future. Pre-hospital antibiotic cover is recommended for all battlefield injuries, as rates of contamination and time to definitive care influence infection rates [14].

Table 1 Tetanus prevention

	Minor non-contaminated wound		Moderate to severe clean or contaminated wound	
Previous vaccination	Tetanus vaccine	Tetanus human Igb*	Tetanus vaccine	Tetanus human Igb
1–2 doses or unknown	Yes	No	Yes	Yes
3 or more	If the last dose was given >10 years ago		If the last dose was given >5 years ago	No

Although compliance is often patchy, this strategy in both orthopedic and torso injuries may reduce infectious complications in combat-related trauma [15].

The use of specific dressings, including those impregnated with antimicrobial agents, silver, or hydro collagen gels for pre-hospital use, is an area under investigation. Even patients with extensive burns seem to derive little benefit from anything other than adequate wound coverage.

5.2 Presumptive Antibiotics in the Emergency Department

Experience with infection reduction in open orthopedic injuries post the advent of antibiotics led to the routine use of antibiotics in most civilian trauma victims. Despite years of interest, however, the evidence for prophylactic antibiotic administration in non-orthopedic injuries is weak. Table 2 provides a common sense–based approach distilled from current evidence [16]. Broadly speaking, non-penetrating injuries anywhere do not require antibiotics; in most other cases, short duration (< 24 h) of antibiotic administration is not inferior to longer courses in preventing infection. In many cases, single doses of antibiotics are sufficient [17].

Clear evidence suggests *no need* for prophylactic antibiotics on admission in the following injury patterns [18]:

- Closed skull fractures including base of skull fractures with or without cerebrospinal fluid leak [19].
- Burn victims [20].
- Blunt injury to thorax (including lung contusions).
- Blunt injury to the abdomen without hollow viscus injury (including pancreatic, liver, spleen, kidney, and bladder injuries).
- Soft tissue injury (including degloving) [21].
- Closed fracture patterns.

Penetrating head injury and frontal and sphenoid sinus fractures (communicating with intracranial contents) require at least 24 h of broad-spectrum antibiotics, and it is an accepted practice to treat for 5 days in the presence of bullet injuries and those with significant tissue or bone fragment contamination. Although the ideal antibiotic choice is unclear, antibiotics which are known to cross the blood-brain barrier such as ampicillin-sulbactam, cefotaxime, or ceftriaxone are logical choices in the absence of allergy. Severe traumatic brain injury of the non-penetrating type has been associated with the development of pneumonia in a few small studies; the use of two doses of antibiotics to cover possible aspiration of oro-pharyngeal secretions in quantities enough to cause pneumonia has shown benefit. This practice is not ubiquitous. Facial fractures are considered closed or open depending on whether sinus fractures co-exist; however, more than 24 h of prophylaxis has not been found to affect infection rates regardless of sinus involvement [22]. Chest injuries including rib fractures, lung contusions, and hemo/pneumothorax do not require prophylactic antibiotics. Although the evidence is not definitive, most guidelines suggest a

Table 2 Suggested presumptive antimicrobials in trauma

Type of injury	Recommended antibiotic	Allergy to penicillin (note penicillin to cephalosporin cross-reactivity <2%)	Duration
Penetrating brain injury	Ampicillin-sulbactam 3–4.5 g IV q 6 h OR Ceftriaxone 2 g IV q 12 h ± metronidazole 500 mg IV/PO q6–8 h	Moxifloxacin 400 mg IV/PO q 24 h	Retained bone or soft tissue: Immediately following injury and for 5 days postoperatively No retained material: At least one dose pre-op
Facial fractures (including open, closed, sinus) and open skull fractures	Cefazolin 1–3 g IV q 8 h PLUS metronidazole 500 mg IV every 8 h OR ampicillin-sulbactam 3–4.5 g IV q 6 h	Clindamycin 600–900 mg IV q 8 h	Open fractures/mandibular and multiple: Time of injury to ≤24 h post-op All other facial fractures/open skull fractures/sinus fractures: Single preoperative dose
Chest tube thoracostomy	Cefazolin 1–3 g OR cefuroxime 1.5 g once	Clindamycin 600–900 mg IV q8 h	At time of chest tube insertion
Penetrating abdominal injury or ruptured hollow viscus	Cefazolin 1–3 g (depending on weight and hypermetabolic state) IV q8h PLUS metronidazole 500 mg IV/PO q8h	Clindamycin 600–900 mg IV q8 h PLUS IV gentamicin every 24 h	No HVI: Single preoperative dose HVI: ≤24 h
Human and animal bites	Ampicillin-sulbactam 3–4.5 g IV q 6 h OR amoxicillin/clavulanate 1200 mg PO BID	Doxycycline 100 mg IV/PO BID	3–5 d depending on risk factors and location (see text)
Aquatic exposure	Cefepime 1 g IV q 6 h PLUS doxycycline 100 mg IV/PO BID if seawater	Ciprofloxacin 400 mg IV q 8 h PLUS Doxycycline 100 mg IV/PO BID if seawater	See above

single dose of antibiotics to be administered at the time of tube thoracostomy to reduce infection with gram-positive organisms (*S. aureus* and *Streptococcus*) and occasionally gram-negative organisms. A cephalosporin is usually adequate. Retained hemothorax is a risk factor for subsequent intrathoracic infection, but evidence for the routine use of antibiotics to prevent infection is lacking. A high index of suspicion is required.

Blunt abdominal injury without hollow viscus injury does not require presumptive infection prevention. There is no definitive evidence in the literature about whether antibiotic prophylaxis is required with penetrating or hollow viscus abdominal injury nor for how long they should be given [23, 24]. Presumptive antibiotics are given at the time of laparotomy and should cover anaerobic organisms if hollow viscus injury has occurred. A 24-h course is likely to be adequate and may reduce the rate of colonization and the emergence of multi-drug-resistant organisms in these patients [25]. There is no evidence addressing this issue in the case of either intra or extraperitoneal bladder ruptures—most experts recommend 24 h of gram-negative and gram-positive cover.

There is no need for antibiotics in soft tissue injury if effective local washout has happened. This includes Morel-Lavallée lesions, which are closed degloving injuries where the skin and subcutaneous tissue have sheared off the underlying fascia. Seromas are common and pose a risk for infection, but prophylactic antibiotics have no role [21]. Mammalian bites should be offered oral co-amoxiclav or doxycycline/metronidazole if the skin is broken and the individual has significant comorbidities or the wound occurs in sensitive or low-circulation areas.

Open fractures (including pelvic fractures) should be treated according to the Gustilo-Anderson grade (Table 3) [26]. Prolonged presumptive courses are not required after wound closure, and probably, the most important step is generous washout and debridement of any contaminated tissue, preferably in the operating room.

Gram-positive organisms should be covered immediately as up to 78% of organisms causing infection are from this genus. Consider adding gram-negative cover in type 3 fractures and a course of 72 h [27]. Fluoroquinolones may worsen fracture healing and cause more infections in severe injuries–avoid them unless treating water-related trauma. If fecal contamination is a possibility (agricultural environments), high-dose penicillin should be given, and add anaerobic cover if heavy soil

Table 3 Gustilo-Anderson grading for open fractures with suggested preventative antibiotic cover

Gustilo-Anderson grade	Laceration characteristics	Recommended antibiotic cover	Duration of treatment (within 6 h, preferably 1 h) and after closure
Type I	1 cm in length and without extensive soft tissue damage, flaps, or avulsions	Cefazolin 2 g ivi q8 or vancomycin (allergy)	24 h
Type II	>1 cm in length and without extensive soft tissue damage, flaps, or avulsions	As above	24 h
Type III a-c	>10 cm wound with extensive soft tissue injury or a traumatic amputation. (special categories include gunshot fractures and open fractures caused by farm injuries)	Cefazolin 2 g q8 PLUS gentamicin 5 mg/kg q24 or clindamycin (allergy) PLUS gentamicin	24–72 h or 1 day after wound closure

contamination has occurred; metronidazole is a reasonable choice here. Ceftriaxone may be used as monotherapy. Open fractures in aquatic settings need special cover as outlined earlier.

6 Intensive Care Management of Suspected or Proven Infection in Trauma Victims

The pathophysiology of trauma means that a considerable proportion of severely injured patients will present with SIRS-like signs and fever related to a hyperdynamic state driven by cytokine release. This is particularly evident in burn victims and in severely injured patients. There is some evidence of genomic variation in this presentation. The syndrome may last for up to 72 h. Up to 50% of injured patients will develop a significant infection-range fever (>38.3–39° C) during their stay in the intensive care unit. It is thus incumbent upon the treating physician to thoroughly evaluate the patient for sources and presence of infection. Axillary temperatures tend to be 0.3° below a central fever (as measured with infrared tympanometry or other central measures), and rectal temperatures tend to be 0.3° higher.

6.1 Evaluating for Infection in ICU Trauma Patients—Risk Factors

Fever is common during the first 48–96 h of admission and may indicate a robust immune response and portend a better prognosis [28, 29]. However, it must be remembered that infection should always be excluded as the patient may have been suffering from an intercurrent infection at the time of trauma, and early septic events do occur and contribute to morbidity [30]. In contrast, hypothermia and a muted bone marrow response (leukopenia) may suggest significant immunosuppression [31].

Specific risk factors for infection include the following:

- Trauma itself, with a linear relationship between Injury Severity Score and infection risk (ISS > 13). This risk is present in addition to the breakdown of the protective skin barrier.
- Multiple injuries in different body regions [32].
- Age—a direct correlation between older age and infection/sepsis risk (>60).
- Male gender (for hypotheses, see [33]).
- Comorbidities
- Severe pelvic injury with perineal soft tissue damage (open or closed) [34]. Early fecal diversion is mandatory.
- Blood transfusions per se and the age of transfused blood (>14 days) have been found in some but not all studies to constitute a modifiable risk factor for infection [35].

6.2 Evaluating for Infection—Clinical Approach

The most common infections in trauma patients are pneumonia and urinary tract infections (> 70%), followed by bloodstream infections, skin and soft tissue, and bowel [36]. The practice of blind pan-cultures is no longer encouraged in the evaluation of fever in trauma victims. This is because surgical wounds are commonly colonized with gram-positive organisms, and abdominal wound drain sites often culture gram-negative organisms when hollow viscus injuries have occurred. Prolonged stay and antibiotic use in intensive care are risk factors for fungal colonization with *Candida albicans* and increasingly with other *Candida* species. A non-directed "sepsis-6" approach may lead to confirmation bias in treating physicians. This may further result in missed diagnoses such as deep vein thrombosis (a cause of a considerable proportion of fever in critically ill patients) or wrongly labeling colonization as an infection. There is emerging evidence that the latter leads to unnecessary antibiotic exposure with the resultant emergence of resistant organisms in trauma patients [37, 38]. Positive cultures in the absence of clinical or investigative features of infection cause ongoing therapeutic uncertainty. This is a particular problem with cultures from wound drains, superficial suture sites, and sputum.

Pneumonia is the most common infection in trauma patients, and the presence of traumatic brain injury may or may not be an additional risk factor. It is also an extremely difficult diagnosis to make in the presence of radiological features of possible aspiration pneumonitis or lung contusions. This condition is often over-diagnosed and subsequently over-treated [39]. Meticulous clinical assessment is thus crucial in the trauma ICU; chest X-ray interpretation must be excellent, and changes in sputum composition and volume along with oxygenation deficits are important; changes in inflammatory markers are not definitive (see below). Sputum production may be due to ventilator-associated tracheitis in patients with tracheal tubes, the incidence of which is like other critically ill patients. The presence of hemothorax, especially if a tube thoracostomy has been performed and the blood has not fully cleared post tube removal, should prompt the physician to consider empyema. The characteristics of the aspirated fluid will be helpful, and often, a CT (computed tomography) scan and then early surgical intervention are required. In keeping with other critically ill patients, urinary tract and catheter-related bloodstream infections can be prevented and managed by proactive removal of lines. Moribund trauma victims will often have lines placed emergently, with the concomitant risk of less-than-ideal anti-septic techniques; remove lines at the earliest opportunity. Other sources of infection in trauma patients occur with the same frequency as in other critically ill patients, including sinusitis and acalculous cholecystitis.

Meningitis deserves special mention, particularly if interventions such as external ventricular drain (EVD) and intracranial pressure monitoring (ICP) have occurred or where the skull has been surgically breached. These conditions, along with cases where the inner table of the frontal sinus or the sphenoid sinus has been broken, should raise suspicion of infection. Diagnosing meningitis in a comatose traumatic brain injury victim is challenging, as central fevers and paroxysmal sympathetic hyperactivity (PSH) may both masquerade as sepsis [40].

Diagnostic tools exist to help in these cases. A driving principle is that infection should be presumed, and central fevers are considered diagnoses of exclusion. The decision to perform a diagnostic lumbar puncture in TBI patients with *and* without a history of neurosurgical interventions should always be taken in conjunction with the treating neurosurgeon; paradoxical herniation has been reported even in lumbar puncture post decompressive craniectomy.

Burn patients are at substantial risk of infection, but the hyperdynamic stress response may persist for a long time; American Burn Association (ABA) sepsis criteria (2007) and the Mann-Salinas novel burn-specific sepsis predictors (2013), as well as the sepsis-3 criteria, are useful in helping make the diagnosis. Sepsis-3 criteria are the most useful, followed by ABA and then Mann-Salinas. A major gram-negative concern is colonization and subsequent infection with *Pseudomonas aeruginosa*.

Orthopedic metalwork may become infected, and pin sites of external fixator devices need to be examined daily. The possibility of osteomyelitis is higher in open pelvic fractures and with devastating perineal soft tissue injuries.

The use of intravascular devices for hemorrhage control is ubiquitous, particularly the non-selective embolization of major pelvic vessels. These can lead to ischemia, necrosis, and infection of surrounding tissues such as gluteal muscles, and these areas should be actively excluded as sources of infection through careful examination and imaging if necessary. Device insertion sites, such as for the REBOA catheter, are often in the groin area and should be removed as soon as possible (within 24–48 h) to prevent thrombosis and infection.

Abdominal trauma of the penetrating kind, or where extensive peritoneal soiling occurred following hollow viscus perforation, poses a significant risk for subsequent infection. Although prolonged prophylaxis is not recommended (see above), early investigation and treatment of suspected sepsis are key in reducing morbidity and mortality. The presence and duration of abdominal packs should be noted as late removal constitutes infection risk. The reader is referred to the specific chapter for more detailed information.

Splenectomized patients require a vaccination history review to fill any gaps in their schedule of routine childhood or seasonal vaccine jabs; they are further at increased risk for certain infections and should receive the following 10–14 days post-surgery.

- The pneumococcal vaccine (13 or 23-valent)
- The *H. influenzae* type b vaccine (Hib)
- The quadrivalent meningococcal conjugate vaccine
- The monovalent meningococcal serogroup B vaccine

Although fungal infections per se are not more common in trauma patients, patients with multiple laparotomies (>2) or with an open abdomen are at particular risk. Appropriate investigation and empiric treatment are discussed elsewhere.

6.3 *Evaluating for Infection—Biomarkers*

As stated previously, the hyperdynamic response to extensive burns and severe trauma can make the diagnosis of infection and sepsis challenging. There has been considerable scientific endeavor at identifying a gold-standard biomarker to assist clinical decision-making; sadly, no candidate has yet emerged with enough specificity and sensitivity to serve this purpose [41].

White blood cell count elevation within 24 h of trauma is extremely common and recrudesces early; there is compelling evidence that other processes such as surgery, procedures, and other ICU-related events can elevate the white blood count during trauma ICU stay, making this a notoriously unreliable marker [42]. The utility of the neutrophil-to-lymphocyte ratio remains unclear.

C-reactive protein (CRP) has been widely studied and is sensitive, but non-specific for infection or sepsis, regardless of level of elevation. Trends and re-induction of a previously lowering CRP may be useful, but concerns about specificity must be borne in mind [43].

Procalcitonin (PCT) has been studied in trauma with equivocal results. At present, it has a definite role in the early discontinuation of antibiotics in trauma ICU patients; a day-3 reduction to 25–50% of the peak should prompt consideration of antibiotic cessation [44]. It has a less clear role in the *diagnosis* of infection and sepsis; early elevations do occur in some individuals purely because of trauma (and may be predictive of susceptibility to sepsis in trauma). It may take up to 24 h to elevate after infection or sepsis, so a severely injured patient who previously had a negative PCT and develops PCT elevation is likely to have sepsis. However, a negative PCT at the time of sepsis suspicion is not sensitive enough to exclude sepsis [45]. Trends may thus be of greatest benefit [46].

Cytokine elevations (interleukin-6 has been most extensively studied) have not been sufficiently evaluated in trauma populations but may show promise as an early specific marker of trauma-related sepsis [47]. Delayed lactate clearance is also associated with higher rates of sepsis.

There are currently more than 100 biomarkers under evaluation for risk prediction, diagnosis, or prognosis of sepsis, with very small sample sizes in trauma populations. At present, the clinical method of history-taking, examination, focused cultures, and ancillary biomarker testing with PCT (and to a lesser extent CRP and IL-6) still serves as the optimum model for sepsis evaluation in trauma. Indeed, several prediction scores using a combination of clinical features and biomarkers in trauma are under evaluation and may soon be useful in trauma-related sepsis [48].

7 Management of Sepsis in Trauma—Special Considerations

In general, the management of severe sepsis follows the standards set by the Surviving Sepsis Campaign [49]. Although it is likely that the initial hyperdynamic state is *not* due to sepsis, a high index of suspicion and low threshold for treatment in this cohort are warranted, considering the risk of mortality; it is better to start antibiotics and stop them early than miss the opportunity to treat devastating sepsis. In patients who have otherwise stabilized, any change in their condition (new fever, hemodynamic change, tachycardia, etc.) mandates a clinical hunt for sepsis, with appropriate cultures and biomarkers.

The contribution of extensive volume loss in terms of hemorrhage or third-space losses from open abdomens, crush injuries, and burns makes the contribution of hypovolemia to the shock state difficult to quantify. The use of effective modalities to measure stroke volume variation (such as invasive cardiac output monitoring) and ultrasound modalities such as IVC (inferior vena cava) dynamics and VTI (velocity time integral) in response to fluid challenges can be helpful. A dedicated intensivist observing clinical response to fluids and vasopressors in the context of time is essential. Akin to other forms of sepsis, fluid overload in trauma intensive care may contribute to worse outcomes; as soon as there is reasonable clinical evidence that bleeding has stopped, a restrictive fluid strategy is wise. The vasopressor of choice in trauma patients with sepsis remains norepinephrine, and upper limits are often not set, considering potential recovery in this usually young cohort [50]. Steroids in doses used to treat refractory septic shock should be considered in the presence of traumatic brain injury despite concerns about their use [51]. It is possible that the benefits may outweigh the potential risks [52]. Infectious disease/antimicrobial management team involvement for this high-risk cohort should be considered a standard of care and may reduce costs, improve appropriate antibiotic use, and improve outcomes [53].

8 Future Directions

The next decades are likely to see advancements in identifying patients at risk of sepsis based on both genotype and phenotype patterns. Several single nucleotide polymorphisms (SNPs) have been identified, and a combination of these may provide infection susceptibility patterns [54]. Trauma sepsis prediction tools based on a combination of clinical features and biomarkers such as procalcitonin and interleukin-6 may provide help [55]. Many novel biomarkers are currently being investigated. However, patient and injury heterogeneity may require more data to ensure accurate prediction [56]. Different clinical patterns of both enhanced and suppressed inflammatory responses are emerging as potentially unique targets for therapies in other forms of sepsis—blood purification technologies have been reported as being successful in modifying specific cytokine responses in severely injured patients, with enhanced survival. Machine learning algorithms may be able to sift through a host of variables and provide a definitive answer to the question: is this severely injured patient suffering an inflammatory cytokine storm, or do they have an infection with sepsis and septic shock? Once sepsis has occurred, cytokine modulation may emerge as a viable strategy to reduce immunosuppression [57], and rapid organism identification techniques will enable the narrowing of the currently broad approach to empiric antimicrobial therapy in suspected trauma-related sepsis.

9 Summary

Sensible presumptive antibiotic strategies with reduced duration courses are key for effective antimicrobial stewardship. Conversely, suspected sepsis must be evaluated, recognized, and aggressively managed with broad-spectrum antibiotics to prevent a rapid "failure to rescue" scenario. Thorough clinical evaluation and focused diagnostics, with thoughtful interpretation of biomarkers, will inform treatment plans; the most promising biomarker in trauma seems to be PCT, with an early elevation being possibly predictive of sepsis and trends in PCT values providing more information than once-off readings.

The following quotations, three decades apart, exemplify the paradox and competing priorities in the management of infection and sepsis in victims of trauma:

> Prophylactic antibiotic therapy probably has no place in the treatment of the traumatized patient. Such patients should be given aggressive therapy aimed at restoring pulmonary, hemodynamic, and metabolic functions. The authors believe that maintenance of circulatory and metabolic homeostasis lessens the likelihood of infection in the posttraumatic patient. M. Allgöwer, M. Dürig, G. Wolff, 1980 [58]

> Since the participation of sepsis in polytrauma has been recognized as a key event in the outcome of patients at the ICU, the ability of the specialist to early recognize a septic process has become a key feature to reduce mortality and improve clinical prognosis. Fernando Mas-Celis, 2021

References

1. Rau CS, Wu SC, Kuo PJ, Chen YC, Chien PC, Hsieh HY, Hsieh CH. Polytrauma defined by the new Berlin definition: a validation test based on propensity-score matching approach. Int J Environ Res Public Health. 2017;14(9):1045. https://doi.org/10.3390/ijerph14091045. PMID: 28891977; PMCID: PMC5615582.
2. Singer M, Deutschman CS, Seymour CW, Shankar-Hari M, Annane D, Bauer M, Bellomo R, Bernard GR, Chiche JD, Coopersmith CM, Hotchkiss RS, Levy MM, Marshall JC, Martin GS, Opal SM, Rubenfeld GD, van der Poll T, Vincent JL, Angus DC. The third international consensus definitions for sepsis and septic shock (Sepsis-3). JAMA. 2016;315(8):801–10. https://doi.org/10.1001/jama.2016.0287. PMID: 26903338; PMCID: PMC4968574.

3. Rossiter ND. Trauma-the forgotten pandemic? Int Orthop. 2022;46(1):3–11. https://doi.org/10.1007/s00264-021-05213-z. Epub 2021 Sep 14. PMID: 34519840; PMCID: PMC8438546.
4. Mas-Celis F, Olea-López J, Parroquin-Maldonado JA. Sepsis in trauma: a deadly complication. Arch Med Res. 2021;52(8):808–16. https://doi.org/10.1016/j.arcmed.2021.10.007. Epub 2021 Oct 25. PMID: 34706851.
5. Jin H, Liu Z, Xiao Y, Fan X, Yan J, Liang H. Prediction of sepsis in trauma patients. Burns Trauma. 2014;2(3):106–13. https://doi.org/10.4103/2321-3868.135479. PMID: 27602370; PMCID: PMC5012019.
6. Eguia E, Bunn C, Kulshrestha S, Markossian T, Durazo-Arvizu R, Baker MS, Gonzalez R, Behzadi F, Churpek M, Joyce C, Afshar M. Trends, cost, and mortality from sepsis after trauma in the United States: an evaluation of the National Inpatient Sample of hospitalizations, 2012-2016. Crit Care Med. 2020;48(9):1296–303. https://doi.org/10.1097/CCM.0000000000004451. PMID: 32590387; PMCID: PMC7872079.
7. Raju R. Immune and metabolic alterations following trauma and sepsis—an overview. Biochim Biophys Acta Mol Basis Dis. 2017;1863(10 Pt B):2523–5. https://doi.org/10.1016/j.bbadis.2017.08.008. Epub 2017 Aug 24. PMID: 28842148; PMCID: PMC5861733.
8. Osuka A, Ogura H, Ueyama M, Shimazu T, Lederer JA. Immune response to traumatic injury: harmony and discordance of immune system homeostasis. Acute Med Surg. 2014;1:63–9. https://doi.org/10.1002/ams2.17.
9. Alexander JW. Prophylactic antibiotics in trauma. Am Surg. 1982;48(2):45–8. PMID: 7041727.
10. https://www.nice.org.uk/guidance/NG37/chapter/recommendations.
11. Chaudhary S, Sen RK, Saini UC, Soni A, Gahlot N, Singh D. Use of gentamicin-loaded collagen sponge in internal fixation of open fractures. Chin J Traumatol. 2011;14(4):209–14. Fuchs T, Stange R, Schmidmaier G, Raschke MJ. The use of gentamicin-coated nails in the tibia: preliminary results of a prospective study. Arch Orthop Trauma Surg 2011;131(10):1419–1425.
12. Havers FP, Moro PL, Hunter P, et al. Use of tetanus toxoid, reduced diphtheria toxoid, and acellular pertussis vaccines: updated recommendations of the Advisory Committee on Immunization Practices—United States, 2019. MMWR Morb Mortal Wkly Rep. 2020;69:77. Liang JL, Tiwari T, Moro P, et al. Prevention of Pertussis, Tetanus, and Diphtheria with Vaccines in the United States: Recommendations of the Advisory Committee on Immunization Practices (ACIP). MMWR Recomm Rep 2018; 67:1.
13. Fabian TC. Infection in penetrating abdominal trauma: risk factors and preventive antibiotics. Am Surg. 2002;68(1):29–35.
14. Schauer SG, Fisher AD, April MD, Stolper KA, Cunningham CW, Carter R 3rd, Fernandez JRD, Pfaff JA. Prehospital Administration of Antibiotic Prophylaxis for open combat wounds in Afghanistan: 2013-2014. J Spec Oper Med. 2018;18(2):53–6. https://doi.org/10.55460/ZRIK-EOE3.
15. Collopy K, Zimmerman L, Westmoreland AM, Powers WF 4th. Prehospital Administration of Cefazolin in trauma patients. Air Med J. 2022;41(5):447–50. https://doi.org/10.1016/j.amj.2022.06.001. Epub 2022 Jul 9
16. Hopkins TL, Daley MJ, Rose DT, Jaso TC, Brown CV. Presumptive antibiotic therapy for civilian trauma injuries. J Trauma Acute Care Surg. 2016;81(4):765–74. https://doi.org/10.1097/TA.0000000000001164.
17. Velmahos GC, Jindal A, Chan L, Kritikos E, Vassiliu P, Berne TV, Demetriades D. Prophylactic antibiotics after severe trauma: more is not better. Int Surg. 2001;86(3):176–83. [Yuan, KC., Huang, HC. Antimicrobial Prophylaxis in Patients with Major Trauma. Curr Trauma Rep 3, 292–299 (2017). https://doi.org/10.1007/s40719-017-0104-0.
18. Poole D, Chieregato A, Langer M, Viaggi B, Cingolani E, Malacarne P, Mengoli F, Nardi G, Nascimben E, Riccioni L, Turriziani I, Volpi A, Coniglio C, Gordini G, Trauma Update Working Group. Systematic review of the literature and evidence-based recommendations for antibiotic prophylaxis in trauma: results from an Italian consensus of experts. PLoS One. 2014;9(11):e113676. https://doi.org/10.1371/journal.pone.0113676. PMID: 25412442; PMCID: PMC4239082.

19. Ratilal BO, Costa J, Pappamikail L, Sampaio C. Antibiotic prophylaxis for preventing meningitis in patients with basilar skull fractures. Cochrane Database Syst Rev. 2015;4:CD004884. https://doi.org/10.1002/14651858.CD004884.pub4.
20. Barajas-Nava LA, López-Alcalde J, Roquéi Figuls M, Solà I, Bonfill Cosp X. Antibiotic prophylaxis for preventing burn wound infection. Cochrane Database Syst Rev. 2013;6:CD008738.
21. Lane JC, Mabvuure NT, Hindocha S, Khan W. Current concepts of prophylactic antibiotics in trauma: a review. Open Orthop J. 2012;6:511–7. https://doi.org/10.2174/1874325001206010511. Epub 2012 Nov 30. PMID: 23248721; PMCID: PMC3522105.
22. Zosa BM, Elliott CW, Kurlander DE, Johnson F, Ho VP, Claridge JA. Facing the facts on prophylactic antibiotics for facial fractures: 1 day or less. J Trauma Acute Care Surg. 2018;85(3):444–50. https://doi.org/10.1097/TA.0000000000002009.
23. Brand M, Grieve A. Prophylactic antibiotics for penetrating abdominal trauma. Cochrane Database Syst Rev. 2019;12(12):CD007370. https://doi.org/10.1002/14651858.CD007370.pub4. PMID: 31830316; PMCID: PMC6907398.
24. Herrod PJ, Boyd-Carson H, Doleman B, Blackwell J, Williams JP, Bhalla A, Nelson RL, Tou S, Lund JN. Prophylactic antibiotics for penetrating abdominal trauma: duration of use and antibiotic choice. Cochrane Database Syst Rev. 2019;12(12):CD010808. https://doi.org/10.1002/14651858.CD010808.pub2. PMID: 31830315; PMCID: PMC6953295.
25. Velmahos GC, Toutouzas KG, Sarkisyan G, et al. Severe trauma is not an excuse for prolonged antibiotic prophylaxis. Arch Surg. 2002;137(5):537–42. https://doi.org/10.1001/archsurg.137.5.537.
26. Gustilo RB, Mendoza RM, Williams DN. Problems in the management of type III (severe) open fractures: a new classification of type III open fractures. J Trauma. 1984;24(8):742–6. https://doi.org/10.1097/00005373-198408000-00009.
27. Sandean D. Open fractures—what is the evidence, and how can we improve? Arch Bone Jt Surg. 2021;9(5):559–66. https://doi.org/10.22038/abjs.2020.53120.2637. PMID: 34692940; PMCID: PMC8503759.
28. Kane WJ, Hassinger TE, Elwood NR, Dietch ZC, Krebs ED, Popovsky KA, Hedrick TL, Sawyer RG. Fever is associated with reduced mortality in trauma and surgical intensive care unit-acquired infections. Surg Infect. 2021;22(2):174–81. https://doi.org/10.1089/sur.2019.352.
29. Mizushima Y, Ueno M, Idoguchi K, Ishikawa K, Matsuoka T. Fever in trauma patients: friend or foe? J Trauma. 2009;67:1062–5. https://doi.org/10.1097/TA.0b013e3181b848fc.
30. Hinson Holly E, Susan R, Cynthia M, Lin Amber L, Schreiber Martin A. Early fever after trauma: does it matter? J Trauma Acute Care Surg. 2018;84(1):19–24. https://doi.org/10.1097/TA.0000000000001627.
31. Choi B, Sung K, Cho J. Fever in trauma patients without brain injury. J Acute Care Surg. 2021;2:6–13.
32. Eguia E, Cobb AN, Baker MS, Joyce C, Gilbert E, Gonzalez R, Afshar M, Churpek MM. Risk factors for infection and evaluation of Sepsis-3 in patients with trauma. Am J Surg. 2019;218(5):851–7. https://doi.org/10.1016/j.amjsurg.2019.03.005. Epub 2019 Mar 8. PMID: 30885453; PMCID: PMC6732249.
33. Choudhry MA, Bland KI, Chaudry IH. Gender and susceptibility to sepsis following trauma. Endocr Metab Immune Disord Drug Targets. 2006;6(2):127–35. https://doi.org/10.2174/187153006777442422.
34. Kataoka Y, Minehara H, Shimada K, Nishimaki H, Soma K, Maekawa K. Sepsis caused by peripelvic soft tissue infections in critically injured patients with multiple injuries and unstable pelvic fracture. J Trauma. 2009;66(6):1548–54. discussion 1554–5. https://doi.org/10.1097/TA.0b013e3181a51b0e.
35. Sowers N, Froese PC, Erdogan M, Green RS. Impact of the age of stored blood on trauma patient mortality: a systematic review. Can J Surg. 2015;58(5):335–42. https://doi.org/10.1503/cjs.011314. PMID: 26384149; PMCID: PMC4599997.

36. Schindler CR, Woschek M, Franz JN, Störmann P, Henrich D, Marzi I. Influence of antibiotic management on microbial selection and infectious complications after trauma. Front Med (Lausanne). 2021;8:678382. https://doi.org/10.3389/fmed.2021.678382. PMID: 34568354; PMCID: PMC8461005.
37. Ceccarelli G, Alessandri F, Moretti S, Borsetti A, Maggiorella MT, Fabris S, Russo A, Ruberto F, De Meo D, Ciccozzi M, Mastroianni CM, Venditti M, Pugliese F, d'Ettorre G. Clinical impact of colonization with Carbapenem-resistant gram-negative bacteria in critically ill patients admitted for severe trauma. Pathogens. 2022;11(11):1295. https://doi.org/10.3390/pathogens11111295. PMID: 36365046; PMCID: PMC9695038.
38. Campbell WR, Li P, Whitman TJ, Blyth DM, Schnaubelt ER, Mende K, Tribble DR. Multi-drug-resistant gram-negative infections in deployment-related trauma patients. Surg Infect. 2017;18(3):357–67. https://doi.org/10.1089/sur.2017.002. Epub 2017 Feb 24. PMID: 29173084; PMCID: PMC5393413.
39. Loftus TJ, Brakenridge SC, Moore FA, Lemon SJ, Nguyen LL, Voils SA, Jordan JR, Croft CA, Smith RS, Efron PA, Mohr AM. Intubated trauma patients receiving prolonged antibiotics for pneumonia despite negative Cultures: predictors and outcomes. Surg Infect (Larchmt). 2016;17(6):766–72. https://doi.org/10.1089/sur.2016.108. Epub 2016 Sep 16. PMID: 27635693; PMCID: PMC5124750.
40. Goyal K, Garg N, Bithal PK. Central fever: a challenging clinical entity in neurocritical care. J Neurocrit Care. 2020;13:19.
41. Jin H, Liu Z, Xiao Y, Fan X, Yan J, Liang H. Prediction of sepsis in trauma patients. Burn Trauma. 2014;2:106–13.
42. Claridge JA, Golob JF, Fadlalla AMA, Malangoni MA, Blatnik J, Yowler CJ. Fever and leukocytosis in critically ill trauma patients: it is not the blood. Am Surg. 2009;75(5):405–10. https://doi.org/10.1177/000313480907500511.
43. Toros A, Grodzinsky E, Karlsson N, Nilsson M, Levander MS. Use of temperature changes and pro-inflammatory biomarkers to diagnose bacterial infections in patients with severe cerebral trauma. J Neurocrit Care. 2022;15(1):21–31. https://doi.org/10.18700/jnc.210031.
44. Bouadma L, Luyt CE, Tubach F, Cracco C, Alvarez A, Schwebel C, Schortgen F, Lasocki S, Veber B, Dehoux M, Bernard M, Pasquet B, Régnier B, Brun-Buisson C, Chastre J, Wolff M, PRORATA trial group. Use of procalcitonin to reduce patients' exposure to antibiotics in intensive care units (PRORATA trial): a multicentre randomised controlled trial. Lancet. 2010;375(9713):463–74. https://doi.org/10.1016/S0140-6736(09)61879-1. Epub 2010 Jan 25.
45. Azzini A, Dorizzi R, Sette P, Vecchi M, Coledan I, Righi E, Tacconelli E. A 2020 review on the role of procalcitonin in different clinical settings: an update conducted with the tools of the evidence based laboratory medicine. Ann Transl Med. 2020;8(9):610. https://doi.org/10.21037/atm-20-1855.
46. Ciriello V, Gudipati S, Stavrou PZ, Kanakaris NK, Bellamy MC, Giannoudis PV. Biomarkers predicting sepsis in polytrauma patients: current evidence. Injury. 2013;44(12):1680–92. https://doi.org/10.1016/j.injury.2013.09.024.
47. Dolin Hallie H, Papadimos Thomas J, Stanislaw S, Xiaohuan C, Pan Zhixing K. A novel combination of biomarkers to herald the onset of sepsis prior to the manifestation of symptoms. Shock. 2018;49(4):364–70. https://doi.org/10.1097/SHK.0000000000001010.
48. Lu HX, Du J, Wen DL, Sun JH, Chen MJ, Zhang AQ, Jiang JX. Development and validation of a novel predictive score for sepsis risk among trauma patients. World J Emerg Surg. 2019;14:11. https://doi.org/10.1186/s13017-019-0231-8.
49. Evans L, Rhodes A, Alhazzani W, Antonelli M, Coopersmith CM, French C, Machado FR, Mcintyre L, Ostermann M, Prescott HC, Schorr C, Simpson S, Wiersinga WJ, Alshamsi F, Angus DC, Arabi Y, Azevedo L, Beale R, Beilman G, Belley-Cote E, Levy M. Surviving sepsis campaign: international guidelines for management of sepsis and septic shock 2021. Intensive Care Med. 2021;47(11):1181–247. https://doi.org/10.1007/s00134-021-06506-y.

50. Auchet T, Regnier MA, Girerd N, Levy B. Outcome of patients with septic shock and high-dose vasopressor therapy. Ann Intensive Care. 2017;7(1):43. https://doi.org/10.1186/s13613-017-0261-x.
51. Edwards P, Arango M, Balica L, Cottingham R, El-Sayed H, Farrell B, Fernandes J, Gogichaisvili T, Golden N, Hartzenberg B, Husain M, Ulloa MI, Jerbi Z, Khamis H, Komolafe E, Laloë V, Lomas G, Ludwig S, Mazairac G, Sanchéz M, RASH Trial Collaborators. Final results of MRC CRASH, a randomised placebo-controlled trial of intravenous corticosteroid in adults with head injury-outcomes at 6 months. Lancet (London, England). 2005;365(9475):1957–9. https://doi.org/10.1016/S0140-6736(05)66552-X.
52. Oliynyk OV, Pereviznyk BO, Yemiashev OV, Shlifirchyk A. The effectiveness of corticosteroid usage in complex therapy for severe sepsis and acute respiratory distress syndrome in cases of severe traumatic brain injury. Adv Clin Exp Med. 2016;25(6):1223–6. https://doi.org/10.17219/acem/61013.
53. Fox BC, Imrey PB, Voights MB, Norwood S. Infectious disease consultation and microbiologic surveillance for intensive care unit trauma patients: a pilot study. Clin Infect Dis. 2001;33(12):1981–9. https://doi.org/10.1086/324083. Epub 2001 Nov 9.
54. Bronkhorst MW, Patka P, Van Lieshout EM. Effects of sequence variations in innate immune response genes on infectious outcome in trauma patients: a comprehensive review. Shock. 2015;44(5):390–6. https://doi.org/10.1097/SHK.0000000000000450.
55. Dolin HH, Papadimos Thomas J, Stanislaw S, Xiaohuan C, Pan Zhixing K. A novel combination of biomarkers to herald the onset of sepsis prior to the manifestation of symptoms. Shock. 2018;49(4):364–70. https://doi.org/10.1097/SHK.0000000000001010.
56. von Groote T, Meersch-Dini M. Biomarkers for the prediction and judgement of sepsis and sepsis complications: a step towards precision medicine? J Clin Med. 2022;11(19):5782. https://doi.org/10.3390/jcm11195782.
57. Bergmann CB, Beckmann N, Salyer CE, Hanschen M, Crisologo PA, Caldwell CC. Potential targets to mitigate trauma- or sepsis-induced immune suppression. Front Immunol. 2021;12:622601. https://doi.org/10.3389/fimmu.2021.622601.
58. Allgöwer M, Dürig M, Wolff G. Infection and trauma. Surg Clin N Am. 1980;60(1):1980.

Intensivist's Role in Epidemic and Pandemic

Umm E Amara, Abdelrahman Balal, Umme Nashrah, Shajahan Idayathulla, Shafee Shaikh, and Nissar Shaikh

Abstract Epidemics and pandemics have occurred in the recent past and will surely occur in the future. To manage patients' suffering and to have a favorable outcome during these epidemics and pandemics, it is essential to have preparedness and use appropriate medical faculty management. The COVID-19 pandemic has shown us the importance of ICU (intensive care unit) beds and the intensivist's role in the management of critically ill patients during these calamities. Each epidemic and pandemic in the past has shown some advancement, which can improve critically ill patients' outcomes, the polio epidemic forced medical specialists to discover and use invasive ventilation patient isolation, and the SARS pandemic has shown us the advantages of using ECMO in the management of critically ill patients with improved outcomes. The Ebola epidemic dictated the importance of the strict following of infectious control and prevention protocols, and the H1N1 influenza pandemic has exposed the importance of ICU bed capacity management. The COVID-19 pandemic apart from showing the importance of intensivist also gave us a good lesson about the management of the "three Ss" the staff, stuff, and space during epidemics and pandemics as well as in natural calamities, with the active participation of intensivists in hospital management and administrative personnel.

Keywords COVID-19 · Epidemics · Ebola · Intensivist · Intensive care unit · Pandemics · Polio · H1N1 · SARS · Space · Staff and Stuff

U. E Amara
Apollo Institute of Medical Sciences, Hyderabad, Telangana, India

A. Balal · S. Idayathulla · S. Shaikh · N. Shaikh (✉)
Surgical Intensive Care Unit, Hamad Medical Corporation, Doha, Qatar
e-mail: smaheboob@hamad.qa

U. Nashrah
Deccan College of Medical Sciences, Hyderabad, Telangana, India

N. Shaikh, A. H. Chanda (eds.), *Applied Microbiology in Intensive Care Medicine*, https://doi.org/10.1007/978-981-97-4006-2_16

1 Introduction

A pandemic is a widespread disease occurrence more than what is normally expected in a geographic region. Usually, it is the spread of disease all over the world. Bubonic plague, cholera, polio, smallpox, and influenza are the most remembered and fatal pandemics in humanity. The smallpox pandemic killed nearly about 500 million people during its existence for 12,000 years. The influenza pandemic also killed around a million people [1].

More recently, the COVID-19 pandemic (Coronavirus pandemic) was a pandemic of coronavirus disease 2019 (COVID-19) that caused a severe acute respiratory syndrome coronavirus 2 (SARS-COV-2). This new virus was first identified in China and soon spread to the world. By mid-2023, COVID-19 has killed more than six million people worldwide, with collateral economic and social losses.

1.1 COVID-19 and Utilization of Intensive Care Units

The COVID-19 pandemic posed greater challenges to intensive care units (ICUs) and intensive care therapy units (ITUs). There were different challenges in the first wave of[2] COVID-19 and the subsequent wave [1].

In one study, bed demand increased during the COVID-19 pandemic, ranging from 52% to 28% of the pre-pandemic bed capacities in the various regions, depending on the disease severity. Initially, the surge in demand caused the cancelation of elective surgeries and the borrowing of beds from the ward and emergency departments [2].

It is well reported by now that the trained intensive care coverage of the ICU will improve the critically ill COVID-19 patients [3].

1.2 Ebola Virus Disease (EVD) 2014 Pandemic

The foreign minister of the West African countries of Liberia and Guinea reported an epidemic of EVD in early 2014. Soon, it spread to nearby countries and overseas, affecting thousands of populations. It is a very contagious disease; hence, it is potentially fatal not only for the patients but equally to the medical and paramedical staff if not trained and have not taken preventive during treatment.

It is demonstrated that the industrialized countries with intensity of resources, staff, and training to care for these patients with EVD had better outcomes than those treated in West African countries. Moreover, in EVD patients previously thought to be futile with critical care admissions and management, interventions such as invasive mechanical ventilation and renal replacement therapy have led to a favorable outcome in these patients with potentially lethal disease [4].

1.3 Influenza A (H1N1) Pandemic 2009

Influenza (H1N1) was reported initially in the southwestern USA and Mexico in early 2009 and soon spread worldwide. By mid-2010, the H1N1 influenza pandemic caused millions of deaths, most of which were due to acute respiratory failure.

H1N1 influenza changed the concept of intensivist and intensive care units in society. It put extra pressure on the hospital system, overcrowding of ICUs, and a significantly higher rate of mortality in patients treated outside the ICU. Even the higher-resource countries such as Australia and New Zealand ICUs were saturated, as there was a five-fold increase in the H1N1 influenza during that winter season, putting the extra workload on the unhealed core body and staff over weeks [5].

In Canada, during this pandemic, 81% of patients required invasive mechanical ventilation over an average of 12 days, with rescue therapy such as neuromuscular blockage in 28%, inhaled nitric oxide in 13%, ECMO in 4%, and prone positioning in 3% of H1N1 influenza patients [6].

1.4 Severe Acute Respiratory Syndrome Associated with Coronavirus Pandemic 2003

This pandemic started in China in 2003 and soon transmitted to the world. In Canada, 60% of nursing staff became ill after coming in contact with SARS patients. Management of SARS was limited by non-specific symptoms and the quick progression to severe disease. Up to 20% of these patients required invasive mechanical ventilatory support. Many hospitals do not have a plan to tackle the sudden surge of patients, patient transfer, and lack of trained critical care staff. Due to a lack of proper infection control prevention protocols, many healthcare workers were infected with SARS [7].

1.5 Polio Pandemic

Although polio was reported in ancient medicine, a major pandemic or epidemic occurred in Europe and the USA in 1900. In 1950, thousands suffered from paralytic poliomyelitis. The iron lung was used for the first time to help them in breathing.

Few of these patients required intubation and mechanical ventilation with a self-inflating bag. There was a huge deficiency of medical and paramedical staff during this pandemic. Thus, this epidemic laid a path for the discovery and invention of invasive mechanical ventilation (Table 1).

Table 1 The advancement and learning from epidemics and pandemics

Pandemic	Advancement
Polio pandemic	Self-inflating bags Invasive ventilation Isolation wards
SARS 2002/2003	Infection control and prevention protocols ECMO (extracorporeal membrane oxygenation)
Influenza H1N1 2009/2010	ICU capacity management position Nitric oxide use
Ebola virus disease 2014–2016	Revised infection prevention and control protocols Use of critical care of invasive ventilation and renal replacement therapy with high-severity illness patient
COVID-19	Intensivists' lead management of critically ill patients had improved outcomes Managing "three Ss": Space, stuff, and staff

2 Intensivist and Intensive Care Planning During the Pandemic

The pandemic due to the infectious disease outbreak creates unique challenges to the hospital and mainly to the intensive care units.

The first isolation ward, intensive respiratory care unit, and invasive mechanical ventilation came into existence. After the 2003 SARS epidemic, the importance of infection control protocol implementation was released.

From H1N1 influenza in 2009, ECMO was a major salvage therapy and a major critical care intervention. Ebola outbreaks share the risk of highly contagious diseases to healthcare workers.

The coronavirus pandemic in 2019 revealed the importance of critical care staff, intensivists, and infection control and prevention procedures which will improve the outcome of critically ill patients during the pandemic and epidemic. All these previous pandemics and epidemics are useful sources for learning and planning future hospitals and ICUs.

The healthcare workers may be unable to reach the hospital due to disaster in the area or get infected due to the epidemic and pandemic. Two factors that affect the infectious disease spread are the virulence of the organism, where a significant portion of the patients are at risk for developing MODS (multi-organ dysfunction syndrome) and may affect a group of patients (elderly or immunocompromised), and the affected percentage of patients may be less.

Another factor is the contagiousness of the disease, where there will be a higher number of patients affected and a higher risk for healthcare workers to get infected as well. The population at risk of exposure generally lacks immunity [8].

3 Critical Care/Intensive Care Therapy Preparation and Planning of Epidemic and Pandemic

While dealing with critically ill patients during epidemics and pandemics, intensivists, hospital administration, and the critical care team should plan according to the "three Ss": stuff, staff, and space.

3.1 *Supply Requirements During and for the Epidemic and Pandemic (Stuff)*

As per the requirements of the Joint Commission International (JCI), the hospital should be able to be autonomous for 96 h of functional reserves without resupply which easily takes care of patients and staff. In spite of these requirements, recurrent epidemics and pandemics face a shortage of routine supply within these 96 h [9]. Disposable items and reusable devices such as mechanical ventilators may be in limited supply. In the USA, there are 20 ventilators per 100,000 population. In the influenza pandemic, with a 30% attack rate, the ventilator demand may be doubled.

When additional supply is needed in these situations, we may consider using a high-flow nasal cannula, non-invasive ventilation, and anesthesia ventilator in selected patients.

Antibiotics, anti-viral, muscle relaxants, sedative agents, and vasopressors are quick to be utilized in epidemics and pandemics. The hospital administrators and clinical and nursing leaders should ensure that these medications are continuously available, if not from routine channels, then from alternative routes. Hospitalists and intensivists should be aware of alternative medicines from the same group of drugs which may be available in the pharmacy, but sure, there will be some differences in action and adverse effects.

The resupply from geographically affected areas during the localized epidemic may be difficult. The Department of Health should maintain the natural stockpile of antibiotics, anti-viral and other emergency and investigational medications, and external devices such as mechanical ventilators.

A highly infectious disease affects patients' needs for dedicated disposable items, ventilators, and laboratory equipment; for example, one EVD patient generates 60 liters of waste product per day. Autoclaving solid waste and decontaminating liquid waste are a must before it is disposed of into the common disposal system [10].

Managing critically ill epidemic and pandemic patients by intensivist-led teams has a better outcome, thus preserving the critical care capabilities of the hospital during epidemics and pandemics, depending on the availability of trained healthcare workers.

Healthcare workers themselves can be victims either bedside or in the community. Safety is very important, and medical and paramedical staff should strictly follow the infection control and prevention protocols; they should also receive vaccination and chemoprophylaxis.

3.2 Staff

Although it is essential to have full force and trained critical care staff, "presenteeism," that is, the act of coming to work while ill, is an important threat to the staff's safety. During the influenza pandemic, around 40% of staff reported to duties while they had respiratory symptoms, thus posing a risk for fellow staff and patients; hence in the epidemic and pandemic, hospitals should maintain the policy that healthcare practitioners with symptoms should avoid working and seek appropriate medical care for themselves [11].

3.3 Space: The Critical Care Without an Intensive Care Unit

It is typical practice to manage a critically ill patient in intensive care units with the availability of trained staff, specific medications, and devices needed to manage potentially ill patients. In a standard hospital surge condition, institutions should be able to manage a 20% increase in critically ill patients with minimal impact. In a situation of epidemic and pandemic where the requirements of clinical care support are by 200%, this needs modification and adaptation of intensive care therapy.

The first modification to intensive care therapy is to manage these acutely ill patients, not needing invasive ventilation, vasopressors, and intensive care monitoring which are managed outside the ICU, such as in PACUs (post-anesthesia care units), monitored step-down units, and temporary ICUs.

Highly contaminant and infectious populations causing epidemics and pandemics require more specific dedicated space and single-patient care, as EVD (Ebola virus disease) causes the loss of six or more standard beds in ICUs; hence, the need for space, and certainly, ICU space will be increased significantly, thus requiring intensivist and critical care teams to take care of acutely ill patients in epidemic and pandemic outside the intensive care units.

In a pandemic and epidemic, the intensivist and critical care teams, in combination with other teams, must do the following tasks in addition to taking care of critically ill patients:

1. Resource allocations and triage.

   Epidemics and pandemics will overwhelm the local capacity of institutions and will need to justify the allocations of scarce resources. There should be coordination if a patient has to transfer to regional or national centers such as ECMO. These decisions of triaging which patient needs intensive care therapy or other specialized services are based on transparent, open, and broadly accepted ethical and medical principles. Critical care admissions may be limited to patients with good chances of survival and use of ventilators. These patients denied ICU care should receive appropriate and compassionate palliative care.

2. Staff optimizations.

   Hospitals should have regular staff training and simulations. There should be appropriate staff trained in infection control and prevention protocols.

3. Equipment, supplies, and space.

   Hospitals and institutions should have plans for additional supplies, apart from 96 hours, by coordinating nearby hospitals and pharmaceuticals. There can be coordination with national and regional stockpiles to replenish the needed supplies. The hospitals should discourage "presentism" and ensure staff vaccinations and chemoprophylaxis.

4. Public affairs.

   Hospitals should be transparent, protecting the privacy and dignity of the patient's community involved in decision-making. Media interaction should be taken by hospital administrative and public affairs personnel in collaboration with public health authorization.

5. Case identification, diagnosis, and tracking of potentially contagious patients should be done in combination with hospital and public authorities.

4 Conclusion

Epidemics and pandemics are frequently occurring due to the mobile and globalized world, and it is impossible to restrict the spread of infectious diseases. The ICU and intensivist have a unique role in epidemics and pandemics, from triaging patients to managing critically ill patients outside the typical intensive care therapy settings. Intensivists also must manage the "three Ss," that is, space, staff, and stuff.

References

1. Arabi YM, Myatra SN, Lobo SM. Surging ICU during COVID-19 pandemic: an overview. Curr Opin Crit Care. 2022;28:638–44.
2. Douin DJ, et al. ICU bed utilization during the coronavirus disease 2019 pandemic in a multistate analysis-march to June 2020. Crit Care Explor. 2021;3:e0361.
3. Oh TK, Kim S, Song I-A. Intensivist coverage and critically ill COVID-19 patient outcomes: a population-based cohort study. J Intensive Care. 2023;11:19.
4. Uyeki TM, et al. Clinical Management of Ebola virus disease in the United States and Europe. N Engl J Med. 2016;374:636–46.
5. ANZIC, Influenza Investigators, et al. Critical care services and 2009 H1N1 influenza in Australia and New Zealand. N Engl J Med. 2009;361:1925–34.
6. Kumar A. Critically ill patients with 2009 influenza A (H1N1) infection in Canada. JAMA. 2009;302:1872.

7. Clinical management and infection control of SARS: lessons learned—PubMed. https://pubmed.ncbi.nlm.nih.gov/23994190/.
8. Intensive care unit preparedness during pandemics and other biological threats—PubMed. https://pubmed.ncbi.nlm.nih.gov/31445608/.
9. Abramson DM, Redlener I. Hurricane Sandy: lessons learned, again. Disaster Med Public Health Prep. 2012;6:328–9.
10. Le AB, et al. A pilot survey of the U.S. medical waste industry to determine training needs for safely handling highly infectious waste. Am J Infect Control. 2018;46:133–8.
11. Chiu S, et al. Working with influenza-like illness: presenteeism among US health care personnel during the 2014-2015 influenza season. Am J Infect Control. 2017;45:1254–8.

GPSR Compliance

The European Union's (EU) General Product Safety Regulation (GPSR) is a set of rules that requires consumer products to be safe and our obligations to ensure this.

If you have any concerns about our products, you can contact us on ProductSafety@springernature.com

In case Publisher is established outside the EU, the EU authorized representative is:

Springer Nature Customer Service Center GmbH
Europaplatz 3
69115 Heidelberg, Germany

Batch number: 10732090

Printed by Printforce, the Netherlands